Opportunities and Challenges of Industrial IoT in 5G and 6G Networks

Poshan Yu
Soochow University, China

Xiaohan Hu
Shanghai University, China

Ajai Prakash
University of Lucknow, India

Nyaribo Wycliffe Misuko
KCA University, Kenya

Gu Haiyue
Shanghai University, China

A volume in the Advances in Wireless Technologies and Telecommunication (AWTT) Book Series

Published in the United States of America by
IGI Global
Information Science Reference (an imprint of IGI Global)
701 E. Chocolate Avenue
Hershey PA, USA 17033
Tel: 717-533-8845
Fax: 717-533-8661
E-mail: cust@igi-global.com
Web site: http://www.igi-global.com

Library of Congress Cataloging-in-Publication Data

Names: Yu, Poshan, editor.
Title: Opportunities and challenges of industrial IoT in 5G and 6G networks / editors: Poshan Yu, Xiaohan Hu, Ajai Prakash, Nyaribo Misuko and Gu Haiyue.
Description: Hershey, PA : Information Science Reference, an imprint of IGI Global, [2022] | Includes bibliographical references and index. | Summary: "This book covers a variety of relevant chapters to the presently under-development and deployment technologies such as 5G and industrial Internet of Things, and upcoming 6G technology and their needs, covering this important part of modernization in communication standards"-- Provided by publisher.
Identifiers: LCCN 2022001286 (print) | LCCN 2022001287 (ebook) | ISBN 9781799892663 (h/c) | ISBN 9781799892670 (s/c) | ISBN 9781799892687 (ebook)
Subjects: LCSH: Internet of things--Industrial applications. | 5G mobile communication systems.
Classification: LCC TK5105.8857 .O67 2022 (print) | LCC TK5105.8857 (ebook) | DDC 004.67/8--dc23/eng/20220324
LC record available at https://lccn.loc.gov/2022001286
LC ebook record available at https://lccn.loc.gov/2022001287

This book is published in the IGI Global book series Advances in Wireless Technologies and Telecommunication (AWTT) (ISSN: 2327-3305; eISSN: 2327-3313)

British Cataloguing in Publication Data
A Cataloguing in Publication record for this book is available from the British Library.

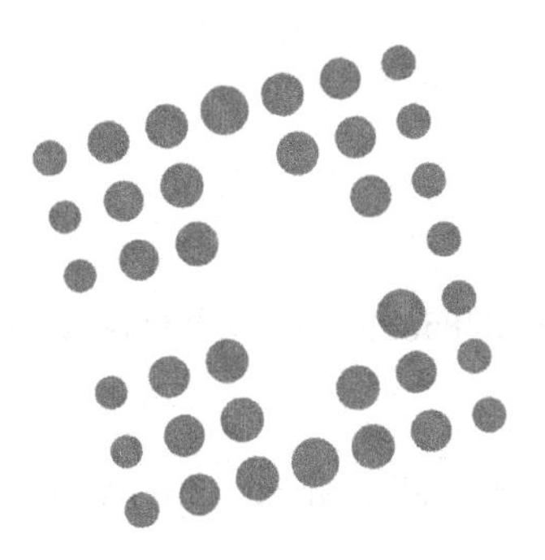

Advances in Wireless Technologies and Telecommunication (AWTT) Book Series

Xiaoge Xu
University of Nottingham Ningbo China, China

ISSN:2327-3305
EISSN:2327-3313

Mission

The wireless computing industry is constantly evolving, redesigning the ways in which individuals share information. Wireless technology and telecommunication remain one of the most important technologies in business organizations. The utilization of these technologies has enhanced business efficiency by enabling dynamic resources in all aspects of society.

The **Advances in Wireless Technologies and Telecommunication Book Series** aims to provide researchers and academic communities with quality research on the concepts and developments in the wireless technology fields. Developers, engineers, students, research strategists, and IT managers will find this series useful to gain insight into next generation wireless technologies and telecommunication.

Coverage

- Digital Communication
- Virtual Network Operations
- Mobile Communications
- Mobile Technology
- Broadcasting
- Radio Communication
- Network Management
- Wireless Broadband
- Mobile Web Services
- Cellular Networks

IGI Global is currently accepting manuscripts for publication within this series. To submit a proposal for a volume in this series, please contact our Acquisition Editors at Acquisitions@igi-global.com or visit: http://www.igi-global.com/publish/.

The Advances in Wireless Technologies and Telecommunication (AWTT) Book Series (ISSN 2327-3305) is published by IGI Global, 701 E. Chocolate Avenue, Hershey, PA 17033-1240, USA, www.igi-global.com. This series is composed of titles available for purchase individually; each title is edited to be contextually exclusive from any other title within the series. For pricing and ordering information please visit http://www.igi-global.com/book-series/advances-wireless-technologies-telecommunication/73684. Postmaster: Send all address changes to above address.

Titles in this Series

For a list of additional titles in this series, please visit: www.igi-global.com/book-series

Designing and Developing Innovative Mobile Applications
Debabrata Samanta (Rochester Institute of Technology, Kosovo)
Information Science Reference • copyright 2023 • 426pp • H/C (ISBN: 9781668485828) • US $225.00 (our price)

Multidisciplinary Applications of Computer-Mediated Communication
Hung Phu Bui (University of Economics Ho Chi Minh City, Vietnam) and Raghvendra Kumar (GIET University, India)
Information Science Reference • copyright 2023 • 322pp • H/C (ISBN: 9781668470343) • US $250.00 (our price)

Role of 6G Wireless Networks in AI and Blockchain-Based Applications
Malaya Dutta Borah (National Institute of Technology, Silchar, India) Steven A. Wright (Georgia State University, USA) Pushpa Singh (GL Bajaj Institute of Technology and Management, Greater Noida, India) and Ganesh Chandra Deka (Ministry of Skill Development and Entrepreneurship, Government of India, India)
Information Science Reference • copyright 2023 • 285pp • H/C (ISBN: 9781668453766) • US $250.00 (our price)

Innovative Smart Materials Used in Wireless Communication Technology
Ram Krishan (Mata Sundri University Girls College, Mansa, India) Manpreet Kaur (Yadavindra Department of Engineering, Guru Kashi College, Punjabi University, Talwandi Sabo, India) and Shilpa Mehta (Auckland University of Technology, New Zealand)
Information Science Reference • copyright 2023 • 369pp • H/C (ISBN: 9781668470008) • US $250.00 (our price)

Economic and Social Implications of Information and Communication Technologies
Yilmaz Bayar (Bandirma Onyedi Eylul University, Turkey) and Lina Karabetyan (Independent Researcher, Turkey)
Information Science Reference • copyright 2023 • 318pp • H/C (ISBN: 9781668466209) • US $235.00 (our price)

Modelling and Simulation of Fast-Moving Ad-Hoc Networks (FANETs and VANETs)
T.S. Pradeep Kumar (Vellore Institute of Technology, India) and M. Alamelu (Kumaraguru College of Technology, India)
Information Science Reference • copyright 2023 • 251pp • H/C (ISBN: 9781668436103) • US $240.00 (our price)

Challenges and Risks Involved in Deploying 6G and NextGen Networks
A.M. Viswa Bharathy (GITAM University, Bengaluru, India) and Basim Alhadidi (Al-Balqa Applied University, Jordan)
Information Science Reference • copyright 2022 • 258pp • H/C (ISBN: 9781668438046) • US $250.00 (our price)

701 East Chocolate Avenue, Hershey, PA 17033, USA
Tel: 717-533-8845 x100 • Fax: 717-533-8661
E-Mail: cust@igi-global.com • www.igi-global.com

Editorial Advisory Board

Table of Contents

Detailed Table of Contents

This chapter aims to explore the importance of the convergence and collaboration of the internet of things and blockchain in enabling the development of smart cities. It will capture the strong link between the burgeoning advanced technology and China's fast-growing smart cities. In addition, the authors used the CiteSpace bibliometric tool to analyse the core journals related to smart cities, IoT, and blockchain in WoS and CNKI operations and visualise the relationships between the literature as a scientific knowledge map. The research topic in this chapter focuses on the keywords with the highest frequency and determines the research frontier and trend. This chapter will further explore the roles of the government in promoting the integration of the internet of things and blockchain in the development agenda of industrial infrastructure and smart cities through policies (i.e., institutions).

From the first generation to the third generation (3G), data communications such as "i-mode" and multimedia information such as photos, music, and video could be communicated using mobile devices. From the fourth generation (4G), smart phones have been explosively popularized by high-speed communication technology exceeding 100 Mbps using the long-term evolution (LTE), and multimedia communication services have appeared. 4G technology continues to evolve in the form of LTE-Advanced and has now reached a maximum communication speed close to 1 Gbps. 5G is expected to provide new value as a basic technology supporting future industry and society, along with artificial intelligence (AI) and the

The internet of things characterizes a unified structure of internet-supported objects that can gather and send data through a wireless network with no intervention of humans. By this structure, the present business world is experiencing remarkable changes due to the irresistible potential derivable from the internet of things. This technology is already bringing significant changes across the industries, of which the digital marketing sector has the maximum benefits. This technology collects various data from the consumers through different forms of digital marketing platforms such as social media marketing, online marketing, electronic mail marketing, pay-per-click advertising. With these types of data, marketers can generate some meaningful insights, develop interactions with customers, communicate with sellers and customers, and can also forecast the behavior and lifestyle of the customers. With this backdrop, the chapter has made an attempt to explain the relevance of the internet of things in digital marketing.

Cities are crucial carriers of economic prosperity and social development. With the approaching Society 5.0, the digital twin city has become the mainstream model for the construction of new smart cities. Digital twins, along with the internet of things, fifth-generation wireless systems, and artificial intelligence technologies, offer great potential in the transformation of the current urban governance paradigm toward smart cities. This chapter will explore the imperative role of digital twin in accelerating digital transformation of smart cities. From the perspective of the overall policy, the industry, and application of the said technology, this chapter gives insight into the development trend of digital twin cities. Additionally, through case studies of the two most outstanding digital pilot regions in China, the chapter investigates the role of digital twin in accelerating digital transformation of smart cities such as the technical approaches on city information modeling (CIM) and building information modeling (BIM).

The chapter provides information as a first step for individuals who are thriving to get a bird's eye view of the aspects underlying the optical networking in the context of 5G technology. Apart from capacity requirement challenges targeted by 5G coverage, it requires a lot of fibers to be successfully provisioned to achieve formidable performance goals of 5G such as diversified capacity requirements, availability, and coverage issues. The goals could be achieved by the underlying optical network with a greater number of interconnected fiber paths. In 5G, the requirements of reliable and ultra-low latency services required at the access side of a network shape up the research and evolution of underlying optical segments spanning from core to access part of the network. The reconfigurability and security issues of the present mode of optical communication need to be addressed, and the proposals given by the researchers are summed

Preface

With the arrival of 5G, we have entered a new era of connectivity, and it is expected that the upcoming 6G network will bring more and more opportunities and challenges to the industrial Internet of Things. The continuous digital transformation led by Industry 4.0 has driven the demand for more advanced Internet of Things systems that can handle massive data, high traffic, and diverse applications while ensuring low latency, high reliability, and security.

The integration of the Internet of Things and 5G/6G networks provides significant opportunities for optimizing production processes, improving operational efficiency, promoting automation, and implementing new business models. Real-time data analysis, advanced communication, and automation capabilities will transform industries such as manufacturing, healthcare, agriculture, logistics, and transportation.

However, deploying the Industrial Internet of Things in 5G/6G networks poses some challenges, including issues related to standardization, interoperability, network security, data privacy, and regulatory compliance. The dynamic and complex nature of these systems requires the collaboration of industry participants, policymakers, and academia to develop powerful solutions that ensure seamless integration, safe operation, and effective results.

This edited book on the opportunities and challenges of industrial IoT in 5G and 6G networks aims to provide a comprehensive and timely overview of the most advanced research, technology, and applications in this evolving field. This book includes the contributions of top experts and researchers from academia and industry, covering the following chapters:

The first chapter is on "Integration of the Internet of Things and Blockchain to Promote Collaboration in Smart Cities: A Case Study in China." This chapter aims to explore the importance of the convergence and collaboration of the Internet of Things and blockchain in enabling the development of smart cities. It captures the strong link between the burgeoning advanced technology and China's fast-growing smart cities. CiteSpace bibliometric tool was used to analyze the core journals related to smart cities, IoT and blockchain in WoS and CNKI operations and visualize the relationships between the literature as a scientific knowledge map. The focus of the research topic in this chapter focuses on the keywords with the highest frequency and determines the research frontier and trend. This chapter further explored the roles of the government in promoting the integration of the Internet of Things and blockchain in the development agenda of industrial infrastructure and smart cities through policies

The second chapter is called "Relevant Technologies to 6G." The chapter highlights the various generations from the first generation to the third generation (3G), data communications such as "i-mode" and multimedia information such as photos, music, and video could be communicated using mobile devices. From the fourth generation (4G), smart phones have been explosively popularized by high-speed communication technology exceeding 100 Mbps using the Long Term Evolution (LTE) and multimedia communication services have appeared. 4G technology continues to evolve in the form of LTE-Advanced, and has now reached a maximum communication speed close to 1 Gbps. 5G is expected to provide new value as a basic technology supporting future industry and society, along with Artificial Intelligence (AI) and the Internet of Things (IoT). The 5G is expected to evolve and the sixth generation (6G) technology will support industry and society in the 2030s. It shows how the 6G will revolutionize the role of technology by achieving several requirements such as ultra-high-speed, high-capacity, and low-latency connectivity.

The third chapter is called "Recent and Emerging Technologies in Industrial IoT." The focus is on the Industrial Internet of Things (IIoT) and how it is evolving through remote monitoring, cognitive analytics, and industrial process control. The chapters show the benefits of IIot through product customization, intelligent monitoring applications for production floor shops and machines that health, and predictive and preventive maintenance of industrial equipment, the primary goal of IIoT is to achieve high operational efficiency, increased productivity, and better management of industrial assets and processes. However, there is a caution that the industrial sector should be conscious in implementing full-stack development solutions with IIoT, there is a need to deal with the problems that are emerging. The IIOT keeps industrial equipment, machines, and cloud-based applications connected. The advantages of IIOT are highlighted, which is one of the chapter's three key themes, which includes emerging IIOT hardware and software technologies followed by challenges of IIOT. Further, this chapter focuses on recent trends and technologies in industrial IOT and challenges in IIOT.

The fourth chapter of this book is about "Making Cities Smarter: IoT and SDN Applications, Challenges, and Future Trends." The Internet of Things (IoT) entails all devices that can get onto the internet, this is mainly because of the technological advancement. This exponential growth of IoT increases on the dense nodes with a huge data volume on the network that affect the collision and network congestion probabilities. This chapter presents a comprehensive description of the central and supporting innovations that are used to make cities smarter, focusing on the fifth generation (5G) IoT paradigm from a software-based network viewpoint. Furthermore, the main initiatives of international significance are discussed. Also, the chapter presents Software Defined Networking (SDN), IoT, and Network Function Virtualization (NFV) challenges as it relates to the user privacy and security, IoT security, energy consumption, integration of IoT with subsystems, and architecture design. A segment of the top five future trends that are made and will make cities smarter is conclusively included.

The fifth chapter is called "Innovative Model of Internet of Things for Industrial Applications." The chapter starts by explaining why the Internet of Things is one of the most significant innovations today. In this chapter, the dual probability-based energy estimation model in the wireless sensor network is proposed. The dual probability-based function measures the expected value of energy for the transmission of data. This function creates a subgroup of networks based on energy function and carries out the operation of energy management in the context sensor node data processing. This function also integrates cloud-based services with the sensor networks. The benefit of this function is demonstrated in that it increases the throughput of network and quality of service. The proposed model was simulated in MATLAB R-2014a environment and the results were obtained using different scenarios of network density. Finally, the performance is analysed with respect to the following metrics: data utility, energy consumptions, and data reconstruction error.

"5G Network Programmability Enabling Industry 4.0 Transformation" is the sixth chapter. It shows how the 5G system aims at enabling innovative services of vertical industries by utilizing network programmability to its full extension. The 3rd Generation Partnership Project (3GPP) has already established the foundations to provide third parties with 5G Core's capabilities by introducing the Common API Framework (CAPIF) and Service Enabler Layer Architecture (SEAL) on top of the services offered by the Network Exposure Function (NEF). However, a scalable, robust and secure ecosystem should be set for the verticals that exploit those capabilities. In this context, cloud native approach is becoming a key enabler for that ecosystem, taking advantage of the inherent cloud-native characteristics of the 5G Service Based Architecture (SBA). The chapter presents the capabilities that 5G provides to verticals as well as the ecosystem that is built around the exploitation of those capabilities. As a matter of better justification and exemplification, Industry 4.0 vertical is targeted while developments related to Vertical Application Enablers (VAEs) for the factory of the future are provided.

The seventh chapter is about "Internet of Things and Its Relevance to Digital Marketing." The chapter starts by demonstrating how the Internet of Things characterizes a unified structure of internet-supported objects that can gather and send data through a wireless network with no intervention of humans. By this structure, the present business world is experiencing remarkable changes due to the irresistible potential derivable from the Internet of Things. Further, it shows what this technology is already bringing across the industries, of which the digital marketing sector has the maximum benefits. This technology is able to collect various data from consumers through different forms of digital marketing platforms such as social media marketing, online marketing, electronic mail marketing, pay-per-click advertising. The benefits accruing from these types of data, marketers can generate some meaningful insights, develop interactions with customers, communicate with sellers and customers, and can also forecast the behavior and lifestyle of the customers. With this backdrop, the present chapter has made an attempt to explain the relevance of the internet of Things in digital marketing.

The eighth chapter is called the "Role of Digital Twin in Accelerating Digital Transformation of Smart Cities: Case Studies in China." It is a case study from China on smart cities. Cities are crucial carriers of economic prosperity and social development. With the approaching Society 5.0, the digital twin city has become the mainstream model for the construction of new smart cities. Digital twins, along with the Internet of Things, fifth-generation wireless systems, and artificial intelligence technologies, offer great potential in the transformation of the current urban governance paradigm toward smart cities. This chapter explores the imperative role of digital twin in accelerating digital transformation of smart cities. From the perspective of the overall policy, the industry, and application of the said technology, this chapter gives insight into the development trend of digital twin cities. Additionally, through case studies of the two most outstanding digital pilot regions in China, the chapter investigates the role of digital twin in accelerating digital transformation of smart cities such as the technical approaches on City Information Modeling (CIM) and Building Information Modeling (BIM).

The ninth chapter is about "Optical Networking Technologies for 5G Services." The article is to help provide information as a first step for individuals who are thriving to get a bird's eye view of the aspects relying the optical networking in the context of 5G technology. First, the chapter explores the challenges of 5G technology. Apart from capacity requirement challenges targeted by 5G coverage, it requires lot of fibers to be successfully provisioned to achieve formidable performance goals of 5G such as diversified capacity requirements, availability, and coverage issues. The goals could be achieved by the underlying optical network with a greater number of interconnected fiber paths. In 5G, the requirements of reliable and ultra-low latency services required at the access side of a network shape up the research and evolution of underlying optical segments spanning from core to access part of the network. The reconfigurability, security issues of the present need of optical communication needs to be addressed and the proposals given by the researchers are summed up. The article includes a general framework and theoretical concepts behind machine learning and software defined networking paradigms.

The tenth chapter is called "A Summary on 5G and Future 6G Internet of Things." This chapter comprehensively surveys the six aspects of the 5G and future 6G Internet of Things (IoT). First, most of the 5G- and 6G-IoT usage scenarios and key performance indicators are summarized in the form of tables, pictures, and diagrams to facilitate readers to understand and compare current and future IoT technologies more easily and quickly. Second, 5G- and 6G-IoT access networks, protocols, and standards were briefly analyzed and compared, such as coverage, transfer data speed, energy consumption, operating frequency, and the number of device connectivity. Third and fourth, the impact of 6G-IoT on society's daily life and industry operation, as well as its underlying research were described. Fifth, five types of 6G-IoT challenges were analyzed and discussed in detail in this chapter, namely transmission path loss at THz, wireless network coverage, transfer data rate, and latency, security, and privacy protection, and energy-efficient and reliable devices/services. Finally, the latest nine IoT business models are described and summarized in tabular form

The final chapter is called "IoT Blockchains for Digital Twins." Digital Twins (DTs) have emerged as a critical concept in cyberspace infrastructure. DTs are virtual representations of physical things including model smart structures or environments, manufacturing processes, humans, and a variety of other things. The value provided by DTs relies on their fidelity in representation. Blockchains provide trust assurance mechanisms, particularly where multiple parties are involved. The expected life cycle operations of the IoT, Blockchain and DT need to be considered to develop economically useful Blockchain Digital Twin (BDT) models. BDTs do not exist in isolation, but rather within a DT Environment (DTE). A DTE may include multiple DTs of different objects, to enable interactions between these objects to be evaluated in both virtual reality and mixed reality cases. To populate DTEs with multiple DTs requires industrialized tooling to support the rapid creation of DTs. The industrialization of DT creation requires frameworks, architectures and standards to enable interoperability between DTs and DTEs.

We hope that this book will become a valuable resource for researchers, practitioners, policymakers, and postgraduate students interested in exploring the potential and challenges of the industrial Internet of Things in the 5G and 6G era.

Poshan Yu
Soochow University, China

Xiaohan Hu
Shanghai University, China

Ajai Prakash
University of Lucknow, India

Nyaribo Wycliffe Misuko
KCA University, Kenya

Gu Haiyue
Shanghai University, China

Chapter 1
Integration of the Internet of Things and Blockchain to Promote Collaboration in Smart Cities:
A Case Study in China

Poshan Yu
Soochow University, China & Australian Studies Centre, Shanghai University, China & EBU Luxembourg, Luxembourg

Zixuan Zhao
Independent Researcher, China

Emanuela Hanes
Independent Researcher, Austria

ABSTRACT

This chapter aims to explore the importance of the convergence and collaboration of the internet of things and blockchain in enabling the development of smart cities. It will capture the strong link between the burgeoning advanced technology and China's fast-growing smart cities. In addition, the authors used the CiteSpace bibliometric tool to analyse the core journals related to smart cities, IoT, and blockchain in WoS and CNKI operations and visualise the relationships between the literature as a scientific knowledge map. The research topic in this chapter focuses on the keywords with the highest frequency and determines the research frontier and trend. This chapter will further explore the roles of the government in promoting the integration of the internet of things and blockchain in the development agenda of industrial infrastructure and smart cities through policies (i.e., institutions).

INTRODUCTION

With the massive application of Internet of Things (IoT) terminal devices in the infrastructure, the shortcomings of low rate, low connection density and time latencies make it difficult for the traditional 4G

DOI: 10.4018/978-1-7998-9266-3.ch001

network to support the future needs of smart city development and construction. The development of 5G networks has become a common choice to enhance urban development, innovate industrial development and realise the overall interconnection. 5G technology is closely related to national development, economic growth and industrial breakthroughs, and the 5G communication standard is not only a technical standard, but also of strategic importance to China's industrial development. From the perspective of connectivity development, the 2G era is like a hanging chime, 3G era is showing its head, 4G era is catching up. In the 5G era, China's goal has changed from being a runner to a leader. This will greatly promote the development of China's communications industry and national economy.

New technologies such as 5G and smart city construction show a complementary relationship, just like the organs that support the operation of the human body, emerging technologies provide technical support for the efficient operation of smart cities, while smart city construction promotes the application and continuous development of new technologies. 5G network technology has three main characteristics: ultra-high speed, ultra-large connectivity and ultra-low latency, and combined with future network slicing services and edge computing capabilities, it can layout 5G. Emerging information technology convergence can give full play to the key enablers of data, algorithms, computing power and distributed networks to promote the rapid development of smart city transportation, finance, government, environment and healthcare industries (Yu, Xu, Cheng & Sampat, 2023; Trivedi et al., 2021; Yu, Lu, Hanes & Chen, 2022; Yu, Lu, Sampat, Li & Ahuja, 2022; Yu, Ge, Mandizvidza & Mulli, 2023; Yu, Zhang, Sampat & Chen, 2023; Yu, Xue & Mahendran, 2022). Only by establishing a credible and secure information environment can smart cities operate efficiently, transparently, safely and credibly.

LITERATURE REVIEW

1. IoT & Smart City

The term IoT was first coined by Kevin Ashton in 1999 with reference to the supply chain management (Ashton, 2009). The concept of IoT revolves around the word ''smartness'', which means ''an ability to independently obtain and apply knowledge'' (Ahmed et al., 2016). Therefore, IoT refers to the ''things or devices and sensors'' that are smart, uniquely addressable based on their communication protocols, and are adaptable and autonomous with inherent security (Bhabad & Bagade, 2015).

With the continuous development of science and technology, IoT technology is appearing more and more in our daily life, and is also receiving more and more attention from all walks of life. Especially in recent years, the IoT has provided great opportunities for the development of various industries (Yu, Liu, Hanes & Mumtaz, 2022), especially in smart city construction (Kim et al., 2017). Smart cities, as well as smart governance, rely on collecting, analysing and processing large amounts of granular real-time data, which can only be achieved with the help of IoT sensors (Cirillo et al., 2020). IoT sensors and cameras can continuously collect detailed data in various forms and in real time, and by using this data, government agencies can make quick decisions on the allocation of different resources and assets (Lee & Lee, 2015).

2. Blockchain & Smart City

"Data" has become a resource in the new context of flourishing information technology, especially with the development of cities, which have the potential to accumulate a large volume of data. However, due to the competitive and exclusive nature of data, it is difficult to achieve an efficient, orderly and low-cost flow of urban data across levels, regions, systems and departments, and it is difficult to avoid data leakage when data is exchanged, making "data islands" and data security issues a constraint to the construction of new smart cities (Singh et al., 2020).

Blockchain is a peer-to-peer distributed ledger technology which records transactions, agreements, contracts, and sales (Christidis et al., 2016). Originally developed to support crypto-currency, blockchain can be utilized for any form of transactions without an intermediary. The benefit of blockchain is that an attacker has to compromise 51% of the systems to surpass the hashing power of the target network (Biswas & Muthukkumarasamy, 2016). Thus, it is computationally impractical to launch an attack against the blockchain network.

With the distributed storage, decentralization and peer-to-peer transmission functions of blockchain, each operation and maintenance management unit of the city can become a node, and the data generated can be sent directly to the designated distributed database without data processing through the centre, so as to achieve direct data transmission and thus solve the problem of difficult data sharing (Xie et al., 2019). At the same time, all data and information of these nodes are open, transparent and traceable, and data cannot be falsified or tampered with during data transmission, as tampering requires the approval of all nodes and leaves a tracking record of data information changes, which is conducive to the tracing of original data information. This makes blockchain technology effective in ensuring guaranteed data security (Li, 2018). Therefore, blockchain technology is ideally suited to solving the difficulties of sharing big data as it is decentralised, distributed, preventive of tampering, secure and anonymous.

Specifically, in the governance of smart cities, the advantages of blockchain are mainly reflected in the following three aspects (Bhushan et al., 2020):

1) blockchain technology reduces the cost of storing governance data in cities, improves the flexibility of data processing and provides feasible conditions for the collaboration of urban governance
2) blockchain technology can simplify the governance process, clarify the subjects of responsibility and obligation, thus promoting rapid responses of urban governance and improving the efficiency of the collaboration of urban governance (Mackenzie et al., 2018);
3) blockchain technology can enhance the transparency of urban governance and strengthen the protection of data privacy at the same time, guaranteeing the orderly operation of the collaborative system of urban governance (Yu, Gong & Sampat, 2022).

3. Big Data, Machine Learning, AI & Smart City

The rapid growth of urban populations worldwide imposes new challenges on citizens' daily lives, including environmental pollution, public security, road congestion, etc.

In order to manage these challenges, new technologies have been developed that support the development of smarter cities (Atitallah et al., 2020). Integrating the Internet of Things (IoT) in citizens' lives enables the innovation of new intelligent services and applications that serve sectors around the city, including healthcare, surveillance, agriculture, etc. IoT devices and sensors generate large amounts of data that can be analyzed to gain valuable information and insights that help to enhance citizens' quality of life (Chin et al., 2017; Yu, Liu & Hanes, 2022; Yu, Chen & Ahuja, 2022).

The IoT is the most important and vital component of most smart city applications and is responsible for generating large amounts of data. With such large and complex data, it is difficult to precisely decide on the most accurate and effective action. Using advanced technologies such as artificial intelligence (AI), machine learning and deep reinforcement learning (DRL), big data can be analysed in order to reach optimal decisions (Atitallah et al., 2020; Sharma et al., 2021).

In smart cities projects, various sectors like Intelligent transportation, cyber-security, smart grids, and UAVs-assisted next-generation communication (5G and 6G), etc. are playing a vital role. All the preceding sectors of smart cities are highly based on big data analytics and effective use of AI, machine learning, and DRL-based techniques that can enhance their efficiency and scalability in a smart city project (Alahakoon et al., 2020). For instance, modern intelligent transportation systems are highly influenced by machine learning and DRL-based techniques to realize self-driving vehicles, ensure security of connected vehicles, and safe travels (Ameer & Shah, 2022).

4. Sustainability & Smart Cities

The process of urbanisation has brought about a high concentration of population, resources and industries, but also a high concentration of resource and energy consumption and carbon emissions. To address climate change, protect the ecological environment and enhance sustainable development, low-carbon energy efficiency and green development have become the new focus of smart cities. The "double carbon" goal[1] has stimulated technological innovation and scenario application in smart cities, and smart city construction is becoming a strong base for the "double carbon" strategy to be fully implemented (Smart Cities White Paper, 2022). On one hand, the "source, network, storage and control" green energy architecture can be used to upgrade the energy supply, improve the efficiency of traditional energy-consuming infrastructure, and support the low-carbon transformation of the management and operation of smart cities. On the other hand, by actively promoting the application of clean energy such as photovoltaic, hydrogen and wind power in production and daily life circumstances such as parks and transportation, optimising the efficiency of city operations with new technologies such as vehicle networking, and adding carbon emissions to the monitoring indicators, we can also promote the upgrading of the city's energy structure and the optimal use of resources (Wang et al., 2019; Yu, Jiao & Sampat, 2022).

China is building a green, low-carbon and sustainable smart city system, mainly in three aspects: a) relying on the "low carbon emission intensity and high emission reduction capacity" of information technology itself, through paperless and sharing economy and other new ways to promote production and lifestyle from "high energy consumption, high material consumption, high pollution and high emissions" to "green, low-carbon and efficient". b) the use of information technology to empower traditional industries and promote energy saving and emission reductions through smart grids, smart buildings, smart logistics and other means (Li et al., 2019; Yu, Liu & Sampat, 2023; Yu, Chang & Mandizvidza, 2023). c) make use of smart terminals in smart cities to monitor and analyse urban environmental data in real time through big data, so as to fight the battle against urban pollution (Sodhro, 2019).

5. 5G & Smart City

5G is short for the 5th generation of mobile communication technology. 5G communication technology is definitely not just about the faster internet speeds that people feel, but also extends to numerous smart devices such as smart homes, smart healthcare, smart transportation, etc. (Rao & Prasad, 2018). The 5G core network minimises the investment in construction costs compared to 4G by building on top of the 4G core network, i.e. on the premise of non-independent networking. Nevertheless, due to the more extensive placement of 5G base stations and their larger power consumption, the operational costs are also several times higher than before. Therefore, it is imperative to bring about higher network resource utilisation by enhancing the user experience and providing a larger number of machine connections (Skouby & Lynggaard, 2014). The network resource utilisation is further optimised in the building configuration.

In the process of smart city construction, IoT is the detector of various intelligent industries, while the 5G network is the "highway" of information transmission. The integration of 5G technology and IoT technology enhances the intelligence and information level of urban management and further promotes the construction of smart cities. It is expected that the 5G will provide $12.3 trillion in goods and services across all industries in 2035 and will support up to 22 million jobs in the 5G value chain globally (Gohar & Nencioni, 2021).

5G, with its advanced features as described in the earlier section, is well positioned to address the demands of the Smart City. The 5G features required for smart cities that enables IoT adoption on a large scale are (Skouby & Lynggaard, 2014):

- Device connectivity serving a huge increase in the number of devices like sensors, cameras, actuators etc. connected to wireless networks. These are required in homes, streets, traffic junctions, public places like bus stands, railway stations, airports etc. This will support smart traffic systems, smart homes, public safety, security and surveillance requirements of smart cities (Minoli & Occhiogrosso, 2019).
- Very large bandwidth is required to support both uplink and download of video rich services over wireless networks and to support high data volumes (Lynggaard et al., 2015). This is supported by 5G.
- Ultra-low latency is required for enhanced user experiences potentially including the delivery of 3D images and holograms and for applications such as driverless cars (Skouby & Lynggaard, 2015). The latency of less than 1ms provided by 5G supports this requirement.
- The "always on" connectivity provided by 5G supports services in high mobility environments such as cars and high-speed trains as well as meets ultra-high reliability requirements for driverless cars and traffic monitoring (Skouby & Lynggaard, 2015). 100 per cent geographical coverage is also required to support an intelligent traffic monitoring and management system and driverless cars. Many healthcare services also require "always on" function, as an outage could have life threatening consequences.
- Energy Efficiency The massive increase in connected devices making up a fully formed IoT grid is likely to require better energy efficiencies than currently possible, with some mobile broadband devices required to be on all the time while others will turn on intermittently (Skouby & Lynggaard, 2014).

The highly scalable and contextual proposed nature of 5G networks could support the diversity of IoT and other applications required for smart cities, with differing requirements for pricing, mobility, latency, network reliability and resilience.

CITESPACE ANALYSIS

CiteSpace is a Java-based application (Chen, 2004) that can analyze connections between authors, institutions, countries, keywords, journals, or references in the scientifific literature. There are many software tools for visualizing knowledge domains (Cobo et al., 2011), but CiteSpace is one of the more balanced and powerful packages. One of the important tools in the CiteSpace package helps identify betweenness centrality between pivotal points in the scientifific literature. Basically, it functions as a measurement to indicate the importance of nodes in a network. It is calculated by identifying the number of shortest paths between all nodes in the network that pass through our node of interest. This measure is used widely in network science for many different applications (Grubesic et al., 2008).

The analysis of keyword co-occurrence is an effective way to show emerging trends and track topics of research over time because keywords provide a concise and precise high-level summarization of a document. The changes in research topics have become an important research issue, which can help researchers to gain deeper insights into the development of a particular research field. Figures 1-5 in this section were generated by CiteSpace, using title terms and a log-likelihood ratio (LLR) weighting algorithm to label the clusters. LLR is an algorithm to calculate and determine each type of labels, which presents core concept of each cluster with professional words.

Figure 1 shows the keyword collaboration network which consisted of 6961 core journals in WoS and their internal links between 2000 and 2022 and Table 1 the corresponding list of keywords based on their frequency.

The author frames the scope of the journals in WoS related to smart cities and IoT. Among the 6192 documents that were screened for the two core terms, we can find that the top ten valid keywords in

Figure 1. A visualization of the keyword collaboration network of core journals in WoS related to smart city and IoT
Source: WoS, 2000-2022.

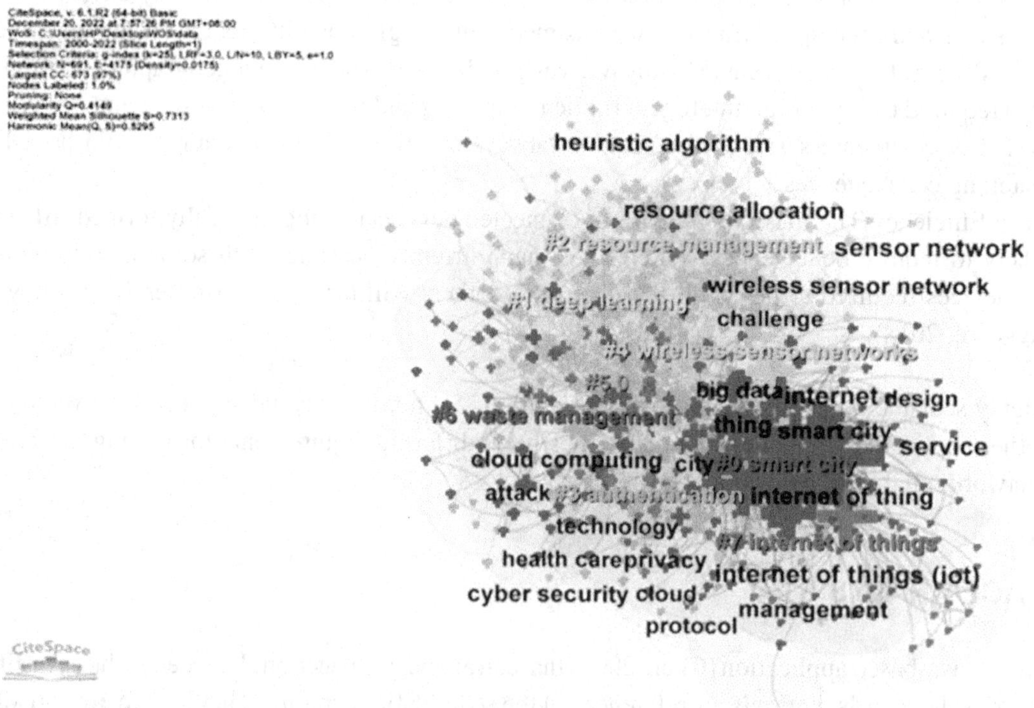

Table 1. Top 25 keywords based on frequency

Count	Year	Keywords
1076	2014	Smart city
977	2013	Internet of thing
321	2016	System
232	2014	Big data
231	2014	City
228	2015	Challenge
224	2013	Wireless sensor network
219	2014	Network
217	2013	Management
196	2017	Security
181	2014	Cloud computing
179	2016	Framework
166	2013	Model
151	2014	Architecture
149	2016	Edge computing
148	2018	Machine learning
139	2015	Algorithm
137	2015	Design
124	2018	Deep learning
116	2016	Technology
94	2018	AI
81	2018	Resource allocation
72	2015	Energy efficiency
48	2020	Blockchain
28	2020	5G mobile communication

Source: WoS, 2000-2022.

terms of frequency of occurrence are deep learning, big data, management, wireless sensor networks, authentication, network, management, security, cloud computing and machine learning.

Figure 2 shows the keyword collaboration network which consisted of 1152 core journals in WoS and their internal links between 2000 and 2022 and Table 2 the corresponding list of keywords based on their frequency.

The author frames the scope of the journals in WoS related to smart cities and blockchain. Among the 1152 documents that were screened for the two core terms, we can find that the top ten valid keywords in terms of frequency of occurrence are smart contract, system, security, management, architecture, framework, technology, privacy, cloud computing and big data.

Figure 3 shows the keyword collaboration network which consisted of 3389 core journals in WoS and their internal links between 2000 and 2022 and Table 3 the corresponding list of keywords based on their frequency. The author frames the scope of the journals in WoS related to smart cities and sustain-

Figure 2. A visualization of the keyword collaboration network of core journals in WoS related to smart city and blockchain
Source: WoS, 2000-2022.

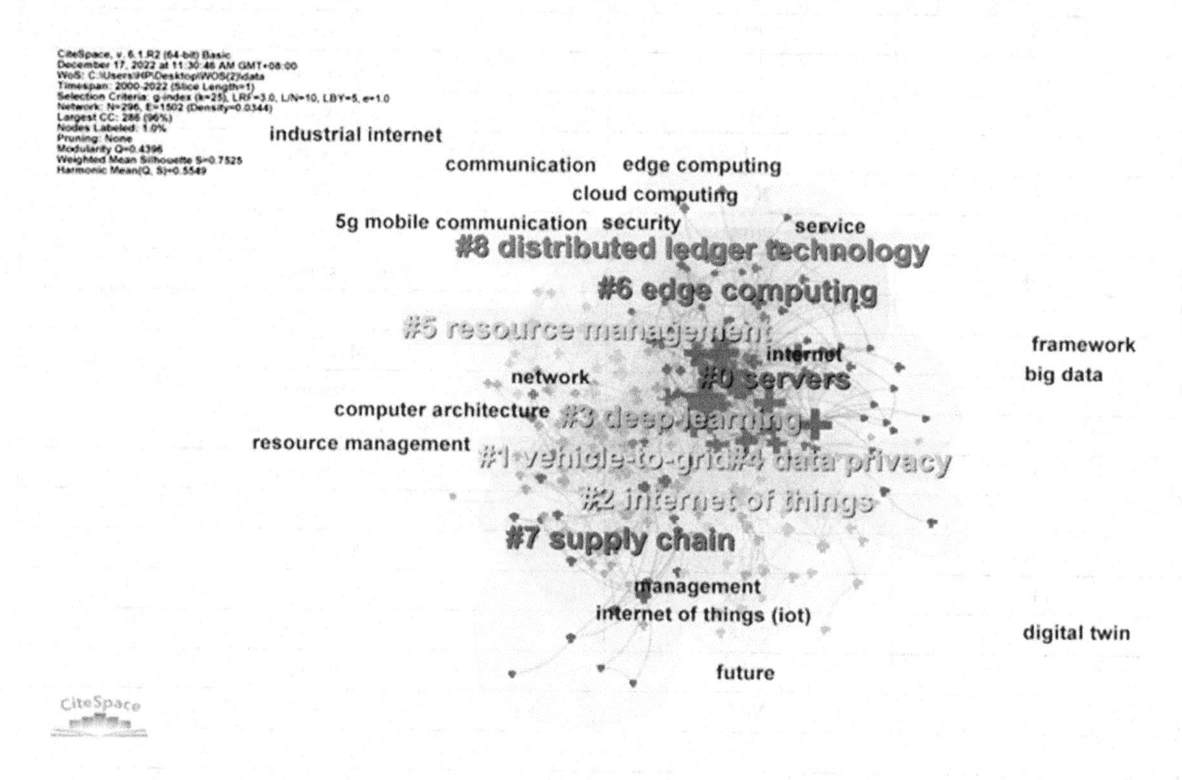

ability. Among the 3389 documents that were screened for the two core terms, we can find that the top ten valid keywords in terms of frequency of occurrence are system, management, impact, framework, technology, big data, policy, governance, innovation and China.

Figure 4 shows the keyword collaboration network which consisted of 6050 core journals in *CNKI* and their internal links between 2000 and 2022 and Table 4 the corresponding list of keywords based on their frequency. The author frames the scope of the journals in *CNKI* related to smart cities and IoT. Among the 6050 documents that were screened for the two core terms, we can find that the top ten valid keywords in terms of frequency of occurrence are cloud computing, big data, intelligent community, AI, smart transportation, digitalization, e-government, information technology, edge computing and digital twin.

As there are only 182 core CNKI documents related to smart cities and blockchain, which does not constitute the basic requirements for citespace cluster mapping analysis, so they are not included in the literature comparison.

Figure 5 shows the keyword collaboration network which consisted of 863 core journals in CNKI and their internal links between 2000 and 2022 and Table 5 the corresponding list of keywords based on their frequency. The author frames the scope of the journals in *CNKI* related to smart cities and sustainability. Among the 863 documents that were screened for the two core terms, we can find that the top ten valid keywords in terms of frequency of occurrence are informatization, big data, cloud computing, urban governance, smart transportation, development, evaluation system, digital twin, public services and policy.

Table 2. Top 25 keywords based on frequency

Count	Year	Keywords
187	2018	Smart city
168	2018	Internet
111	2018	Smart contract
105	2019	Internet of thing
86	2018	Blockchain
73	2019	Challenge
72	2018	System
61	2018	Security
60	2019	Management
55	2018	Architecture
54	2019	Framework
49	2020	Technology
44	2019	Privacy
39	2018	Cloud computing
37	2018	Big data
35	2020	Scheme
30	2019	Network
25	2019	Machine learning
25	2018	Edge computing
25	2018	Design
24	2019	AI
22	2018	Access control
21	2020	Efficient
20	2018	Service
14	2020	5G mobile communication

Source: WoS, 2000-2022.

Comparing the three groups from WoS and the two groups from CNKI, we can conclude that the intersection of core keywords in first place with the highest frequency in core journals related to smart city & IoT, smart city & blockchain, smart city & sustainability are the keywords big data, 5G, clouding computing, digitalization, management, policy, system and governance. The secondary intersection core keywords are smart transportation, sustainable development, digital twin, public services, technology and China. Based on these keywords the authors can summarise the hot areas under the research topics of smart cities, blockchain, IoT and sustainability, and identify the research frontiers and trends, so that the authors can better develop the key research directions and targeted contents below.

In the next sections, we will use this basis and answer the following questions:

1. What is the roles of govenment in promoting 5G to develop smart cities?

Figure 3. A visualization of the keyword collaboration network of core journals used in WoS related to smart city and sustainability
Source: WoS, 2000-2022.

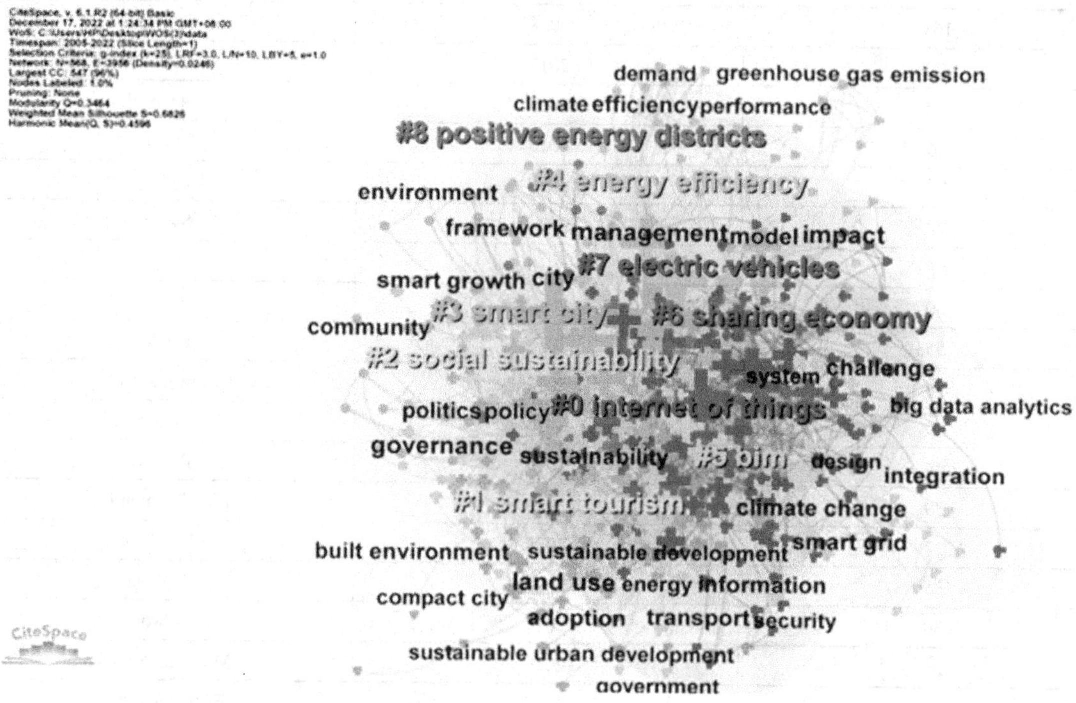

2. What is the added value of integrating IoT, blockchain from stakeholders (including regulators, management, citizen and investors) perspective?
3. A case study discussion in order to analyse the benefits/challenges of using AI/IoT to develop smart cities. The Chinese case studies will also be discussed in relation to the hot keywords that have been summarised using CiteSpace.

GOVERNMENT ROLES IN SHAPING SMART CITIES IN CHINA

China's urbanization level exceeded 60% in 2019, which showed that China had fully entered the middle and late stages of urbanization. At the same time, it also exposed the imbalance, incoordination, and unsustainability of China's urbanization. According to international experience, the "mid and late urbanization" is an important period for resolving the accumulated contradictions in the early urbanization period (Wang et al., 2020). In addition, with the rapid development of information technology, informatization has impacted traditional urban production and lifestyles, and put forward new requirements for the optimization of urban governance. Smart cities, as the most cutting-edge exploration of urban development, will become the mainstay of China's new urbanization in the future (Huang et al., 2021).

Development of smart city construction has been developed in China for more than ten years in three stages and has experienced a budding period, a period of exploration and development, and a new type

Table 3. Top 25 keywords based on frequency

Count	Year	Keywords
703	2013	Smart city
196	2015	System
187	2008	Management
181	2008	Sustainability
150	2010	Impact
136	2009	Framework
131	2011	Internet
118	2014	Technology
117	2015	Big data
117	2008	Challenge
99	2013	Design
98	2009	Policy
95	2008	Performance
91	2016	Internet of thing
90	2010	Governance
89	2008	Sustainable development
89	2014	Innovation
75	2015	Sustainable city
73	2017	Future
69	2012	Climate change
62	2013	Urban sustainability
50	2012	Smart grid
43	2013	Energy efficiency
40	2018	China
40	2014	AI

Source: WoS, 2000-2022

of wisdom. Before 2010, the construction and promotion of smart cities in China lacked top-level design and unified planning, and was mainly driven by technology. During this period, the focus on smart city development was on digital construction. A large part of urban information infrastructures such as networks and data have been established for the normal operation of the city. On this basis, remote sensing technology, geographic information technology and global positioning systems were used for collecting and monitoring information (Bartholomew, 2020). Since 2010, China has successively issued relevant development plans for smart cities, and the construction of smart cities has entered a period of exploration and development, and pilot cities have been announced batch by batch for construction and testing. During this period, the policies have mainly been focusing on planning schemes, guidance opinions, and project management methods. In the preliminary exploration stage, a series of problems such as information islands and repeated construction were gradually exposed (Xu & Jia, 2021). In 2016, the state clearly pointed out that in order to build a new type of smart city, the focus of smart city development

Figure 4. A visualization of the keyword collaboration network of core journals used in CNKI related to smart city and IoT
Source: CNKI, 2000-2022.

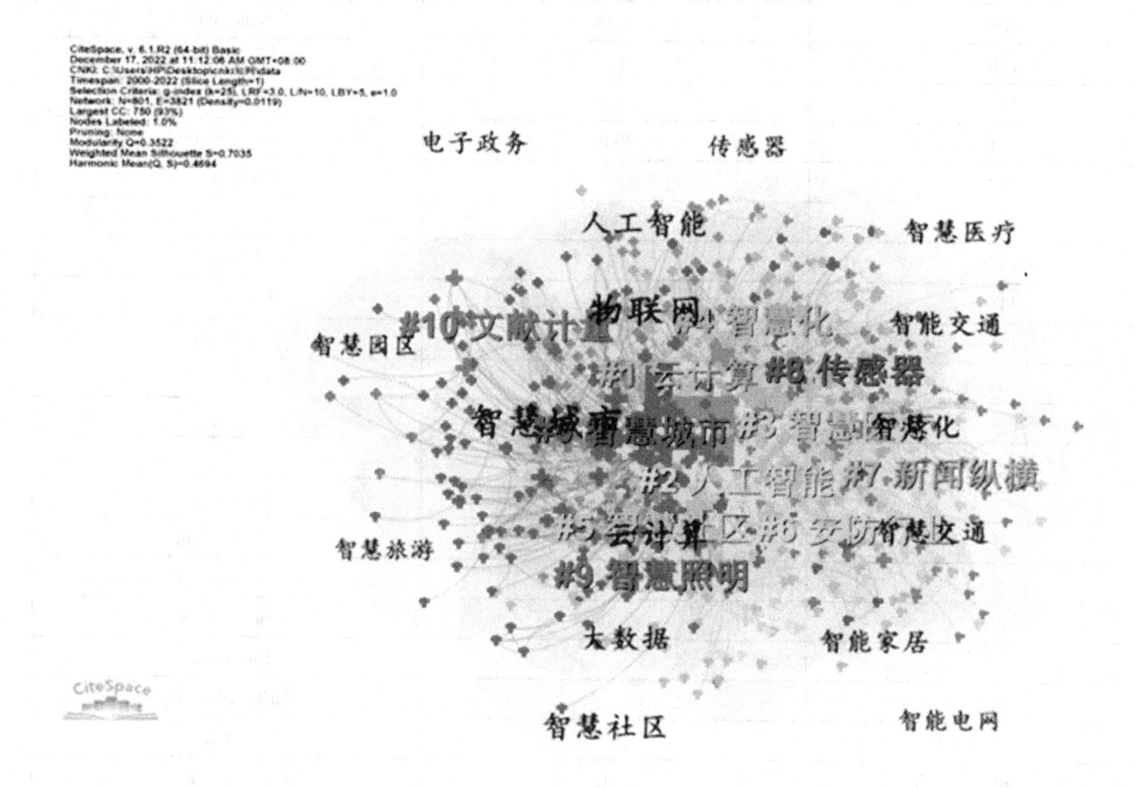

will shift from concept popularization to implementation in the "13th Five-Year Plan". The national level has successively released a series of smart city-related evaluation models and national standards for the field of smart cities. The standard system gradually formed. The application of 5G technology will bring further earth-shaking changes to the development of smart cities.

Table 6 clearly shows that the Chinese government is vigorously pursuing the use of 5G technology in smart cities (Xu & Jia, 2021). The most significant results achieved so far are:

1. The top-level design of industrial policies has been gradually improved. The "2021 Government Work Report" proposed to increase the construction of 5G networks and gigabit optical networks and enrich application scenarios, and "the 14th Five-Year Plan for National Economic and Social Development of the People's Republic of China and the Outline of Vision 2035" proposed to build 5G-based application scenarios and industrial ecology. The Ministry of Industry and Information Technology and nine other departments jointly issued the "5G Application 'Sailing' Action Plan (2021-2023)", a programmatic document to promote the application of 5G, proposing to focus on promoting the integration of 5G in 15 industries in the three major areas of information consumption, real economy and livelihood services (Chen et al., 2020).
2. Cross-sectoral collaboration to promote key sectors. Several Chinese Ministries and Commissions have jointly issued implementation plans for industry applications, organized pilot demonstration projects, and encouraged localities to build pilot zones and demonstration bases to guide the devel-

Table 4. Top 25 keywords based on frequency

Count	Year	Keywords
2281	2009	智慧城市 (Smart city)
1668	2000	物联网 (IOT)
444	2010	云计算 (Cloud computing)
267	2013	大数据 (Big data)
186	2012	智慧社区 (Smart community)
163	2012	人工智能 (AI)
82	2012	智慧交通 (Smart transportation)
70	2010	数字城市 (Digital city)
70	2011	信息化 (Informatization)
68	2012	智能家居 (Smart home)
65	2014	智慧园区 (Smart park)
65	2012	智慧旅游 (Smart tourism)
54	2017	智慧消防 (Smart fire)
51	2010	传感器 (Sensors)
51	2012	云平台 (Cloud platform)
51	2011	安防行业 (Security industry)
49	2010	智慧医疗 (Smart healthcare)
48	2012	电子政务 (e-government)
46	2012	信息技术 (Information technology)
44	2000	边缘计算 (Edge computing)
40	2019	数字孪生 (Digital twin)
37	2019	顶层设计 (Top-Level design)
34	2010	信息安全 (Information security)
34	2017	城市治理 (Urban governance)
28	2018	区块链 (Blockchain)

Source: CNKI, 2000-2022.

opment of 5G convergence applications in key industry sectors and enrich 5G application scenarios. For example, it has organized pilot projects on "5G+Healthcare" and "5G+Wisdom Education", released the "Implementation Plan for 5G Applications in the Energy Sector", and promoted the construction of pilot zones for 5G+industrial Internet convergence applications and Telematics pilot zones in many places.

3. Local governments promote the implementation of specialised applications according to local conditions. Each province (municipality or region) has introduced local industrial policies in combination with industrial advantages, and has focused on the development of 5G specialised application landing promotion through special policies, investment funds and demonstration projects in order to achieve large-scale application.

Figure 5. A visualization of the keyword collaboration network of core journals used in CNKI related to smart city and sustainability
Source: CNKI, 2000-2022.

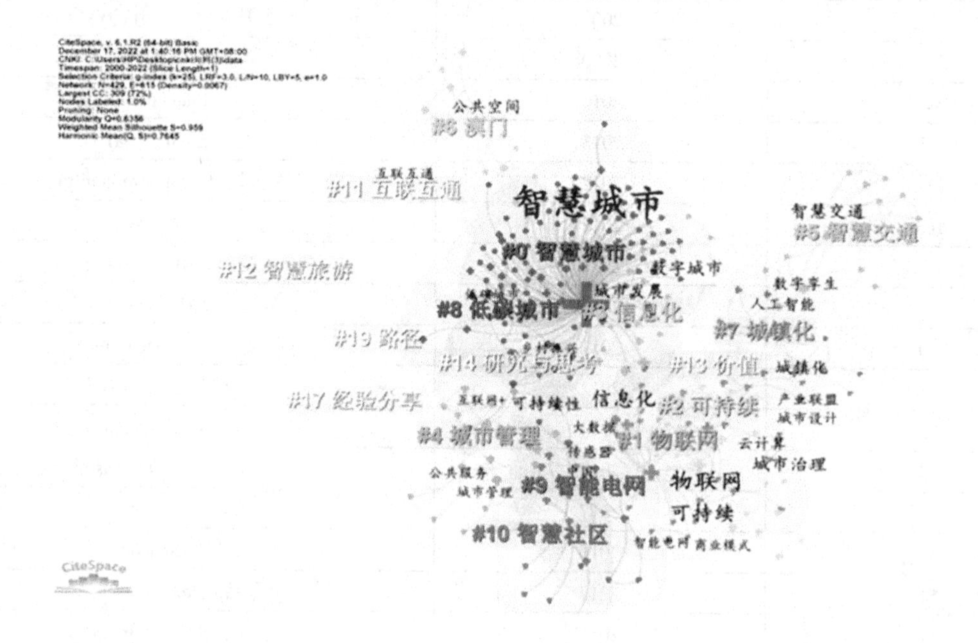

Thanks to the active industrial policies of the central and local governments and the concerted efforts of all parties in the industry (Yu et al., 2023), China's 5G has been commercially available for two years at the time of writing and its applications are covering many areas in primary, secondary and tertiary industries, with some key areas having entered the commercial implementation stage. For example, in the industrial sector, 5G+ quality inspection, remote operation and maintenance, multi-machine collaborative operation and other typical applications have been formed around R&D and design, manufacturing, operation and management, and product services. In the medical industry, 5G+ emergency medical services, remote consultation, remote diagnosis and health management have been applied, effectively improving the level of medical services and management efficiency. In the media and cultural tourism industries, 5G applications have enhanced the visitor experience and improved the intelligent management and service level of scenic spots and venues (Xu & Jia, 2021).

THE ADDED VALUE OF INTEGRATING IoT, BLOCKCHAIN, 5G, BIG DATA, DIGITAL TWIN, AI FROM STAKEHOLDERS (INCLUDING REGULATORS, MANAGEMENT, CITIZEN, AND INVESTORS) PERSPECTIVES

China has officially proposed the vision of "carbon peaking" by 2030 and "carbon neutrality" by 2060 in 2020, and "green" and "low carbon" are also the direction of exploration for the construction and development of new smart cities in China (Wang, 2021). As can be seen in Figure 6, the specific path of digital technology enabling "double carbon" can be divided into four major processes: data mapping, scenario prediction, clear path, and implementation adjustment. Among them, 5G and IoT technologies

Table 5. Top 25 keywords based on frequency

Count	Year	Keywords
414	2011	智慧城市 (smart city)
36	2012	物联网 (IoT)
16	2011	信息化 (Informatization)
14	2015	大数据 (Big data)
14	2012	云计算 (cloud computing)
11	2017	城市治理 (Urban governance)
10	2015	智慧交通 (Smart transportation)
9	2013	顶层设计 (Top-level design)
8	2013	可持续发展 (Sustainable development)
7	2019	评价体系 (Evaluation system)
7	2020	数字孪生 (Digital twins)
6	2012	公共服务 (Public services)
6	2014	对策 (Countermeasures)
6	2014	智慧旅游 (Smart tourism)
6	2015	低碳城市 (Low carbon cities)
5	2012	城镇化 (Urbanisation)
5	2015	中国 (China)
5	2018	人工智能 (AI)
5	2016	智慧建筑 (Smart buildings)
5	2012	创新驱动 (Innovation driven)
5	2012	信息技术 (Information technology)
5	2013	以人为本 (People-oriented)
4	2013	电子政务 (E-government)
4	2014	智慧社区 (Smart communities)
3	2021	碳中和 (Carbon neutral)

Source: CNKI, 2000-2022.

play the role of data collection and transmission in the monitoring, counting, accounting and verification of carbon emission data, but also provide the communication basis and sensory control capability in the specific implementation process. In addition, Integrating IoT with blockchain brings numerous advantages (Atlam et al., 2020). For example, employing decentralized and distributed attributes of blockchain technology can manage issues of security and a single point of failure associated with the centralized IoT architecture, as there is no need for a central server to control IoT devices and their communications with each other (Allam, 2022). Furthermore, blockchain delivers better security and privacy, since blockchain utilizes sophisticated cryptography algorithms, hash functions and timestamps, which provide a secure computing environment. Besides, the blockchain provides tamper-proof and immutable ledger to safeguard data against harmful attacks such that any data change cannot be stored in the ledger, except if the majority of contributing users validate it (Karafiloski & Mishev, 2017). This, in turn,

Table 6. Important policies related to 5G adopted for smart cities in China since 2016

Issue Date	Policy/Conference	Authority	Core Contents
2018.3	The Government Work Report	/	Proposed to accelerate the construction of a manufacturing power and promote the development of the fifth generation mobile communication and other industries
	Key Points of National Radio Management in 2018	Ministry of Industry and Information Technology	1.It is proposed to speed up the frequency planning of 5G systems and formulate radio frequency technical indicators for mid-band radio equipment 2.The frequency planning scheme of partial millimeter wave frequency band is proposed
2018.4	5G development prospect and policy orientation	Ministry of Industry and Information Technology	It mentioned that 5G in China will be initially ready for commercial use in the second half of 2019
2018.5	Opinions on the Implementation of the 2018 Special Action to Increase Internet Speed and Reduce Internet Fees and Accelerate the Cultivation of New Drivers of Economic Development	Ministry of Industry and Information Technology, State-owned Assets Supervision and Administration Commission	1. Make solid progress in 5G standardization, research and development, application, maturity of the industrial chain, and supporting security 2. Organize the implementation of the major project of "a new generation of broadband wireless mobile communication network", complete the third phase of technology research and development trials, and promote the formation of a unified global 5G standard 3. Organize 5G application competition to promote the integrated development of 5G and vertical industries, and prepare for large-scale 5G networking and application
2018.7	Three-Year Action Plan for Expanding and Upgrading Information Consumption (2018-2020)	Ministry of Industry and Information Technology, Development and Reform Commission	1. Organise and implemente a new generation of information infrastructure construction projects 2. Propose to accelerate 5G standards research, technology trials, promote 5G scale network construction and application demonstration projects, and ensure the launch of 5G commercialization
2018.10	Implementation Plan for Improving the Institutional Mechanism for Promoting Consumption (2018 - 2020)	State Council (PRC)	Accelerate the commercialization of 5G technology and cultivate a number of strong digital innovation enterprises
2019.5	Notice on the Special Action to Deeply Promote Broadband Network Speed Up and Reduce Charges to Support High Quality Economic Development 2019	Ministry of Industry and Information Technology, State-owned Assets Supervision and Administration Commission	1. Guide localities to do a good job in planning 5G base station sites and other work, a optimize the 5G development environment 2. Continue to promote the development and industrialisation of 5G technology, into the further maturation of the system, chip, terminal and other industrial chains
2019.6	Implementation Plan for Promoting the Renewal and Upgrading of Key Consumer Products to Smooth the Recycling of Resources (2019-2020)	Development and Reform Commission	Accelerate the commercial application of 5G mobile phones
2019.11	"5G+Industrial Internet" 512 Project Promotion Programme	Ministry of Industry and Information Technology	1. Enhance the industrial capacity, innovative application capacity and resource supply capacity of "5G+Industrial Internet" network key technologies 2. Strengthen the promotion of guidance and experience
2019.12	Outline of the Yangtze River Delta Regional Oneness Development Plan	State Council (PRC)	80% 5G network coverage and basic infrastructure interconnection by 2025

continued on following page

Table 6. Continued

Issue Date	Policy/Conference	Authority	Core Contents
2020.3	Notice from the Ministry of Industry and Information Technology on Promoting the Accelerated Development of 5G	Ministry of Industry and Information Technology	1. Accelerate the construction of 5G networks and support increased base station site resources 2. Strengthen power and frequency security, promote network sharing and cross-network roaming
	Implementation Opinions on Promoting the Expansion and Quality of Consumption to Accelerate the Formation of a Strong Domestic Market	National Development and Reform Commission and twenty-three other departments	Accelerate the construction and commercialisation of information infrastructure such as 5G networks
	Standing Committee of the Political Bureau of the CPC Central Committee	/	Accelerate the construction of new infrastructure such as 5G networks and data centres
2020.11	Notice on the Organisation of Pilot Projects for 5G+ Healthcare Applications	Ministry of Industry and Information Technology	1. Focus on enriching the application scenarios of 5G technology in the healthcare industry, recruiting and selecting a group of key units to collaborate on research and development 2. Focus on forming a number of 5G+ healthcare landmark applications with advanced technology, superior performance and obvious results to set a benchmark and direction for the innovative development of 5G+ healthcare 3. Cultivate the main force in the development of 5G smart healthcare innovation in China
2021.2	Notice from the Ministry of Industry and Information Technology on Improving the Quality of 5G Services	Ministry of Industry and Information Technology	1. Develop and improve the 5G service standards of this enterprise and increase the service assessment of front-line windows such as physical business halls and customer service hotlines 2. Make 5G service quality an important element of front-line window performance assessment
2021.3	Notice of 5G High-Tech Video Series Standard System (Version 2021)	National Radio and Television Administration	1. Organise the development of four standards system documents for interactive video, immersive video, VR video and cloud gaming 2. Promote high quality and innovative development of the radio, television and online audiovisual industries
2021.7	5G Application "Sailing" Action Plan (2021-2023)	Ministry of Industry and Information Technology and other ten departments	1. Propose general objectives for the development of 5G applications in China 2. Achievie a double breakthrough in the depth and breadth of 5G applications in key areas
2022.3	2022 Government Work Report	State Council (PRC)	1. Strengthen the overall layout of digital China construction 2. Build digital information infrastructure and gradually construct a national integrated big data centre system 3. Promote the large-scale application of 5G to facilitate the digital transformation of industry 4. Develop smart cities and digital villages

Source: www.gov.cn; Yu et al. (2022).

delivers a trusted system where the participating IoT devices are the only objects to accept or discard a transaction based on their consent (Reyna et al., 2018). Artificial intelligence and blockchain technology support the trend analysis of energy saving and emission reduction actions and decision optimisation for the implementation of the dual carbon strategy. The digital twin, big data and cloud computing are inextricably linked in enabling the implementation of the dual carbon strategy (Hang & Kim, 2019).

Figure 6. How digital technology can help develop sustainable smart cities in China
Source: Beijing Software and Information Service Industry Association.

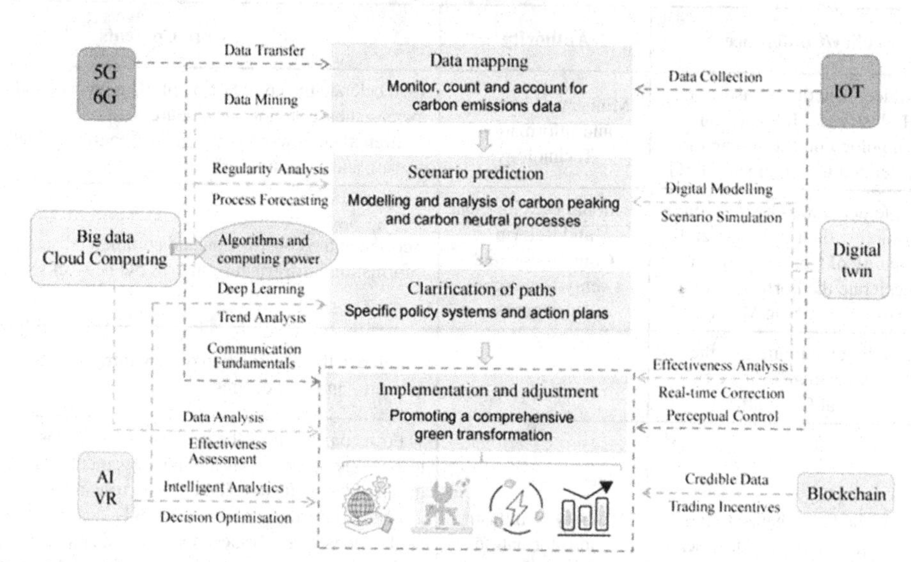

These three technological capabilities play an important role in the four stages of data mapping, scenario prediction and implementation adjustment, and are the most deeply involved of all new technologies in achieving China's dual carbon goals.

There is no doubt that digital technology is advancing the transformation of the economy and society to a low-carbon and green direction (Yu et al., 2022). The combination of digital technology and artificial intelligence with traditional industries has produced new industrial forms such as smart energy, smart agriculture, smart manufacturing, smart transportation, smart healthcare and smart cities that create new value while effectively integrating various factor resources, realising precise input of factors, precise control of production processes, efficient matching of supply and demand, and playing a key role in energy conservation, efficiency and recycling (Lee et al., 2021; Yu, Liu, Mahendran & Lu, 2022; Yu, Chen, Sampat, & Misuko, 2022; Yu, Yu, & Sampat, 2022).

CASE STUDIES

After more than a decade of exploration and practice, China has explored a road of smart city construction in line with Chinese characteristics.In the new era, the demand for urban development is constantly changing. The construction of smart cities in China are highlighting several key points and strengthen smart foundations. At the same time, the construction of smart cities should fully take into account the basic and supporting general infrastructure of the city, such as 5G, IoT, blockchain, industrialization degree, smart devices, intelligent manufacturing, digitalization, and business integration. Therefore, the construction of smart city needs to be based on the local conditions and construction characteristics of each province and city, highlighting their own advantages.

After the introduction of the new smart city strategy, the pilot and construction of smart cities in China have shown the development trends of graded construction, multifocal development, and improved

Figure 7. The distribution of smart pilot cities in China: Characteristics of the three selected cases
Source: China Ministry of Housing and Construction, China Internet Information Office, and official government websites across China.

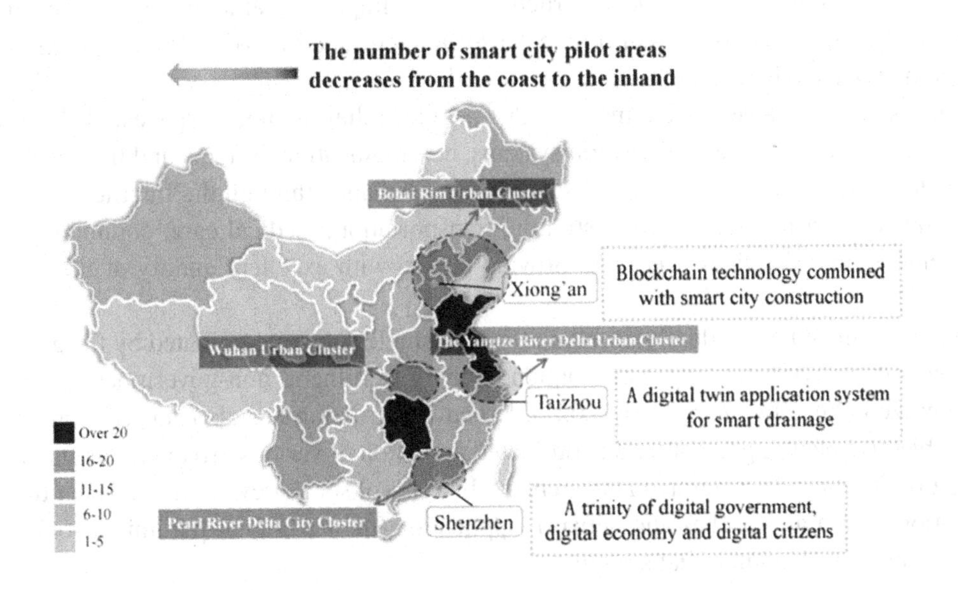

quality and efficiency. Among the cities and regions selected as national smart city pilots, most of them are located along the Bohai Sea Rim and the Yangtze River Delta city cluster (Ifeng, 2021). As we can see, under the vigorous promotion of Chinese government policies, the construction of smart cities has already started in first-tier cities and developed second-tier cities. However, the level of development and the degree of informatization also varies widely from city to city. For developed cities, it will focus on the development of livelihood-related smart city construction to improve urban innovation and competitiveness with smart cities; while for medium-sized cities, it will focus more on the combination of smart cities and local tourism, ports and other resources (Tencent, 2021). In the following, we provide three case studies of Chinese smart cities as reference and application of the analysis above.

Xiong'an New Area: Blockchain Technology Combined With Smart City Construction

Xiong'an New Area focuses on exploring the application of the underlying blockchain technology in government integration, data sharing and cross-chain interaction. Fintech companies give full play to the core technology advantages of banks in big data, blockchain, digital twin technology and other fields, and lead the rapid development of new fintech industries in the Xiong'an New Area.The application of innovative technology ensures the overall management and efficient operation of all kinds of funds in the new district, which not only improves government supervision, but also realizes data sharing and process transparency, and finally helps the implementation of the construction concept of "digital Xiong'an" and "intelligent Xiong'an" concepts in the new district.

According to Oldenburg (1999) and Norberg-Schulz (2019), "People's activity locations are divided into three categories . . . including family residence, study workplace and other public places". The construction of "smart city" in Xiongan New Area does not blindly emphasize the application of high-end technology, but focuses on: 1) meeting people's various needs with the assistance of technology, 2) solving practical problems in urban construction, and 3) improving and making people's lives in the city easier. Smart cities is not only a concept, but also an important form of dynamic operations of urban space based on concrete information.

In the new smart city development model, "City people" relies on mobile space and place and on the technical infrastructure such as information systems, communication systems and transportation lines. The goal is to express and realize citizens' interests and demands through the Internet network as well as to implement smart municipal administration, transportation, medical care, community, education and other public smart services needed to provide citizen with excellent quality of life (Yu, Zhao & Sampat, 2023).

Urban management has gradually developed from a single subject dominated by the government to a diversified subject with the joint participation of the government, non-governmental organizations and urban citizen (Yu, Zhao & Sampat, 2023). Urban construction has also shifted from the original artificial subjective planning to realize a coordinated operation of various urban system functions based on information sharing and integrated management. In terms of social development, under the influence of information-driven approaches and network application, the traditional pyramidal social hierarchy structure begins to evolve into a flat structure.

Taizhou: Constructing A Digital Twin Application System for Smart Drainage

Since April 2021, Taizhou City has relied on advanced technologies such as digital twin, 5G and augmented reality (AR) to show the operational situation of above-ground and underground drainage facilities in three dimensions and in real time, forming a set of infrastructure management library for the safe operation of urban drainage facilities (Zj.gov.cn, 2022).

In the preliminary research and investigation stage, more than ten experts set up a research and mapping group to manage major problems in depth: difficulty in regulating side sewage, permanent sewage treatment overload as well as sewage flow obstructions during rainy days. So far the existing 15 scattered drainage management systems, 32 rainwater and sewage pumping stations and more than 90 river gate wells and gates along the river have been examined (Zj.gov.cn, 2022). Using online hydraulic models, digital twins, building information model, AR and other technologies, the drainage system in Taizhou has successfully changed from a fuzzy analysis based on empirical judgments to a refined approach of quantitative analysis and predictive forecasting. Through digital twin application system, risks of overloading and sewage overflow are monitored and visible in advance. The system simulates the distribution and extent of the risk of impending flooding and implements a flood risk alert mechanism (Zj.gov.cn, 2022).

The practical effectiveness of the digital twin application system can be seen in following improvements:

- 20% reduction in sludge disposal costs
- 15% reduction in pipe network operation and maintenance expenditure
- Information coverage of sewage system facilities increased from less than 30% to over 95%
- Closed-loop disposal of wastewater operational risk events: efficiency doubled

Besides, the government departments not only provide 30 million project fund guarantee, but combine local enterprises and talents from all over the country to set up a technical research and development team, which realizes the sustainable development of drainage industry in Taizhou (Zj.gov.cn, 2022).

Digital Shenzhen: Creating a Trinity of Digital Government, Digital Economy, and Digital Citizens

1. Intelligent Transportation System to Improve Road Congestion

With the city's road mileage of only 600 kilometers, Shenzhen has a population of more than 20 million people, 3.52 million vehicles, and the average density of 530 vehicles per kilometer, ranking first in the country. Shenzhen has effectively improved congestion on some key roads in Shenzhen with the help of Huawei Cloud, while improving the enforcement efficiency of Shenzhen traffic police (Sz.gov.cn, 2022).

For example, in the control of signal lights, the data from all cameras will be connected and based on Huawei CloudStack, through AI technology, big data and edge computing, the overall regulation and intelligent real-time management of signal lights. Through the cooperation with Huawei Cloud, Shenzhen traffic police will launch a red light timing solution based on AI, big data and other technologies, and deploy it to about 200 intersections in the city.

Shenzhen's intelligent transportation has promoted the city's transformation from an empirical judgement-based to a data analysis-based traffic management, further liberating the traffic police while facilitating daily life for the public and becoming a new label for intelligent Shenzhen.

2. Government and Business Are Working Together to Build and Share the City of the Future

Data chimneys and information silos are another problem for many cities in the construction of smart cities. In order to promote data availability and sharing, Shenzhen has held the Shenzhen Open Data Application Innovation Competition every year since 2019, focusing on the economic and social development of the Greater Bay Area, using government data availability for traction, promoting the integration of government data and social data, and promoting the realization of real scenario applications (Sz.gov.cn, 2022).

By building a smart airport, Shenzhen has fully implemented facial recognition boarding and intelligent allocation of aircraft slots. Due to intelligent scheduling, Shenzhen sees a reduction in numbers of ferry rides by 2.6 million per year.

With the support of intelligent healthcare, Shenzhen has achieved interoperability and mutual recognition of medical testing and imaging results in all hospitals, allowing patients to save an average of about 40 minutes of queuing time.

The smart traffic solution has greatly optimised the waiting time at main road junctions. Previously, drivers watched the lights and read the seconds to pass, but now the lights watch the cars and read the number of cars to let pass, resulting in a 17.7% reduction in the average waiting time at key junctions.

As an international and innovative city, Shenzhen has nurtured many leading technology companies with international competitive advantages, such as Huawei, and these companies have provided the driving system and technical support for Shenzhen's developed digital economy. In the process of promoting the construction of a new type of smart city, the Shenzhen government and enterprises are making concerted efforts to fully take advantage of the concentration of high-tech enterprises and Internet enterprises, and continue to move towards the goal of promoting business and industry, benefiting the people and good governance.

CONCLUSION AND RECOMMENDATIONS

In the following paragraphs we conclude by giving recommendations on how the integration of IoT, 5G, AI, digital twin and blockchain can transform or enhance smart cities and the economy as a whole, and how this change can be related to the issue of sustainability.

The combination of AI and IoT is highly valuable for enabling enterprise carbon management. It can effectively solve the problems of system information islands, low real-time information, low carbon verification, insufficient data analysis, unclear management systems, and lack of intelligent approaches in energy management and carbon emission management. It can also provide more comprehensive, more energy-efficient and more intelligent system services, bringing 3-10% long-term energy-saving benefits, and improve the carbon emission readiness of enterprises. At the same time, based on the reliability of IoT and blockchain, technology can be used to provide enterprises with a long-term way to interact with their consumers while measuring carbon emissions.

In the future, blockchain will also play a great role in solar panels, solar power plants and other fields of the sustainability industry. Solar power plants, wind power plants, and hydroelectric power plants have the common feature of being located in more remote areas. Investors worry about the efficiency of the regional governance structure for their remote projects. Based on digital twin and block chain technology, effective management of clean energy power plant operation can be implemented for such remote areas. Blockchain can provide complete transparency to these renewable power generation assets and make them more credible and reliable, thus lowering the risk threshold for investors and financial institutions.

In the energy sector, 5G can help to build a smarter power grid to better adapt to the development of renewable energies and distributed generations, and achieve energy conservation and efficiency. 5G can enable the next generation of smart grids, with more robust, more responsive and larger capacity grids expected to reduce gas and electricity consumption by 12%, thus effectively responding to the growing energy demand.

In addition, the low-carbon industries need the participation of regulation and financial policy stakeholders, financial markets and government agencies. With this nexus not only the overall environment can be saved but ultimately the life cycle of the "forest" of sustainable smart cities will also become more sustainable.

ACKNOWLEDGMENT

The authors extend sincere gratitude to:

• Our colleagues from Soochow University, The Australian Studies Centre of Shanghai University, The European Business University of Luxembourg and Krirk University as well as the independent research colleagues who provided insight and expertise that greatly assisted the research, although they may not agree with all of the interpretations/conclusions of this chapter.

• China Knowledge and Belt and Road Blockchain Association for supporting our research.

• The Editor and the International Editorial Advisory Board (IEAB) of this book who initially desk reviewed, arranged a rigorous double/triple blind review process and conducted a thorough, minute and critical final review before accepting the chapter for publication.

• All anonymous reviewers who provided very constructive feedbacks for thorough revision, improvement and fine tuning of the chapter.

REFERENCES

Ahmed, E., Yaqoob, I., Gani, A., Imran, M., & Guizani, M. (2016). Internet-Of-Things-Based Smart Environments: State of the Art, Taxonomy, and Open Research Challenges. *IEEE Wireless Communications, 23*(5), 10–16. doi:10.1109/MWC.2016.7721736

Alahakoon, D., Nawaratne, R., Xu, Y., De Silva, D., Sivarajah, U., & Gupta, B. (2020). Self-Building Artificial Intelligence and Machine Learning to Empower Big Data Analytics in Smart Cities. *Information Systems Frontiers*, 1–20. doi:10.100710796-020-10056-x

Allam, Z., Bibri, S. E., Jones, D. S., Chabaud, D., & Moreno, C. (2022). Unpacking the '15-Minute City' via 6G, IoT, and Digital Twins: Towards a New Narrative for Increasing Urban Efficiency, Resilience, and Sustainability. *Sensors (Basel), 22*(4), 1369. doi:10.339022041369 PMID:35214271

Ameer, S., & Shah, M. A. (2018, August). Exploiting Big Data Analytics for Smart Urban Planning. In *2018 IEEE 88th Vehicular Technology Conference (VTC-Fall)* (pp. 1-5). IEEE. 10.1109/VTCFall.2018.8691036

Atitallah, S. B., Driss, M., Boulila, W., & Ghézala, H. B. (2020). Leveraging Deep Learning and IoT Big Data Analytics to Support the Smart Cities Development: Review and Future Directions. *Computer Science Review, 38*, 100303. doi:10.1016/j.cosrev.2020.100303

Atlam, H. F., Azad, M. A., Alzahrani, A. G., & Wills, G. (2020). A Review of Blockchain in Internet Of Things and AI. *Big Data and Cognitive Computing, 4*(4), 28. doi:10.3390/bdcc4040028

Bartholomew, C. (2020). China and 5G. *Issues in Science and Technology, 36*(2), 50–57. https://www.jstor.org/stable/26949108

Bhabad, M. A., & Bagade, S. T. (2015). Internet Of Things: Architecture, Security Issues and Countermeasures. *International Journal of Computers and Applications, 125*(14).

Bhushan, B., Khamparia, A., Sagayam, K. M., Sharma, S. K., Ahad, M. A., & Debnath, N. C. (2020). Blockchain for Smart Cities: A Review of Architectures, Integration Trends and Future Research Directions. *Sustainable Cities and Society, 61*, 102360. doi:10.1016/j.scs.2020.102360

Biswas, K., & Muthukkumarasamy, V. (2016, December). Securing Smart Cities Using Blockchain Technology. In *2016 IEEE 18th international conference on high performance computing and communications; IEEE 14th international conference on smart city; IEEE 2nd international conference on data science and systems (HPCC/SmartCity/DSS)* (pp. 1392-1393). IEEE. 10.1109/HPCC-SmartCity-DSS.2016.0198

Chen, C. (2004). Searching for Intellectual Turning Points: Progressive Knowledge Domain Visualization. *Proceedings of the National Academy of Sciences of the United States of America, 101*(suppl_1), 5303–5310. doi:10.1073/pnas.0307513100 PMID:14724295

Chen, H., Yuan, L., & Jing, G. (2020, October). 5G Boosting Smart Cities Development. In *2020 2nd International conference on artificial intelligence and advanced manufacture (AIAM)* (pp. 154-157). IEEE. 10.1109/AIAM50918.2020.00038

Chin, J., Callaghan, V., & Lam, I. (2017, June). Understanding and Personalising Smart City Services Using Machine Learning, the Internet-Of-Things and Big Data. In *2017 IEEE 26th international symposium on industrial electronics (ISIE)* (pp. 2050-2055). IEEE. 10.1109/ISIE.2017.8001570

Christidis, K., & Devetsikiotis, M. (2016). Blockchains and Smart Contracts for the Internet of Things. *IEEE Access : Practical Innovations, Open Solutions, 4*, 2292–2303. doi:10.1109/ACCESS.2016.2566339

Cirillo, F., Gómez, D., Diez, L., Maestro, I. E., Gilbert, T. B. J., & Akhavan, R. (2020). Smart City IoT Services Creation Through Large-Scale Collaboration. *IEEE Internet of Things Journal, 7*(6), 5267–5275. doi:10.1109/JIOT.2020.2978770

Cobo, M. J., López-Herrera, A. G., Herrera-Viedma, E., & Herrera, F. (2011). Science Mapping Software Tools: Review, Analysis, and Cooperative Study Among Tools. *Journal of the American Society for Information Science and Technology, 62*(7), 1382–1402. doi:10.1002/asi.21525

Gohar, A., & Nencioni, G. (2021). The Role of 5G Technologies in a Smart City: The Case for Intelligent Transportation System. *Sustainability, 13*(9), 5188. doi:10.3390u13095188

Grubesic, T. H., Matisziw, T. C., Murray, A. T., & Snediker, D. (2008). Comparative Approaches for Assessing Network Vulnerability. *International Regional Science Review, 31*(1), 88–112. doi:10.1177/0160017607308679

Hang, L., & Kim, D. H. (2019). Design and Implementation of an Integrated IoT Blockchain Platform for Sensing Data Integrity. *Sensors (Basel), 19*(10), 2228. doi:10.339019102228 PMID:31091799

Hang, L., & Kim, D. H. (2019). Design and Implementation of an Integrated IoT Blockchain Platform for Sensing Data Integrity. *Sensors (Basel), 19*(10), 2228. doi:10.339019102228 PMID:31091799

Huang, K., Luo, W., Zhang, W., & Li, J. (2021). Characteristics and Problems of Smart City Development in China. *Smart Cities, 4*(4), 1403–1419. doi:10.3390martcities4040074

Ifeng. (2021). *Super Smart City 2.0, Artificial Intelligence Leads the Way to a New Trend.* https://www.ifeng.com/

Karafiloski, E., & Mishev, A. (2017, July). Blockchain Solutions for Big Data Challenges: A Literature Review. In *IEEE EUROCON 2017-17th International Conference on Smart Technologies* (pp. 763-768). IEEE. 10.1109/EUROCON.2017.8011213

Kim, T. H., Ramos, C., & Mohammed, S. (2017). Smart City and IoT. *Future Generation Computer Systems*, *76*, 159–162. doi:10.1016/j.future.2017.03.034

Lee, D., Lee, S. H., Masoud, N., Krishnan, M. S., & Li, V. C. (2021). Integrated Digital Twin and Blockchain Framework to Support Accountable Information Sharing in Construction Projects. *Automation in Construction*, *127*, 103688. doi:10.1016/j.autcon.2021.103688

Lee, I., & Lee, K. (2015). The Internet of Things (IoT): Applications, Investments, and Challenges for Enterprises. *Business Horizons*, *58*(4), 431–440. doi:10.1016/j.bushor.2015.03.008

Li, S. (2018, August). Application of Blockchain Technology in Smart City Infrastructure. In *2018 IEEE International Conference on Smart Internet of Things (SmartIoT)* (pp. 276-2766). IEEE. 10.1109/SmartIoT.2018.00056

Li, X., Fong, P. S., Dai, S., & Li, Y. (2019). Towards Sustainable Smart Cities: An Empirical Comparative Assessment and Development Pattern Optimization in China. *Journal of Cleaner Production*, *215*, 730–743. doi:10.1016/j.jclepro.2019.01.046

Lynggaard, P., & Skouby, K. E. (2015). Deploying 5G-technologies in Smart City and Smart Home Wireless Sensor Networks with Interferences. *Wireless Personal Communications*, *81*(4), 1399–1413. doi:10.100711277-015-2480-5

Mackenzie, B., Ferguson, R. I., & Bellekens, X. (2018, December). An Assessment of Blockchain Consensus Protocols for the Internet of Things. In *2018 International Conference on Internet of Things, Embedded Systems and Communications (IINTEC)* (pp. 183-190). IEEE. 10.1109/IINTEC.2018.8695298

Minoli, D., & Occhiogrosso, B. (2019). Practical Aspects for the Integration of 5G Networks and IoT Applications in Smart Cities Environments. *Wireless Communications and Mobile Computing*, *2019*, 1–30. Advance online publication. doi:10.1155/2019/5710834

Nakamoto, S. (2008). Bitcoin: A Peer-to-Peer Electronic Cash System. *Decentralized Business Review*, *21260*. Advance online publication. doi:10.2139srn.3440802

Notice of the State Council on the Issuance of "Made in China 2025". (n.d.). https://www.gov.cn/

Quaye, I., Amponsah, O., Azunre, G. A., Takyi, S. A., & Braimah, I. (2022). A Review of Experimental Informal Urbanism Initiatives and their Implications for Sub-Saharan Africa's Sustainable Cities' Agenda. *Sustainable Cities and Society*, *103938*, 103938. Advance online publication. doi:10.1016/j.scs.2022.103938

Rao, S. K., & Prasad, R. (2018). Impact of 5G technologies on Smart City Implementation. *Wireless Personal Communications*, *100*(1), 161–176. doi:10.100711277-018-5618-4

Reyna, A., Martín, C., Chen, J., Soler, E., & Díaz, M. (2018). On Blockchain and Its Integration with IoT. Challenges and Opportunities. *Future Generation Computer Systems*, *88*, 173–190. doi:10.1016/j.future.2018.05.046

Sharma, R., Mehta, K., & Sharma, O. (2021). Exploring Deep Learning to Determine the Optimal Environment for Stock Prediction Analysis. *2021 International Conference on Computational Performance Evaluation,* 148–152. 10.1109/ComPE53109.2021.9752138

Singh, S., Sharma, P. K., Yoon, B., Shojafar, M., Cho, G. H., & Ra, I. H. (2020). Convergence of Blockchain and Artificial Intelligence in IoT Network for the Sustainable Smart City. *Sustainable Cities and Society, 63*, 102364. doi:10.1016/j.scs.2020.102364

Skouby, K. E., & Lynggaard, P. (2014, November). Smart Home and Smart City Solutions Enabled by 5G, IoT, AAI and CoT Services. In *2014 International Conference on Contemporary Computing and Informatics (IC3I)* (pp. 874-878). IEEE. 10.1109/IC3I.2014.7019822

Statistical Bulletin of the People's Republic of China on National Economic and Social Development in 2021. (2022). Chinese Government.

Sz.gov.cn. (2022). *The 14th Five-Year Plan for digital government and smart city of Shenzhen was released.* http://www.sz.gov.cn/

Tencent. (2021). *Smart City Policy in China and 31 Provinces and Cities in 2021.* https://news.qq.com/

The State Council of The People's Republic of China. (2015). *Notice of the State Council on Printing and Distributing "Made in China 2025".* www.gov.cn/zhengce/content/2015-05/19/content_9784.htm

Trivedi, S., Mehta, K., & Sharma, R. (2021). Systematic Literature Review on Application of Blockchain Technology in E-Finance and Financial Services. *Journal of Technology Management & Innovation, 16*(3), 90–102. doi:10.4067/S0718-27242021000300089

Wang, K., Lin, C., & Wu, C. (2020). Trends and Planning Choices After the 60% Urbanization Rate in China. *City Planning Review, 12*, 9–17.

Wang, Y., Guo, C. H., Chen, X. J., Jia, L. Q., Guo, X. N., Chen, R. S., ... Wang, H. D. (2021). Carbon Peak and Carbon Neutrality in China: Goals, Implementation Path and Prospects. *China Geology, 4*(4), 720–746. doi:10.31035/cg2021083

Wang, Y., Ren, H., Dong, L., Park, H. S., Zhang, Y., & Xu, Y. (2019). Smart Solutions Shape for Sustainable Low-Carbon Future: A Review on Smart Cities and Industrial Parks in China. *Technological Forecasting and Social Change, 144*, 103–117. doi:10.1016/j.techfore.2019.04.014

Wątróbski, J., Bączkiewicz, A., Ziemba, E., & Sałabun, W. (2022). Sustainable Cities and Communities Assessment Using the DARIA-TOPSIS Method. *Sustainable Cities and Society, 103926*. Advance online publication. doi:10.1016/j.scs.2022.103926

Wilkins, G., & Stiff, A. (2019). Hem Realities: Augmenting Urbanism Through Tacit and Immersive Feedback. *Architecture and Culture, 7*(3), 505–521. doi:10.1080/20507828.2019.1673545

Xie, J., Tang, H., Huang, T., Yu, F. R., Xie, R., Liu, J., & Liu, Y. (2019). A Survey of Blockchain Technology Applied to Smart Cities: Research Issues and Challenges. *IEEE Communications Surveys and Tutorials, 21*(3), 2794–2830. doi:10.1109/COMST.2019.2899617

Xu, Y., & Jia, S. (2021, April). Research on Application of 5G to Smart City. *IOP Conference Series: Earth and Environmental Science, 760(1), 012014*. doi:10.1088/1755-1315/760/1/012014

Yu, P., Chang, X., & Mandizvidza, K. (2023). Development of New Energy Vehicles in Entrepreneurial Ecosystem Under the Carbon Neutrality Policy in China. In B. Marco-Lajara, J. Martínez-Falcó, & L. Millán-Tudela (Eds.), *Corporate Sustainability as a Tool for Improving Economic, Social, and Environmental Performance* (pp. 55–84). IGI Global. doi:10.4018/978-1-6684-7422-8.ch004

Yu, P., Chen, D., & Ahuja, A. (2022). Smart and Sustainable Economy: How COVID-19 Has Acted as a Catalyst for China's Digital Transformation. In S. Kautish & G. Kaur (Eds.), *AI-Enabled Agile Internet of Things for Sustainable FinTech Ecosystems* (pp. 106–146). IGI Global. doi:10.4018/978-1-6684-4176-3.ch006

Yu, P., Chen, J., Sampat, M., & Misuko, N. (2022). The Digital Transformation of Rural Agricultural Business Management: A Case Study of China. In S. Bilgaiyan, J. Singh, & H. Das (Eds.), *Empirical Research for Futuristic E-Commerce Systems: Foundations and Applications* (pp. 23–52). IGI Global. doi:10.4018/978-1-6684-4969-1.ch002

Yu, P., Ge, Y., Mandizvidza, K., & Mulli, J. (2023). How Can Small and Medium Enterprises in the Chinese Market Achieve Sustainable Development Goals Through Blockchain? In D. Taleb, M. Abdelli, A. Khalil, & A. Sghaier (Eds.), *Examining the Vital Financial Role of SMEs in Achieving the Sustainable Development Goals* (pp. 52–85). IGI Global. doi:10.4018/978-1-6684-4834-2.ch004

Yu, P., Gong, R., & Sampat, M. (2022). Blockchain Technology in China's Digital Economy: Balancing Regulation and Innovation. In P. Tehrani (Ed.), *Regulatory Aspects of Artificial Intelligence on Blockchain* (pp. 132–157). IGI Global. doi:10.4018/978-1-7998-7927-5.ch007

Yu, P., Gu, H., Zhao, Y., & Ahuja, A. (2022). Digital Transformation Driven by Internet Data Center: Case Studies on China. In D. Piaggesi, H. Landazuri, & B. Jia (Eds.), *Cases on Applying Knowledge Economy Principles for Economic Growth in Developing Nations* (pp. 203–230). IGI Global. doi:10.4018/978-1-7998-8417-0.ch011

Yu, P., Jiao, A., & Sampat, M. (2022). The Effect of Chinese Green Transformation on Competitiveness and the Environment. In P. Ordóñez de Pablos, X. Zhang, & M. Almunawar (Eds.), *Handbook of Research on Green, Circular, and Digital Economies as Tools for Recovery and Sustainability* (pp. 257–279). IGI Global. doi:10.4018/978-1-7998-9664-7.ch014

Yu, P., Liu, X., Mahendran, R., & Lu, S. (2022). Analysis and Comparison of Business Models of Leading Enterprises in the Chinese Hydrogen Energy Industry. In R. Felseghi, N. Cobîrzan, & M. Raboaca (Eds.), *Clean Technologies and Sustainable Development in Civil Engineering* (pp. 179–216). IGI Global. doi:10.4018/978-1-7998-9810-8.ch008

Yu, P., Liu, Z., & Hanes, E. (2022). Supply Chain Resiliency, Efficiency, and Visibility in the Post-Pandemic Era in China: Case Studies of MeiTuan Waimai, and Ele.me. In Y. Ramakrishna (Ed.), *Handbook of Research on Supply Chain Resiliency, Efficiency, and Visibility in the Post-Pandemic Era* (pp. 195–225). IGI Global. doi:10.4018/978-1-7998-9506-0.ch011

Yu, P., Liu, Z., Hanes, E., & Mumtaz, J. (2022). Integration of IoT and Blockchain for Smart and Secured Supply Chain Management: Case Studies of China. In S. Goyal, N. Pradeep, P. Shukla, M. Ghonge, & R. Ravi (Eds.), *Utilizing Blockchain Technologies in Manufacturing and Logistics Management* (pp. 179–207). IGI Global. doi:10.4018/978-1-7998-8697-6.ch010

Yu, P., Liu, Z., & Sampat, M. (2023). Enhancing the Resilience of Food Cold Chain Logistics Through Digital Transformation: A Case Study of China. In I. Masudin, M. Almunawar, D. Restuputri, & P. Sud-On (Eds.), *Handbook of Research on Promoting Logistics and Supply Chain Resilience Through Digital Transformation* (pp. 200–224). IGI Global. doi:10.4018/978-1-6684-5882-2.ch014

Yu, P., Lu, S., Hanes, E., & Chen, Y. (2022). The Role of Blockchain Technology in Harnessing the Sustainability of Chinese Digital Finance. In P. Swarnalatha & S. Prabu (Eds.), *Blockchain Technologies for Sustainable Development in Smart Cities* (pp. 155–186). IGI Global. doi:10.4018/978-1-7998-9274-8.ch009

Yu, P., Lu, S., Sampat, M., Li, R., & Ahuja, A. (2022). How AI-Enabled Agile Internet of Things Can Enhance the Business Efficiency of China's FinTech Ecosystem. In S. Kautish & G. Kaur (Eds.), *AI-Enabled Agile Internet of Things for Sustainable FinTech Ecosystems* (pp. 190–223). IGI Global. doi:10.4018/978-1-6684-4176-3.ch009

Yu, P., Shen, X., & Hanes, E. (2023). Promoting Responsible Research and Innovation in China's Hi-Tech Zones: Based on Case Studies of Zizhu Hi-Tech Zone, East Lake Hi-Tech Zone, and Guangzhou Hi-Tech Zone. In B. Marco-Lajara, J. Martínez-Falcó, & L. Millán-Tudela (Eds.), *Corporate Sustainability as a Tool for Improving Economic, Social, and Environmental Performance* (pp. 222–245). IGI Global. doi:10.4018/978-1-6684-7422-8.ch012

Yu, P., Weng, Y., & Ahuja, A. (2022). Carbon Financing and the Sustainable Development Mechanism: The Case of China. In A. Rafay (Ed.), *Handbook of Research on Energy and Environmental Finance 4.0* (pp. 301–332). IGI Global. doi:10.4018/978-1-7998-8210-7.ch012

Yu, P., Xu, S., Cheng, Z., & Sampat, M. (2023). Does the Development of New Energy Vehicles Promote Carbon Neutralization?: Case Studies in China. In A. Pego (Ed.), *Climate Change, World Consequences, and the Sustainable Development Goals for 2030* (pp. 109–131). IGI Global. doi:10.4018/978-1-6684-4829-8.ch006

Yu, P., Xue, W., & Mahendran, R. (2022). The Development and Impact of China's Digital Transformation in the Medical Industry. In M. Rodrigues & J. Proença (Eds.), *Impact of Digital Transformation on the Development of New Business Models and Consumer Experience* (pp. 97–128). IGI Global. doi:10.4018/978-1-7998-9179-6.ch006

Yu, P., Yu, M., & Sampat, M. (2022). Smart Management for Digital Transformation in China. In B. Barbosa, S. Filipe, & C. Santos (Eds.), Handbook of Research on Smart Management for Digital Transformation (pp. 411-438). IGI Global. doi:10.4018/978-1-7998-9008-9.ch019

Yu, P., Zhang, Y., Sampat, M., & Chen, Y. (2023). Research on Cross-Industry Digital Transformation Under the New Normal: A Case Study of China. In B. Marco-Lajara, J. Martínez-Falcó, & L. Millán-Tudela (Eds.), *Corporate Sustainability as a Tool for Improving Economic, Social, and Environmental Performance* (pp. 246–277). IGI Global. doi:10.4018/978-1-6684-7422-8.ch013

Yu, P., Zhao, Z., & Sampat, M. (2023). How Digital Twin Technology Promotes the Development of Smart Cities: Case Studies in China. In I. Vasiliu-Feltes (Ed.), *Impact of Digital Twins in Smart Cities Development* (pp. 198–227). IGI Global. doi:10.4018/978-1-6684-3833-6.ch008

Zj.gov.cn. (2022). *Taizhou City "Three Roles" to Build Intelligent Drainage Digital Twin Application System*. http://jst.zj.gov.cn/

ENDNOTE

[1] The "double carbon" goal refers to the goals of "carbon peak" and "carbon neutrality", which is China's long-term development strategy for low greenhouse gas emissions in the 21st century.

Chapter 2
Relevant Technologies to 6G

Subramaniam Meenakshi Sundaram
GSSS Institute of Engineering and Technology for Women, India

Tejaswini R. Murgod
NITTE Meenakshi Institute of Technology, Bengaluru, India

Madhu M. Nayak
GSSS Institute of Engineering and Technology for Women, India

Usha Rani Janardhan
GSSS Institute of Engineering and Technology for Women, India

Usha Obalanarasimhaiah
GSSS Institute of Engineering and Technology for Women, India

ABSTRACT

From the first generation to the third generation (3G), data communications such as "i-mode" and multimedia information such as photos, music, and video could be communicated using mobile devices. From the fourth generation (4G), smartphones have been explosively popularized by high-speed communication technology exceeding 100 Mbps using the long-term evolution (LTE), and multimedia communication services have appeared. 4G technology continues to evolve in the form of LTE-Advanced and has now reached a maximum communication speed close to 1 Gbps. 5G is expected to provide new value as a basic technology supporting future industry and society, along with artificial intelligence (AI) and the internet of things (IoT). 5G is expected to evolve, and the sixth generation (6G) technology will support industry and society in the 2030s. The objectives of 6G technology include simultaneous achievement of several requirements such as ultra-high-speed, high-capacity, and low-latency connectivity.

INTRODUCTION

We are moving toward a society of fully automated and remote management systems. The very rapid development of various emerging applications, such as Artificial Intelligence (AI), Virtual Reality (VR), three-dimensional

DOI: 10.4018/978-1-7998-9266-3.ch002

(3D) media, and the Internet of Everything (IoE), has led to a massive volume of traffic (Mumtaz, 2017). This statistics clearly depicts the importance of the improvement of communication systems. Autonomous systems are becoming popular in every sector of society, such as industry, health, roads, oceans, and space. To provide a smart life and automated systems, millions of sensors will be embedded into cities, vehicles, homes, industries, foods, toys, and other environments. Hence, a high-data-rate with reliable connectivity will be required to support these applications. In certain parts of the world, fifth-generation (5G) wireless networks have already been deployed. By 2020, it is expected that 5G will be fully deployed worldwide.

5G networks will not have the capacity to deliver a completely automated and intelligent network that provides everything as a service and a completely immersive experience (Nawaz S, 2019). Although, the 5G communication systems that are going to be released very soon will offer significant improvements over the existing systems, they will not be able to fulfill the demands of future emerging intelligent and automation systems after 10 years (Giordani, 2020). The 5G network will provide new features and provide better Quality of Service (QoS) as compared with fourth-generation (4G) communications (Shafi, 2017; Zhang, 2016; Jaber, 2016; Andrews, 2014). The 5G technology will include several new additional techniques, such as new frequency bands (e.g., the millimeter wave (mmWave) and the optical spectra), advanced spectrum usage and management, and the integration of licensed and unlicensed bands (Giordani, 2020). Nevertheless, the fast growth of data-centric and automated systems may exceed the capabilities of 5G wireless systems. Certain devices, such as Virtual Reality (VR) devices need to go beyond 5G (B5G) because they require a minimum of 10 Gbps data rate (Mumtaz, 2017). Hence, with 5G reaching its limits in 2030, the design goals for its next step are already being explored in literature. To overcome the constraints of 5G for supporting new challenges, a sixth-generation (6G) wireless system will need to be developed with new attractive features.

The key drivers of 6G will be the convergence of all the past features, such as network densification, high throughput, high reliability, low energy consumption, and massive connectivity. The 6G system would also continue the trends of the previous generations, which included new services with the addition of new technologies. The new services include AI, smart wearables, implants, autonomous vehicles, computing reality devices, sensing, and 3D mapping (Saad.W). The most important requirement for 6G wireless networks is the capability of handling massive volumes of data and very high-data-rate connectivity per device (Mumtaz, 2017).

The 6G system will increase performance and maximize user QoS several folds more than 5G along with some exciting features. It will protect the system and secure the user data. It will provide comfortable services. The 6G communication system is expected to be a global communication facility. It is envisioned that the per-user bit rate in 6G will be approximately 1 Tb/s in many cases (Mumtaz, 2017; David, 2018). The 6G system is expected to provide simultaneous wireless connectivity that is 1000 times higher than 5G. Moreover, ultra-long-range communication with less than 1-ms latency is also expected (F. Tariq). The most exciting feature of 6G is the inclusion of fully supported AI for driving autonomous systems. Video-type traffic is likely to be dominant among various data traffic systems in 6G communications. The most important technologies that will be the driving force for 6G are the terahertz (THz) band, AI, optical wireless communication (OWC), 3D networking, unmanned aerial vehicles (UAV), and wireless power transfer.

In this section we describe how 6G communication systems can be developed; we also describe the expected 6G technologies, and the research issues required to address the needs of future smart networks. Section A presents the 6G enabled technologies. The possible network architectures with the applications of future 6G communication systems is presented in Section B. The expected service requirements and the network characteristics in 6G communication systems are presented in Section C. The possible key technologies for the development of 6G system are briefly discussed in Section D. In Section E, we present the various research activities on 6G. We provide the key challenges and the research directions to reach the goal of 6G in Section F. Finally, we draw our conclusions in Section G.

A. 6G Enabling Technologies

6G technologies will integrate three aspects namely physical, biological, and digital world. The 6G networks along with the typical radio frequency communication will include robots, digital twins, artificial intelligence, and emotion-driven devices, smart communicating surfaces, communication through brain implanted chips or brain-machine interface (Chowdhury, 2019) to enable all-round cyber-physical-biological communication experience. Consequently, 6G will be much more than the present smart connected networks, where the network components will largely integrate intelligence to bring in a paradigm shift from smart to intelligent networks. Some of the enabling technologies for 6G include artificial intelligence, THz communication, 3D connectivity, Visible light communication, block chain, quantum communication, intelligent surfaces and digital twins. Figure 1 shows the possible 6G communication architecture scenario.

Figure 1. Possible 6G communication architecture scenario

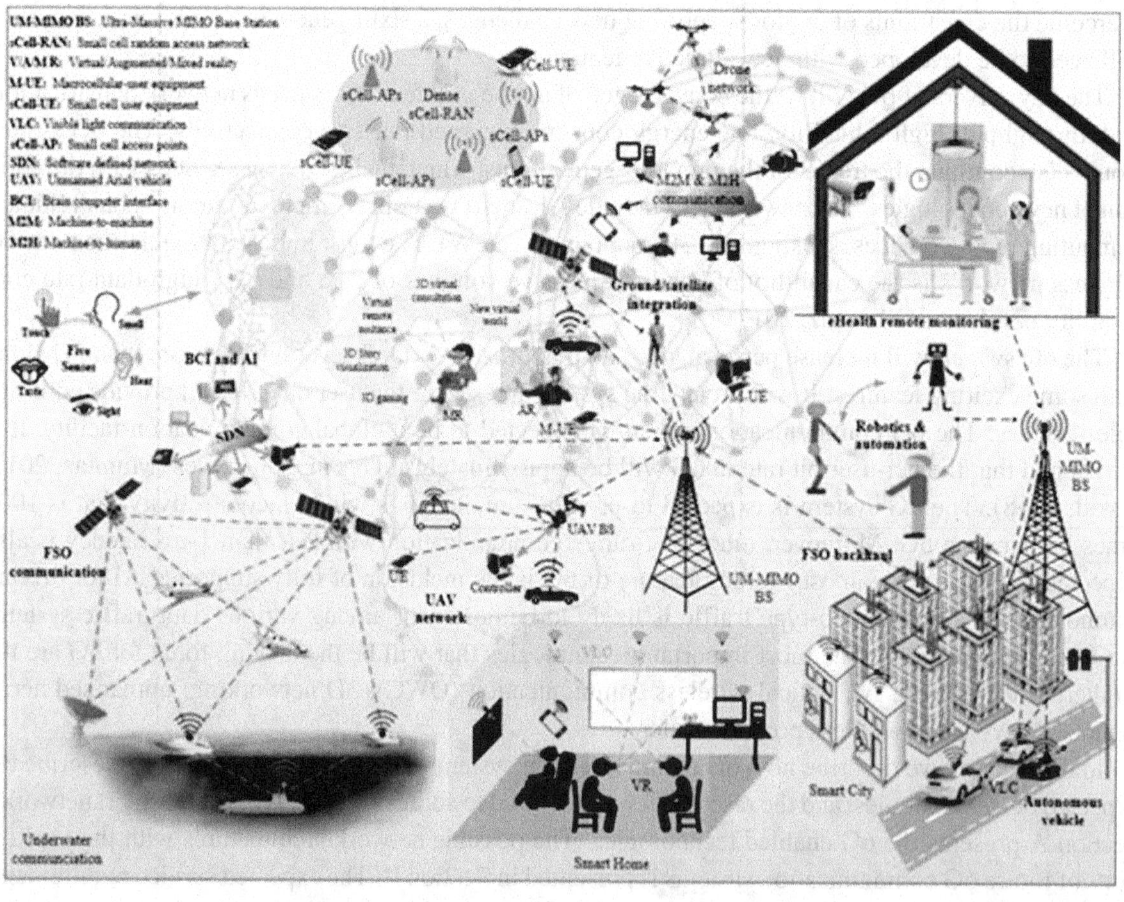

B. Architectures For 6G

A few researchers have proposed new architectures of 6G networks. These include:

1. Cyber Twin Architecture

The authors proposed two architectures for 6G networks, namely, (1) cloud-centric internet model and (2) a decoupled Radio Access Network (RAN) model (Yu, 2020). In the cloud-centric internet model, the existing IP architecture is slightly retained with certain modifications. First, the users are connected to the RAN, and the data from the RAN enters the edge cloud layer where cyber-twins accept the data. These cyber-twins act as data loggers, assent owners, or virtual representation of the user. Edge clouds, in turn, are connected to the cloud layer, where multiple clouds are interconnected to form the center of the network architecture. The cloud layer consists of resource scheduler, orchestrator, communication, computing, and caching functions. This cloud layer enables applications to provide services in the edge and cloud at reasonable cost and QoS.

2. Decoupled RAN Architecture

Here are distinct APs for handling the *control* and *user plane* data different from the earlier cellular generations. The control plane BS will be a macro BS, which a user shall connect to exchange the control information. These control BSs will be connected to the user plane (data) BSs for high level signaling. In addition, the uplink and downlink traffic are also decoupled and handled by separate BSs to the users. Here, the uplink BS can be a micro-cell dense deployment near the user to collect the user data at low power. The uplink and downlink base stations will have internal coordination for efficient communication (Yu, 2020).

3. Generalized Architecture for 6G

A generalized 6G architecture for IoT and vehicular networks is presented in (She, 2020). It consists of three levels, namely, *user level* composed of smart devices with caching ability. These devices send the sensed data to the base station level, where base stations or APs have edge servers to perform scheduling and resource allocation. Finally, a central server will do the slicing, handover actions at the network level. This architecture will reduce the delay for critical services (autonomous cars) by processing and storing at the edge level. However, due to its partially centralized control plane, some non-critical tasks still use cloud processing.

4. A High-Level Architecture of 6G

A three-level 6G network architecture consists of AI plane, user plane, and control plane (Tataria, 2020) In this case, storage, compute, and networking are done at the same level (in user plane) to eliminate the hierarchy. The user plane is that and it is defined between an access network and the internet. The control plane and AI planes are distributed and virtualized for various services. Further, the transport network is virtualized and isolated from the rest by software defined virtualization. The core network functions are made as micro-services and accessible by server-less systems.

5. Multi-Level Architecture for 6G

A three-tier architecture of 6G consists of smart users, edge devices, and the cloud (She, 2020). The users implement intelligent decision-making techniques such as data-driven or model-based techniques to predict mobility and traffic patterns. At edge intelligence level, the mobile edge devices use deep reinforcement learning, or deep neural networks to optimize the scheduler for resource allocation to mobile users based on the CSI. Moreover, at the cloud level, having a high-capacity central control can train the system with a numerical platform that presents the labeled samples for optimization algorithms. These can be used to train deep neural networks and later be implemented in the control plane.

6. Three-Dimensional 6G Architecture

The FG-NET 2030 has envisioned a 3D architecture for 6G networks, covering three key aspects, namely, communication (infrastructure view), intelligence (control view), and management (network view) (Huang, 2019). It has different communication layers at the infrastructure view ranging from terrestrial, underwater, aerial, and satellite to enhance the range of communication in 6G networks. In addition, the authors at the control view presented how 6G networks will include intelligence to control and optimize the overall functions of sensing, spectrum access, communication, storage, and processing with the help of AI, deep learning, and ML. It recommends that the intelligence will be distributed across various network entities. Finally, in the network view, the functions of the overall network have been divided as sub layers. They include the application sub layer, routing, management, spectrum access, and physical medium, which resemble the layered IP stack. A simplified version of the same architecture for 6G has been proposed in Nguyen (2020).

7. Neuroscience-Based 6G Architecture

This architecture is more of a framework that tries to integrate neuro signals to emulate the wireless signals to be applied in 6G networks (Moioli, 2020). Here, human brain's intelligence and radiating properties of wireless signals shall be integrated to enable an intelligent communication between the human and computers. Recent advances in bio-IoT and implantable communication devices have made us bold to envision short range 6G network communication, where wireless modules implanted inside the human brain acquire intelligence from the brain and communicate directly with the outer world wireless devices and base station or with another human with similar capability.

C. Specifications and Requirements of 6G

5G technologies have been associated with trade-offs of several issues such as throughput, delay, energy efficiency, deployment costs, reliability, and hardware complexity. It is very likely that 5G will not be able to meet the market demands after 2030. Then, 6G will fill the gap between 5G and the market demand. Based on the previous trends and predictions of future needs, the main objectives for the 6G systems are (1) extremely high data rates per device, (2) a very large number of connected devices, (3) global connectivity, (4) very low latency, (5) lowering the energy consumption with battery-free IoT devices, (6) ultra-high reliable connectivity, and (7) connected intelligence with machine learning capability. Table 1 shows a comparison of 6G with the 4G and 5G communication systems.

Table 1. Comparison of 6G with 4G and 5G communication systems

Issue	4G	5G	6G
Per device peak data rate	1 Gbps	10 Gbps	1 Tbps
E2E latency	100 ms	10 ms	1 ms
Maximum spectral efficiency	15 bps/Hz	30 bps/Hz	100 bps/Hz
Mobility support	Up to 350 km/hr	Up to 500 km/hr	Up to 1000 km/hr
Satellite integration	No	No	Fully
AI	No	Partial	Fully
Autonomous vehicle	No	Partial	Fully
XR	No	Partial	Fully
Haptic Communication	No	Partial	Fully

1. Service Requirements

The 6G wireless system will have the following key factors:

- Enhanced mobile broadband (eMBB)
- Ultra-reliable low latency communications (URLLC)
- Massive machine-type communication (mMTC)
- AI integrated communication
- Tactile internet
- High throughput
- High network capacity
- High energy efficiency
- Low backhaul and access network congestion
- Enhanced data security

It is estimated that the 6G system will have 1000 times higher simultaneous wireless connectivity than the 5G system. URLLC, which is a key 5G feature, will be a key driver again in 6G communications by providing end-to-end (E2E) delay of less than 1 ms (F.Tariq.et.al). Volume spectral efficiency, as opposed to the often-used area spectral efficiency, will be much better in 6G (F.Tariq.et.al). The 6G system will provide ultra-long battery life and advanced battery technology for energy harvesting. In 6G systems, mobile devices will not need to be separately charged.

2. New Network Characteristics

Satellites integrated network: To provide global mobile collectivity, 6G is expected to integrate with satellites. Integrating terrestrial, satellite, and airborne networks into a single wireless system will be crucial for 6G. Connected intelligence: In contrast to the earlier generation of wireless communication systems, 6G will be transformative, and will update the wireless advancement from "connected things" to "connected intelligence" (Letaief et al., 2019). AI will be introduced in each step of the communication process. The pervasive introduction of AI will produce a new paradigm of communication systems.

Seamless integration of wireless information and energy transfer: The 6G wireless networks will also transfer power to charge battery devices, such as smart phones and sensors. Hence, wireless information and energy transfer (WIET) will be integrated.

Ubiquitous super 3D connectivity: Accessing the network and core network functionalities on drones and very low earth orbit satellites will make the super-3D connectivity in 6G ubiquitous.

3. Few General Requirements in Network Characteristics

Small cell networks: The idea of small cell networks has been introduced to improve the received signal quality as a consequence of throughput, energy efficiency, and spectral efficiency enhancement in cellular systems (Chowdhury, 2018; Zadid, 2018; Mahbas, 2019). As a result, small cell networks are an essential characteristic for the 5G and beyond (5GB) communication systems. Therefore, 6G communication systems also adopt this network characteristic.

Ultra-dense heterogeneous networks: Ultra-dense heterogeneous networks (Zhou, 2018; Andreev, 2019) will be another important characteristic of 6G communication systems. Multi-tier networks consisting of heterogeneous networks will improve the overall QoS and reduce the cost.

High-capacity backhaul: The backhaul connectivity must be characterized by high-capacity backhaul networks to support a huge volume of traffic. High-speed optical fiber and free space optical (FSO) systems are possible solutions for this problem.

Radar technology integrated with mobile technologies: High-accuracy localization with communication is also one of the features of the 6G wireless communication system. Hence, radar systems will be integrated with 6G networks.

Softwarization and virtualization: Softwarization and virtualization are two important features that are the basis of the design process in 5GB networks to ensure flexibility, reconfigurability, and programmability. Moreover, they will allow sharing the billions of devices in a shared physical infrastructure.

D. Key Enabling Technologies Of 6G

The 6G system will be driven by many technologies. A few expected key technologies for 6G are discussed below.

Artificial Intelligence: The most important and newly introduced technology for 6G communication systems is AI (Zanella, 2020; Lopez 2020; Yuan, 2020; Hema, 2020). There was no involvement of AI for 4G communication systems. The upcoming 5G will support partial or very limited AI. However, 6G will be fully supported by AI for automation. Advancements in machine learning will create more intelligent networks for real-time communications in 6G. The introduction of AI in communication will simplify and improve the transport of real-time data. Using numerous analytics, AI can determine the way a complex target job is performed. AI will increase the efficiency and reduce the processing delay of the communication steps. Time-consuming tasks, such as handover and network selection, can be performed promptly by using AI. AI will also play a vital role in M2M, machine-to-human, and human-to-machine communications. It will also prompt communication in the BCI. AI-based communication systems will be supported by meta materials, intelligent structures, intelligent networks, intelligent devices, intelligent cognitive radio, self-sustaining wireless networks, and machine learning.

Terahertz Communications: Spectral efficiency can be increased by increasing the bandwidth; this can be done by using sub-THz communication with wide bandwidths and by applying advanced massive multiple input, multiple output (MIMO) technologies. The RF band has been almost exhausted, and now it is insufficient to meet the high demands of 6G. The THz band will play an important role in 6G communication (Giordani, 2020; Sliwa, 2020). The THz band is intended to be the next frontier of high-data-rate communications. THz waves, also known as sub millimeter radiation, usually refer to the frequency band between 0.1 THz, and 10 THz with the corresponding wavelengths in the 0.03 mm–3 mm range (Kantola, 2020). According to the recommendations of ITU-R, the 275 GHz–3 THz band range is considered to be the main part of the THz band for cellular communications (Kantola, 2020). The capacity of 6G cellular communications will be increased by adding the THz band (275 GHz–3THz) to the mmWave band (30–300 GHz). The band within the range of 275 GHz–3 THz has not yet been allocated for any purpose worldwide; therefore, this band has the potential to accomplish the desired high data rates (Sliwa 2020). When this THz band is added to the existing mmWave band, the total band capacity increases a minimum of 11.11 times. Of the defined THz bands, 275 GHz–3THz, and 275 GHz–300 GHz lie on the mmWave, and 300 GHz–3 THz lie on the far infrared (IR) frequency band. Even though the 300 GHz–3 THz band is part of the optical band, it is at the boundary of the optical band and immediately after the RF band. Hence, this 300 GHz–3 THz band shows similarities with the RF. THz heightens the potentials and challenges of high frequency communications (Huang, 2019). The key properties of THz communications include (1) widely available bandwidth to support very high data rates, (2) high path loss arising from the high frequency (highly directional antennas will most probably be indispensable) (David, 2019). The narrow beam widths generated by the highly directional antennas reduce the interference. The small wavelength of the THz signals allows a much greater number of antenna elements to be incorporated into the devices and BSs operating in this band. This allows the use of advanced adaptive array technologies that can overcome the range limitations (David, 2019).

Optical Wireless Technology: OWC technologies are envisioned for 6G communications in addition to RF-based communications for all possible device-to-access networks; these networks also access network-to-backhaul/fronthaul network connectivity. OWC technologies are already being used since 4G communication systems. However, it will be used more widely to meet the demands of 6G communication systems. OWC technologies, such as light fidelity, visible light communication, optical camera communication, and FSO communication based on the optical band are already well-known technologies (Peltonen, 2020; Katz, 2020; Yang, 2019; evolution, 2020). Researchers have been working on enhancing the performance and overcoming the challenges of these technologies. Communication based on optical wireless technologies can provide very high data rates, low latencies, and secure communications. LiDAR, which is also based on the optical band, is a promising technology for very-high-resolution 3D mapping in 6G communications.

FSO Backhaul Network: It is not always possible to have optical fiber connectivity as a backhaul network because of remote geographical locations and complexities. The FSO backhaul network is very promising for 5GB communication systems (Monserrat, 2020; Akyildiz, 2020). The transmitter and receiver characteristics of the FSO system are similar to those of optical fiber networks. Therefore, the data transfer in the FSO system is comparable with the optical fiber system. Hence, along with the optical fiber networks, FSO is an excellent technology for providing backhaul connectivity in 6G. Using FSO, it is possible to have very long-range communications even at a distance of more than 10,000 km. FSO supports high-capacity backhaul connectivity for remote and non-remote areas, such as the sea, outer space, underwater, isolated islands; FSO also supports cellular BS connectivity.

Massive Multiple Input, Multiple Output Techniques: One key technique to improve spectral efficiency is the application of the MIMO technique. When the MIMO technique improves, the spectral efficiency also improves. Therefore, a massive MIMO technology will be crucial in the 6G system.

Block Chain: Block chain will be an important technology to manage massive data in future communication systems (Sergiou, 2020; 6G Drivers, 2020). Block chains are just one form of the distributed ledger technology. A distributed ledger is a database that is distributed across numerous nodes or computing devices. Each node replicates and saves an identical copy of the ledger. The block chain is managed by peer-to-peer networks. It can exist without being managed by a centralized authority or a server. The data on a block chain is gathered together and structured in blocks. The blocks are connected to one another and secured using cryptography. The block chain is essentially a perfect complement to the massive IoT with improved interoperability, security, privacy, reliability, and scalability (Pouttu, 2020). Therefore, the block chain technology will provide several facilities, such as interoperability across devices, traceability of massive data, autonomic interactions of different IoT systems, and reliability for the massive connectivity of 6G communication systems.

3D Networking: The 6G system will integrate the ground and airborne networks to support communications for users in the vertical extension. The 3D BSs will be provided through low orbit satellites and UAVs (Rajatheva, 2020). The addition of new dimensions in terms of altitude and related degrees of freedom makes 3D connectivity considerably different from the conventional 2D networks.

Quantum Communications: Unsupervised reinforcement learning in networks is promising in the context of 6G networks. Supervised learning approaches will not be feasible for labeling huge volumes of data generated in 6G. Unsupervised learning does not need labeling. Hence, this technique can be used for autonomously building the representations of complex networks. By combining reinforcement learning and unsupervised learning, it is possible to operate the network in a truly autonomous fashion (Tariq, 2019).

Unmanned Aerial Vehicle: UAVs or drones will be an important element in 6G wireless communications. In many cases, high-data-rate wireless connectivity will be provided using the UAV technology. The BS entities will be installed in UAVs to provide cellular connectivity. A UAV has certain features that are not found in fixed BS infrastructures, such as easy deployment, strong line-of-sight links, and degrees of freedom with controlled mobility (Calvanese, 2019). During emergency situations, such as natural disasters, the deployment of terrestrial communication infrastructures is not economically feasible, and sometimes it is not possible to provide any service in volatile environments. UAVs can easily handle these situations. A UAV will be the new paradigm in the field of wireless communication. This technology can facilitate three fundamental requirements of wireless networks that are, eMBB, URLLC, and mMTC (Nguyen, 2020). UAVs can also serve several purposes, such as the enhancement of network connectivity, fire detection, emergency services in disaster, security and surveillance, pollution monitoring, parking monitoring, accident monitoring, and so on. Therefore, UAV technology is recognized as one of the most important technologies for 6G communication.

Cell-Free Communications: The tight integration of multiple frequencies and heterogeneous communication technologies will be crucial in 6G systems. As a result, the user will move seamlessly from one network to another network without the need for making any manual configurations in the device (Tariq, 2019). The best network will be automatically selected from the available communication technology. This will break the limits of the concept of cells in wireless communications. Currently, the user movement from one cell to another cell causes too many handovers in dense networks, and also causes handover failures, handover delays, data losses, and the ping-pong effect. The 6G cell-free communications will overcome all these and provide better QoS. Cell-free communication will be achieved through multi-connectivity and muti-tier hybrid techniques and by different and heterogeneous radios in the devices (Tariq, 2019).

Integration of Wireless Information and Energy Transfer: WIET in communication will be one of the most innovative technologies in 6G. WIET uses the same fields and waves as wireless communication systems. In particular, sensors and smart phones will be charged by using wireless power transfer during communication. WIET is a promising technology for lengthening the lifetime of the battery-charging wireless systems (Samdanis, 2020). Hence, devices without batteries will be supported in 6G communications.

Integration of Sensing and Communication: A key driver for autonomous wireless networks is the capability to continuously sense the dynamically changing states of the environment and exchange the information among different nodes (Business of 6G, 2020). In 6G, the sensing will be tightly integrated with communication to support autonomous systems.

Integration of Access-Backhaul Networks: The density of the access networks in 6G will be huge. Each access network is connected with backhaul connectivity, such as optical fibers and FSO networks. To cope with the very large number of access networks, there will be tight integration between the access and backhaul networks.

Dynamic Network Slicing: Dynamic network slicing permits a network operator to allow dedicated virtual networks to support the optimized delivery of any service toward a wide range of users, vehicles, machines, and industries. It is one of the most important elements for management when a large number of users are connected to a large number of heterogeneous networks in 5GB communication systems.

Holographic Beam Forming: Beam forming is a signal processing procedure by which an array of antennas can be steered to transmit radio signals in a specific direction. It is a subset of smart antennas or advanced antenna systems. The beam forming technique has several advantages, such as a high signal-to-noise ratio, interference prevention, and rejection, and high network efficiency. Holographic beam forming (HBF) is a new method for beam forming that is considerably different from the MIMO systems because it uses software-defined antennas. HBF will be a very effective approach in 6G for the efficient and flexible transmission and reception of signals in multi-antenna communication devices.

Big Data Analytics: Big data analytics is a complex process for analyzing a variety of large data sets or big data. This process uncovers information, such as hidden patterns, unknown correlations, and customer inclinations, to ensure perfect data management. The big data is collected from a wide variety of sources, such as videos, social networks, images, and sensors. This technology will be widely used for handling of huge data in 6G systems.

E. Standardization and Research Activities

The 5G specifications have already been prepared, and even though it has already been launched in some parts of the world, the full phase of 5G will be deployed in 2020. Research activities on 6G are in their initial stages. From 2020, a number of studies will be performed worldwide on the standardization of 6G; 6G communication is still in its infancy. Many researchers have defined 6G as B5G or 5G+. Preliminary research activities have already started in the United States of America. The US president has requested the deployment of 6G in the country. China has already started the concept study for the development and standardization of 6G communications in 2019. The Chinese are planning for active research work on 6G in 2020. Most European countries, Japan, and Korea are planning several 6G projects. The research activities on 6G are expected to start in 2020. In this section, we present a few research activities, and standardization efforts.

Samsung Electronics: Samsung Electronics has opened an R&D center for the development of essential technologies for 6G mobile networks. To accelerate the development of solutions and for the standardization of 6G, Samsung research is conducting extensive research on cellular technologies; they have upgraded a next-generation telecommunication research team to a center. Finnish 6G Flagship program: University of Oulu began the 6G research activities under Finland's flagship program. Research in 6G Flagship is organized into four unified planned research parts: wireless connectivity, distributed computing, services, and applications. Scientific innovations will be developed for important technology components of 6G systems.

International Telecommunication Union: Standardization activities on 5G of the ITU radio communication sector ITU-R was based on IMT-2020. Consequently, ITU-R will probably release IMT-2030, which will summarize the possible requirements of mobile communications in 2030 (i.e., 6G).

6G Wireless Summit: A successful first 6G wireless summit was held in Lapland, Finland, in March 2019. A wide and effective discussion was performed by academicians, industry persons, and vendors from all over the world. Pioneering wireless communication researchers were present at the summit. Moreover, the world's leading telecom companies also attended the summit. This 6G summit initiates the discussions on key issues, such as the motivation behind 6G, the way to move from 5G to 6G, the current industry trends for 6G, and the enabling technologies.

F. 6G: Prospects and Applications of 6G

AI will be integrated into the 6G communication systems. All the network instrumentation, management, physical layer signal processing, resource management, service-based communications, and so on will be incorporated by using AI (Strinati, 2019). It will foster the Industry 4.0 revolution, which is the digital transformation of industrial manufacturing (Stoica, 2019). Some key prospects and applications of 6G wireless communication include: 1) Super-smart society where the superior features of 6G will accelerate the building of smart societies leading to life quality improvements, environmental monitoring, and automation using AI-based M2M communication and energy harvesting (Stoica, 2019). 2) Extended reality services including augmented reality (AR), mixed reality (MR), and VR are very important features of 6G communication systems. 3) Connected robotics and autonomous systems in automotive technology where researchers have been investigating automated and connected vehicles. The 6G systems help in the deployment of connected robots and autonomous systems. 4) Wireless brain–computer interactions which use the Brain Computer Interface (BCI) approach to control the appliances that are used daily in smart societies, especially the appliances used at home and in medical systems (Yu & Zhang, 2019). Haptic communication for nonverbal communication that uses the sense of touch where remote users will be able to enjoy haptic experiences through real-time interactive systems (Soderi, 2020). 6) Smart healthcare where medical health systems will also benefit by the 6G wireless systems because innovations, such as AR/VR, holographic tele presence, mobile edge computing, and AI, will help build smart healthcare systems (Stoica, 2019).

G. 6G Research Worldwide

While the 5G network is being deployed worldwide, the research and development activities on 6G networks are gearing up both in industry and academia. In March 2020, 3GPP completed the 5G standard release 16, which will be followed by release 17 to support all three scenarios in 2021 (Pouttu, 2020; Lu, 2020). Therefore, in the next couple of years, 3GPP will initiate the 6G research. It is anticipated that by 2027, the 5G infrastructure market will increase to 47.75bn USD by 2027. The top five players in the 5G infrastructure market include Huawei, Ericsson, Samsung, Nokia, and ZTE. They have currently shifted their attention toward the 6G network due to the enormous potential and benefit that the 6G network is anticipated to offer over the 5G network. Along the same lines, FCC has decided to promote 6G network research and trials in the THz band by opening a 95 GHz -3 THz frequency band for research (Lu, 2020). Samsung has anticipated that ITU-R will define the official vision of the 6G network in mid-2021, and the initial commercialization of 6G network will begin by the end of 2028. The global telecom companies have even endorsed this. In the following paragraph, we will list the major activities in the research and standardization efforts by various organizations worldwide.

1. Finland

The 6G Flagship program (6Genesis) of the University of Oulu in association with other industries and academic institutes such as Nokia, VTT research center, Business Oulu, and Key sight technologies are at the forefront of research groups. They have initiated two 6G network summits and have released 12 white papers on the key research areas of the 6G network (L. Khan et. al., 2020). Moreover, Mediatek has started its research on 6G chipset along with Nokia in Finland.

2. *China*

The major players in the 6G network in China include Huawei and ZTE. They have their independent research units aside government-sponsored 6GR&D promotion and expert group. Their research mainly includes THz, AI, and block chain for 6G networks along with other operators such as China Mobile and China Unicom.

3. USA

In Feb 2019, US President Donald Trump US telecom companies gave a call to intensify their research to launch 6G network at the earliest. From academia, The New York University wireless center, headed by Prof. T. Rappaport, has highly engaged in developing the THz channel modeling and has achieved a 100 Gbps data rates in its trials (Rappaport et al., 2019).

5. South Korea

One of the key telecom players SK Telecom, has undertaken 6G network research in the areas of THz communication, ultra-massive MIMO, and aerial communication. Further, they have collaborated with Ericsson and Nokia in 6G equipment manufacturing technology development. Samsung has recently released the white paper on 6G vision emphasizing three aspects: holographic communication, truly immersive XR, and digital twins. Moreover, LG and KAIST University have developed a 6G network research laboratory to jointly conduct research in the technology areas of the 6G network.

5. Japan

Japan has focused its research on AI and Cyber physical system to promote 6G networks. In the process, NTT has demonstrated 100 Gbps at 28 GHz band. In addition, Japan Govt., released 2.04 billion USD to promote R&D. Apart from these, Toshiba and Tokyo University have initiated 6G networks research (Lu et al., 2020). Similarly, Sony, Nippon, and Intel have planned to work together in different fields of 6G technology.

6. Europe

In Europe, besides Finland, several universities namely, University of Dresden and Deutsche Telekom (Germany) are involved in the research of Tactile internet, HCI technologies. Next, University of Padua, Italy and NYU wireless group are also involved in the 6G network research.

H. Challenges and Future Research Directions of 6G

Several technical problems need to be solved to successfully deploy 6G communication systems. A few of possible concerns are briefly discussed below.

High Propagation and Atmospheric Absorption of THz: The high THz frequencies provide high data rates. However, the THz bands need to overcome an important challenge for data transfer over relatively long distances because of the high propagation loss, and atmospheric absorption characteristics (K. David, 2018). We require a new design for the transceiver architecture for the THz communication systems. The transceiver must be able to operate at high frequencies, and we need to ensure the full use of very widely available bandwidths. A very small gain and an effective area of the distinct THz band antennas is another challenge of THz communication. Health and safety concerns related to THz band communications also need to be addressed.

Complexity in Resource Management for 3D Networking: The 3D networking extended in the vertical direction. Hence, a new dimension was added. Moreover, multiple adversaries may intercept legitimate information, which may significantly degrade the overall system performance. Therefore, new techniques for resource management and optimization for mobility support, routing protocol, and multiple access are essential. Scheduling needs a new network design.

Heterogeneous Hardware Constraints: In 6G, a very large number of heterogeneous types of communication systems, such as frequency bands, communication topologies, service delivery, and so on, will be involved. Moreover, the access points and mobile terminals will be significantly different in the hardware settings. The massive MIMO technique will be further upgraded from 5G to 6G, and this might require a more complex architecture. It will also complicate the communication protocol and the algorithm design. However, machine learning and AI will be included in communication. Moreover, the hardware design for different communication systems is different. Unsupervised and reinforcement learning may create complexities in hardware implementation as well. Consequently, it will be challenging to integrate all the communication systems into a single platform.

Autonomous Wireless Systems: The 6G system will provide full support to automation systems such as autonomous car, UAVs, and Industry 4.0 based on AI. To make autonomous wireless systems, we need to have the convergence of many heterogeneous sub-systems, such as autonomous computing, interoperable processes, system of systems, machine learning, autonomous cloud, machines of systems, and heterogeneous wireless systems (Letaief, 2019). Thus, the overall system development becomes complex and challenging. For example, developing a fully autonomous system for the driverless vehicle will be much more challenging because 6G researchers need to design fully automated self-driving vehicles that perform better than the human-controlled vehicles.

Modeling of Sub-mmWave (THz) Frequencies: The propagation characteristics of the mmWave and sub-mmWave (THz) is subject to atmospheric conditions; therefore, absorptive and dispersive effects are seen (R. Shatin, 2019). The atmospheric condition is frequently changeable and thus quite unpredictable. Therefore, the channel modeling of this band is relatively complex, and this band does not have any perfect channel model.

Device Capability: The 6G system will provide a number of new features. Devices, such as smart phones, should have the capability to cope with the new features. In particular, it is challenging to support 1 Tbps throughput, AI, XR, and integrated sensing with communication features using individual devices. The 5G devices may not support few of the 6G features, and the capability improvement in 6G devices may increase the cost as well. There will be billions of devices connected to the 5G technology; therefore, we need to ensure that those devices are compatible with the 6G technology also.

High-Capacity Backhaul Connectivity: The access networks in 6G will have a very high density. Moreover, these access networks are diverse in nature and widespread within a geographical location. Each of these access networks will support very high-data-rate connectivity for diverse types of users. The backhaul networks in 6G must handle the huge volume of data for connecting between the access networks and the core network to support high-data-rate services at the user level; otherwise, a bottleneck will be created. The optical fiber and FSO networks are possible solutions for high-capacity backhaul connectivity; therefore, any improvement in the capacity of these networks is challenging for the exponentially growing data demands of 6G.

Spectrum and Interference Management: Due to the scarcity of the spectrum resources and interference issues, it is very important to efficiently manage the 6G spectra including the spectrum-sharing strategies and innovative spectrum management techniques. Efficient spectrum management is important for achieving the maximum resource utilization with QoS maximization. In 6G, researchers have to address concerns, such as how to share the spectrum, and how to manage the spectrum mechanism in the heterogeneous networks that synchronize the transmission at the same frequency. Researchers also need to investigate how the interference can be cancelled using the standard interference cancellation methods, such as parallel interference cancellation, and successive interference cancellation.

Beam Management in THz Communications: Beam forming through massive MIMO systems is promising technology for supporting high-data-rate communications. However, beam management in sub-mmWave, that is, the THz band is challenging because of the propagation characteristics of the sub-mmWave.

CONCLUSION

The mobile communication system has been evolving technically every decade, while the services of mobile communications have changed greatly in cycles of approximately 20 years. Therefore, the "Third

Wave" initiated by 5G is expected to become a larger wave through 5G evolution and the sixth generation (6G) technology, and will support industry and society in the 2030s. The objectives of 6G technology include simultaneous achievement of several requirements such as ultra-high-speed, high-capacity, and low-latency connectivity. The pioneering of new frequency bands including terahertz frequencies and expansion of communication coverage in the sky, at sea, and in space will include provisioning of extremely low energy and low-cost, reliable communications with capabilities for extremely massive connectivity and sensing. The vision of the 6G network and the next-generation network requirements indicate that the 6G network will immensely outperform the 5G network due to its ability to serve the extreme needs of future use cases such as autonomous vehicles, tele healthcare, industry automation, and the rise of several verticals.

Some of the enabling technologies for 6G include artificial intelligence, THz communication, 3D connectivity, Visible light communication, block chain, quantum communication, intelligent surfaces and digital twins. The promising 6G architectures like Cyber-Twin architecture, decoupled RAN architecture, generalized architecture for 6G, a high-level architecture of 6G, a multi-level architecture for 6G, three-dimensional 6G architecture and Neuroscience based 6G architecture have been discussed.

The applications of 6G that includes Super-smart society, Extended reality (XR), Connected robotics and autonomous systems, wireless brain–computer interactions, haptic communication, smart healthcare, automation, and manufacturing, five sense information transfer and internet of everything have been discussed. The 6G technologies research carried out worldwide in Finland, China, USA, South Korea, Japan, and Europe have been highlighted. The future technology explorations of 6G for researchers that include security provision, virtualization of radio access interface, vertical edge caching, mobility and localization, rural connectivity, analysis of meta materials and reflecting surfaces, EI and sustainable goals have been discussed.

REFERENCES

Akyildiz, Kak, & Nie. (2020). 6G and beyond: The future of wireless communications systems. *IEEE Access*, *8*, 133995-134030.

Andreev, S., Petrov, V., Dohler, M., & Yanikomeroglu, H. (2019, June). Future of ultra-dense networks beyond 5G: Harnessing heterogeneous moving cells. *IEEE Communications Magazine*, *57*(6), 86–92. doi:10.1109/MCOM.2019.1800056

Andrews, G., & Zhang, J. (2014). What will 5G be? *IEEE Journal on Selected Areas in Communications*, *32*(6), 1065–1082. doi:10.1109/JSAC.2014.2328098

Calvanese Strinati, Barbarossa, Gonzalez-Jimenez, Ktenas, Cassiau, Maret, & Dehos. (2019). 6G: The next frontier: From holographic messaging to artificial intelligence using subterahertz and visible light communication. *IEEE Veh. Technol. Mag.*, *14*(3), 42-50.

Chowdhury, M., & Jang, Y. (2019). *6G wireless communication systems: Applications, requirements, technologies, challenges, and research directions*. https://arxiv.org/abs/1909.11315 doi:10.1109/ACCESS.2018.2844843

Chowdhury, M. Z., Hossan, M. T., & Jang, Y. M. (2018, June). Interference management based on RT/nRT traffic classification for FFR-aided small cell/macrocell heterogeneous networks. *IEEE Access : Practical Innovations, Open Solutions*, *6*, 31340–31358. doi:10.1109/ACCESS.2018.2844843

David & Berndt. (2018). 6G vision and requirements: Is there any need for beyond 5G? *IEEE Veh. Technol. Mag.*, *13*(3), 72-80.

Giordani, M. (n.d.). *Towards 6G networks: use cases and technologies.* arXiv:1903.12216

Giordani, Polese, Mezzavilla, Rangan, & Zorzi. (2020). Towards 6G networks: Use cases and technologies. *IEEE Commun. Mag., 58*(3), 51-63.

Hewa, Gür, Kalla, Ylianttila, Bracken, & Liyanage. (2020). The role of block chain in 6G: Challenges, opportunities and research directions. *Proc. 2nd 6G Wireless Summit (6G SUMMIT),* 1-5.

Huang, Yang, Wu, Ma, Zhang, & Zhang. (2019). A survey on green 6G network: Architecture and technologies. *IEEE Access*, *7*, 175758-175768.

Jaber, M., & Tukmanov, A. (2016). 5G backhaul challenges and emerging research directions: A survey. *IEEE Access : Practical Innovations, Open Solutions*, *4*, 1743–1766. doi:10.1109/ACCESS.2016.2556011

Kantola, R. (2020). Trust networking for beyond 5G and 6G. Proc. 2nd 6G Wireless Summit (6G SUMMIT), 1-6. doi:10.1109/6GSUMMIT49458.2020.9083917

Katz, M., Matinmikko-Blue, M., & Latva-Aho, M. (2018). 6 Genesis flagship program: Building the bridges towards 6G-enabled wireless smart society and ecosystem. *Proc. IEEE 10th Latin-American Conf. Commun. (LATINCOM)*, 1-9.

Khan & Hong. (2020). 6G wireless systems: A vision, architectural elements, and future directions. *IEEE Access, 8*, 147029-147044.

Letaief, Chen, Shi, Zhang, & Zhang. (2019). The roadmap to 6G: AI empowered wireless networks. *IEEE Commun. Mag.*, *57*(8), 84-90.

Lopez-Lopez, Matinmikko-Blue, Cardenas-Juarez, Stevens-Navarro, Aguilar-Gonzalez, & Katz. (2020). Spectrum challenges for beyond 5G: The case of Mexico. *Proc. 2nd 6G Wireless Summit (6G SUMMIT),* 1-5.

Lu. (2020). Security in 6G: The prospects and the relevant technologies. *J. Ind. Integr. Manage., 5*, 1-24.

Lu & Zheng. (2020). 6G: A survey on technologies, scenarios, challenges, and the related issues. *J. Ind. Inf. Integr., 19*, 1-52.

Mahbas, A. J., Zhu, H., & Wang, J. (2019, February). Impact of small cells overlapping on mobility management. *IEEE Transactions on Wireless Communications*, *18*(2), 1054–1068. doi:10.1109/TWC.2018.2889465

Moioli & Latré. (2020). Neurosciences and 6G: Lessons from and needs of communicative brains. Available: https://arxiv.org/abs/2004.01834 doi:10.1109/TWC.2018.2889465

Monserrat, J. F., Martin-Sacristan, D., Bouchmal, F., Carrasco, O., de Valgas, J. F., & Cardona, N. (2020). Key technologies for the advent of the 6G. *Proc. IEEE Wireless Commun. Netw. Conf. Workshops (WCNCW)*, 1-6. 10.1109/WCNCW48565.2020.9124725

Mumtaz, S. (2017). Terahertz communication for vehicular networks. *IEEE Transactions on Vehicular Technology*, *66*(7), 5617–5625. doi:10.1109/TVT.2017.2712878

Nawaz, S., & Asaduzzaman, M. (2019). Quantum machine learning for 6G communication networks: State-of-the-art and vision for the future. *IEEE Access : Practical Innovations, Open Solutions*, *7*, 46317–46350. doi:10.1109/ACCESS.2019.2909490

Nguyen, T., Tran, N., Loven, L., Partala, J., Kechadi, M.-T., & Pirttikangas, S. (2020). Privacy-aware blockchain innovation for 6G: Challenges and opportunities. Proc. 2nd 6G Wireless Summit (6G SUMMIT), 1-5. doi:10.1109/6GSUMMIT49458.2020.9083832

Peltonen, E., Leppanen, T., & Loven, L. (2020). EdgeAI: Edge-native distributed platform for arti_cial intelligence. Proc. 2nd 6G Wireless Summit (6G SUMMIT), 1-2.

Pouttu, A., Burkhardt, F., Patachia, C., & Mendes, L. (2020). *6G white paper on validation and trials for verticals towards 2030.* 6GFlagship, 6G White Paper.

RajathevaN.White paper on broadband connectivity in 6G. Available: http://arxiv. org/abs/2004.14247

Rappaport & Trichopoulos. (2019). Wireless communications and applications above 100 GHz: Opportunities and challenges for 6G and beyond. *IEEE Access, 7*, 78729-78757.

Saad, W., & Chen, M. (2020). *A vision of 6G wireless systems: applications, trends, technologies, and open research problems.* White Paper on 6G-the Next Hyper Connected Experience for All, Samsung, Seoul, South Korea.

Samdanis & Taleb. (2020). The road beyond 5G: A vision and insight of the key technologies. *IEEE Netw., 34*(2), 135-141.

Sergiou, Lestas, Antoniou, Liaskos, & Pitsillides. (2020). Complex systems: A Communication networks perspective towards 6G. *IEEE Access, 8*, 89007-89030.

Shafi, M., Molisch, A. F., Smith, P. J., Haustein, T., Zhu, P., De Silva, P., Tufvesson, F., Benjebbour, A., & Wunder, G. (2017). 5G: a tutorial overview of standards, trials, challenges, deployment, and practice. *IEEE Journal on Selected Areas in Communications*, *35*(6), 1201–1221. doi:10.1109/JSAC.2017.2692307

Shatin, Liu, Chandrasekhar, Chen, Reed, & Zhang. (2019). Artificial intelligence-enabled cellular networks: A critical path to beyond-5G and 6G. *IEEE Wireless Commun.*, *27*(2), 212-217.

SheC.VuceticB. (2020). A tutorial on ultra-reliable and low-latency communications in 6G: Integrating domain knowledge into deep learning. Available: https://arxiv.org/abs/2009.06010

SheC.VuceticB. (2020). Deep learning for ultra-reliable and low-latency communications in 6G networks. Available: https://arxiv.org/abs/2002.11045

Sliwa, Falkenberg, & Wietfeld. (2020). Towards cooperative data rate prediction for future mobile and vehicular 6G networks. *Proc. 2nd 6G Wireless Summit (6G SUMMIT)*, 1-5.

Soderi. (2020). Enhancing security in 6G visible light communications. *Proc. 6G Wireless Summit*, 1-5.

StoicaR.AbreuF. (2019). 6G: The wireless communications network for collaborative and AI applications. Available: https://arxiv.org/abs/1904.03413

Strinati & Dehos. (2019). 6G: The next frontier: From holographic messaging to artificial intelligence using subterahertz and visible light communication. *IEEE Veh. Technol. Mag., 14*(3), 42-50.

Tariq, F. (n.d.). *A Speculative Study on 6G*. arXiv:1902.06700.

TatariaH.TufvessonF. (2020). 6G wireless systems: Vision, requirements, challenges, insights, and opportunities. Available: https://arxiv.org/abs/2008.03213

YangH.AlphonesA.XiongZ.NiyatoD.ZhaoJ.WuK. (2019). Artificial intelligence-enabled intelligent 6G networks. Available: https://arxiv.org/abs/1912.05744

YuQ.ZhangW. (2019). Cybertwin: An origin of next generation network architecture. Available: https://arxiv.org/abs/1904.11313

Yu, Q., & Zhang, W. (2020). A cybertwin based network architecture for 6G. Proc. 2nd 6G Wireless Summit (6G SUMMIT), 1-5. doi:10.1109/6GSUMMIT49458.2020.9083808

YuanX.ZhangY.-J. A.ShiY.YanW.LiuH. (2020). Reconfigurable intelligent- surface empowered wireless communications: Challenges and opportunities. Available: https://arxiv.org/abs/2001.00364

Yuan, Zhao, Zong, & Parolari. (2020). Potential key technologies for 6G mobile communications. *Sci. China Inf. Sci., 63*(8), 1-19.

Zadid Shifat, A. S. M. (2018, January). Game-based approach for QoS provisioning and interference management in heterogeneous networks. *IEEE Access : Practical Innovations, Open Solutions, 6*, 10208–10220. doi:10.1109/ACCESS.2017.2704094

Zanella, Filgueiras, Valério, Dartora, Mariano, & Cerqueira. (2020). Nano-antenna modelling based on plasmonic charge distribution for THz-based 6G applications. *Proc. 2nd 6G Wireless Summit (6G SUMMIT),* 1-4.

Zhang, D., & Sato, T. (2016). One integrated energy efficiency proposal for 5G IoT communications. *IEEE Internet of Things Journal, 3*(6), 1346–1354. doi:10.1109/JIOT.2016.2599852

Zhou, T., Jiang, N., Liu, Z., & Li, C. (2018). Joint cell activation and selection for green communications in ultra-dense heterogeneous networks. *IEEE Access : Practical Innovations, Open Solutions, 6*, 1894–1904. doi:10.1109/ACCESS.2017.2780818

KEY TERMS AND DEFINITIONS

Artificial Intelligence (AI): Is intelligence—perceiving, synthesizing, and inferring information—demonstrated by machines, as opposed to intelligence displayed by non-human animals and humans.

Example tasks in which this is done include speech recognition, computer vision, translation between (natural) languages, as well as other mappings of inputs.

Augmented Reality: Is an interactive experience that combines the real world and computer-generated content. The content can span multiple sensory modalities, including visual, auditory, haptic, somato, sensory, and olfactory.

Autonomous Vehicles: Is one that can drive itself from a starting point to a predetermined destination in "autopilot" mode using various in-vehicle technologies and sensors, including adaptive cruise control, active steering (steer by wire), anti-lock braking systems (brake by wire), GPS navigation technology, lasers.

Brain-Computer Interface (BCI): Sometimes called a brain–machine interface (BMI) or smart brain, is a direct communication pathway between the brain's electrical activity and an external device, most commonly a computer or robotic limb.

Communication Systems: Is a collection of individual telecommunications networks, transmission systems, relay stations, tributary stations, and terminal equipment usually capable of interconnection and interoperation to form an integrated whole. The components of a communications system serve a common purpose, arc technically compatible, use common procedures, respond to controls, and operate in union.

Cyber Twin Architecture: Helps out in serving stronger communication and also contains several features that help out in assisting communication like maintaining a log record of network data and managing all digital assets like images, audio, videos, etc.

Data Loggers: Are electronic devices which automatically monitor and record environmental parameters over time, allowing conditions to be measured, documented, analyzed, and validated. The data logger contains a sensor to receive the information and a computer chip to store it. Then the information stored in the data logger is transferred to a computer for analysis.

Digital Twins: Is a digital representation of an intended or actual real-world physical product, system, or process (a physical twin) that serves as the effectively indistinguishable digital counterpart of it for practical purposes, such as simulation, integration, testing, monitoring, and maintenance.

Edge Cloud: Refers to a middle ground between cloud computing and the edge of a network. This is often not a precise definition because the area between the edge and the cloud is ambiguous.

Haptic Communication: Is a branch of nonverbal communication that refers to the ways in which people and animals communicate and interact via the sense of touch. Touch is the most sophisticated and intimate of the five senses.

Internet of Everything (IoE): Is the most innovative and Ubiquitous technology advancement which is going to make networked connections more relevant and valuable than ever before. Turning information into action creates new capabilities, richer experiences, and unprecedented economic opportunities for businesses, individuals, and countries.

Internet of Things (IoT): Describes physical objects (or groups of such objects) with sensors, processing ability, software and other technologies that connect and exchange data with other devices and systems over the Internet or other communications networks.

Long-Term Evolution (LTE): Is a fourth-generation (4G) wireless standard that provides increased network capacity and speed for cell phones and other cellular devices compared with third-generation (3G) technology.

Mixed Reality (MR): Is a term used to describe the merging of a real-world environment and a computer-generated one. Physical and virtual objects may co-exist in mixed reality environments and interact in real time.

Mobile Communication: Is the use of technology that allows us to communicate with others in different locations without the use of any physical connection (wires or cables) which makes our life easier as it saves time and effort.

Mobile Edge Computing (MEC): Also known as multi-access computing, is the near-real-time processing of large amounts of data produced by edge devices and applications closest to where it's captured—in other words, extending the edge of your edge network infrastructure.

Optical Wireless Communication (OWC): Is a form of optical communication in which unguided visible, infrared (IR), or ultraviolet (UV) light is used to carry a signal. It is generally used in short-range communication.

Quality of Service (QoS): Is the description or measurement of the overall performance of a service, such as a telephony or computer network, or a cloud.

Radio Access Network (RAN): Is the latest architecture in wireless communication and uses 5G radio frequencies to provide wireless connectivity to devices. A radio access network (RAN) is a key component of a mobile telecommunication system that connects devices like smartphones to a network via a radio link.

Unmanned Aerial Vehicles (UAV): Commonly known as a drone, is an aircraft without any human pilot, crew, or passengers on board.

Virtual Reality (VR): Is a simulated experience that employs pose tracking and 3D near-eye displays to give the user an immersive feel of a virtual world.

Chapter 3
Recent and Emerging Technologies in Industrial IoT

Rekha R. Nair
School of Engineering, Dayananda Sagar University,Bengaluru, India

Tina Babu
School of Engineering, Dayananda Sagar University, Bengaluru, India

Kishore S.
School of Engineering, Dayananda Sagar University, Bengaluru, India

ABSTRACT

The industrial internet of things (IIoT) is evolving through remote monitoring, cognitive analytics, and industrial process control. Through product customization, intelligent monitoring applications for production floor shops and machine health, and predictive and preventive maintenance of industrial equipment, the primary goal of IIoT is to achieve high operational efficiency, increased productivity, and better management of industrial assets and processes. But because the industrial sector is only now beginning to implement full-stack development solutions with IIoT, there is a need to deal with the problems that are emerging. The IIoT keeps industrial equipment, machines, and cloud-based applications connected. The authors emphasize the advantages of IIoT, which is one of the chapter's three key themes, which include emerging IIoT hardware and software technologies followed by the challenges of IIoT. Further, this chapter focuses on recent trends and technologies in industrial IoT and challenges in IIoT.

INTRODUCTION

In the industrial sector, disruptive technologies are creating enormous waves by enabling enterprises to drastically enhance their access to data to help connect people, organizations, and technologies. Organizations can now easily access factory, production, and industrial equipment data from a distance thanks to these revolutionary technologies Behrendt (2019). The Internet of Things (IoT) is the technology that has caused the most disruption in recent years. Two of the key breakthroughs in the digitizing trend are the

DOI: 10.4018/978-1-7998-9266-3.ch003

Internet of Things (IoT) and the Industrial Internet of Things (IIoT). IoT and IIoT can assist businesses in functioning optimally, making better decisions, and opening up new funding sources by linking multiple devices and items of equipment through the Internet Khajenasiri et al. (2017).

Functioning of Internet of Things (IoT)

A network of connected devices known as the Internet of Things (IoT) allows users to access data and exchange messages. IoT devices often feature sensors that allow them to collect data and connect to the Internet. The IoT consists of a wide variety of gadgets Riahi Sfar et al. (2018). With more connected devices being added daily, it can include anything from factory equipment to electrical substations to buildings and infrastructure. The IoT is used by a range of different types of enterprises, including manufacturers, energy providers, local governments, and others.

With the use of IoT technology, you can gather data automatically from many different tasks, such as how much energy a building's lighting uses or how much water is passing through a sewage treatment facility. The data that IoT systems and devices collect can be sent to a centralized system over the Internet Karmakar et al. (2019). In order to make decisions, managers can use this data. It is possible to go further into the data using data analysis techniques in order to make predictions about the future and gain deeper insights.

Functioning of Industrial Internet of Things (IoT)

The development and implementation of the internet of things (IoT) throughout industrial applications and sectors is referred to as the "industrial internet of things" (IIoT). The IIoT helps businesses and industries to operate more efficiently and reliably thanks to its strong emphasis on machine-to-machine (M2M) connectivity, big data, and machine learning Meyer et al. (2018). The IIoT includes software-defined production processes, robotics, and medical advancements.

Beyond the standard networking of physical gadgets and consumer electronics related to the IoT, the IIoT encompasses a wider range of technologies. It differs due to the point at which operational technology (OT) and information technology (IT) converge. Supervisory control, Industrial control systems (ICSs), such as programmable logic controllers, distributed control systems (DCSs), and data acquisition (SCADA) systems, and human machine interfaces (HMIs), are referred to as operational processes when they are networked (PLCs).

A subtype of IoT is IIoT. IoT technology utilized in industrial settings, namely in manufacturing plants, is what the phrase alludes to. Industry 4.0, the subsequent phase of the industrial revolution, heavily relies on the IIoT. Smart technology, data, automation, interconnection, artificial intelligence, and other technologies and capabilities are highlighted by Industry 4.0 Zhu et al. (2018). The management of factories and industrial organizations is being revolutionized by these technologies.

Many of the same applications and advantages as IoT are possible with IIoT. Manufacturing equipment, energy systems, and infrastructure like pipelines and wiring can all use smart sensors. These sensors can assist industrial businesses in increasing their efficiency, production, employee safety, and other factors through the data they collect and the cutting-edge capabilities they provide.

Plant managers receive data from the IIoT that improves machine-to-machine connectivity and gives them a better understanding of how their facility is running. Industrial businesses can keep a better eye on how much energy, water, and other resources they're using, as well as when and how much they're creating, by continuously gathering granular data Aazam et al. (2018). To maximize their operation, operators can then make manual modifications or have equipment that can adjust automatically. Businesses may save a lot of energy, water, and resources by continuously optimizing their processes, while maintaining or even boosting productivity. This is how IIoT may be utilized to support businesses in achieving their lean manufacturing objectives. IIoT can be used for additional commercial objectives, such as accelerating product development and informing predictive maintenance programmes.

Difference Between IoT and IIoT

As you distinguish between IoT and industrial IoT, it is crucial to realise that IIoT is a subset of IoT technology that pertains exclusively to technologies used for industrial operations as you distinguish between IoT and industrial IoT (Sisinni et al., 2018; Liao et al., 2018). The different application areas of IOT and IIOT are illustrated in Figure 1. While there are many similarities between IIoT and traditional IoT applications, there are also notable variances.

1. **Focus on the Market:** IoT covers a wide range of industries and end users. IoT technology may be used by consumers as well as experts in fields including healthcare, business, and the public sector. IoT hence tends to concentrate on broader applications. IIoT technology, in comparison, has a narrower market emphasis because it is exclusively employed in industrial settings by professionals. Power plants, oil and gas refineries, and manufacturing facilities are some of the key locations for IIoT applications.

Figure 1. Application areas of IoT and IIoT
Source: Aazam et al. (2018)

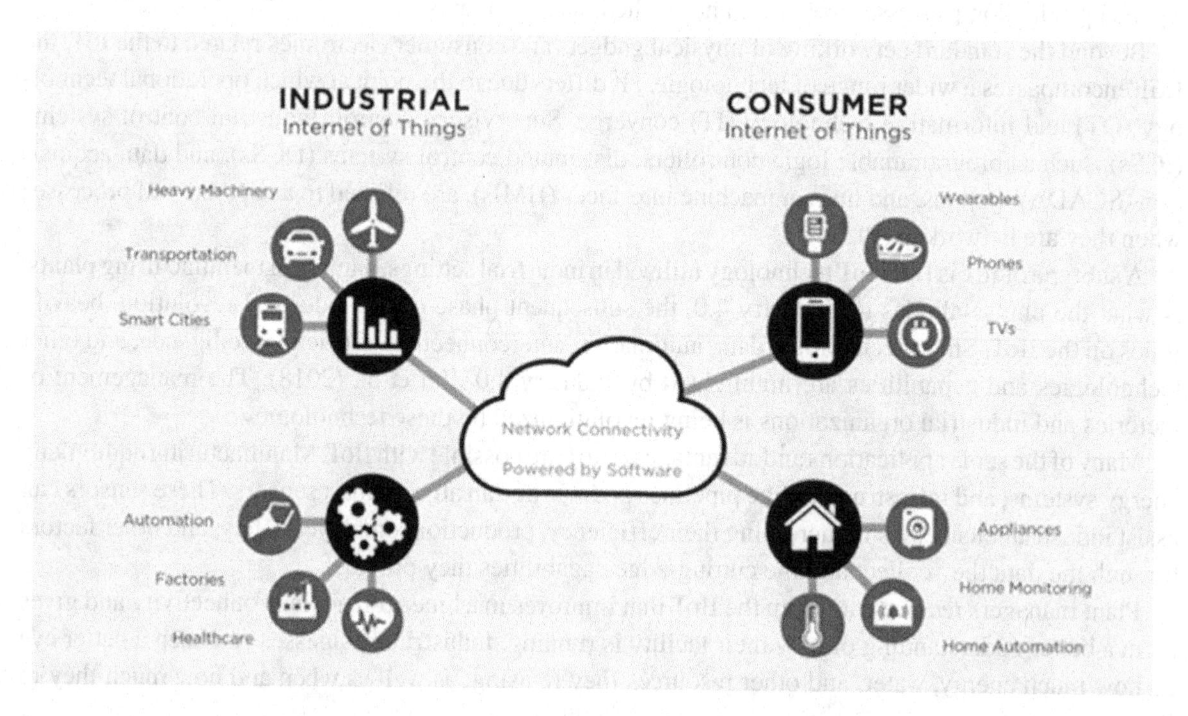

2. **Environmental Criteria:** The architecture of IoT devices is often built to endure the pressures of common settings, including average temperatures. IIoT devices must be more dependable and robust because they are employed in tougher environments like factories, oil refineries, and power plants. In order to guarantee consistent performance, IIoT device makers often build their products to endure dampness, radio interference, and high temperatures.
3. **Goals:** IoT and IIoT users typically have slightly different objectives in mind. The deployment of IoT frequently attempts to boost productivity, enhance health and safety, and offer better experiences. IIoT is less user-centric and often more focused on the first two objectives. IIoT is not used by ordinary people in their daily lives; rather, it is a component of industrial processes.
4. **Compliance With Old Technology:** IoT devices typically don't have to be backwards compatible with legacy systems. Device makers don't have to make these devices backward compatible because they frequently function independently. IIoT devices, on the other hand, frequently need to be interoperable with a variety of legacy machinery and devices used in industrial operations. Many IIoT devices are needed to assist legacy equipment in providing digital data and accepting IT system commands because these plants frequently feature equipment without digital interfaces or capabilities.
5. **End Devices:** IIoT and IoT frequently make use of various equipment because their focuses and objectives fluctuate. Devices in the IIoT are intended to give users information about their equipment, integrating with it rather than operating independently. Controllers and Programmable Logic Controllers are two examples of these gadgets. IoT devices are typically everyday objects that you can use on your own, such as smart assistants, watches, and thermostats. IoT smart sensors and other devices are occasionally utilised to upgrade infrastructure.
6. **Needs for Development:** Companies that create new IoT devices typically want to make users daily lives more convenient. Because there is such a great desire for usability among consumers, development is more heavily weighted toward enhancing comfort. IIoT development normally concentrates on creating new gadgets that can improve client operations. IIoT developers use data metrics to create devices that help businesses reduce costs and boost efficiency because industrial plants need to optimise their processes to stay competitive in their industry.
7. **Failure Possibility:** Since IoT devices are only sometimes used, there is a low danger involved when they malfunction. IoT devices aren't typically employed for critical functions that could endanger life or be hazardous if they go wrong. When compared to IoT, IIoT devices and technology failure can be more hazardous because they are connected via a network and can result in potentially fatal scenarios when a piece of large machinery breaks down.

Benefits of IIOT

In a market that is changing quickly, producers today must move at breakneck speed to stay up with shifting consumer needs, market trends, and international rivals (Zhu et al., 2018); Aazam et al., 2018); ur Rehman et al., 2019). The advantages of the IIoT have an impact on everything, despite the fact that many conventional obstacles—including rising prices, unplanned downtime, quality control (QC), changing business requirements, and ageing infrastructure—remain obstacles to its development.

1. **Minimizing Errors:** Thanks to industrial IoT, manufacturers can digitize almost every aspect of their operations. By reducing the number of manual methods and submissions, manufacturers can reduce the primary risk associated with manual labor, human error. This would go further

than inefficiencies in both production and operations. IIoT technologies can help reduce the risk of cyberattacks and data theft caused by gross negligence. FIn accordance with a poll by Cyber Security Trend, individuals are the primary cause of breaches in cyber security with human error being responsible for 37% of instances. Software and hardware with AI and machine learning capabilities can do most calculations on their own, eliminating the danger of a human making a minor mistake and compromising the manufacturer's data.

2. **Enhance Safety:** Security in the workplace is being improved by the data and sensors required for an IIoT manufacturing process to function properly. In the situation of an accident, every person within the building can be informed of it, services may be halted, and top management can get involved to make sure the incidents and problems are handled. Such mishap might also provide insightful data that can be applied to prevent a repeat from happening in the future. A more recent alternative that some manufacturers are implementing among their personnel is wearable technology. Wearables have been a part of the IoT from the outset, but industrial IoT applications have only lately started to leverage them. By enabling management to keep an eye on things like employee posture and outside noise levels, wearables can enhance the working environment and perhaps even productivity. They could also alert staff when proper workplace safety procedures aren't being followed.
3. **Boost Performance:** The main benefit of IIoT is the ability for manufacturers to computerise, increasing their operational performance. Robotics and computerized machinery can work better accurately and efficiently, boosting revenue and helping businesses streamline existing procedures. Furthermore, sensors which constantly productivity factors can link hardware and software. As a result, manufacturers have a better awareness of the firm profitability of both individual pieces of equipment and whole navies. Using IIoT-enabled information systems, manufacturers can improve process efficiency, which include:
    - Digitizing manual operations and substituting digital ones for them.
    - Relying on data to arrive at decisions regarding all aspects of manufacturing.
    - Keeping track of performance from anywhere, either on the production floor or thousands of kilometers away.
4. **Prevention-Based Maintenance:** The greatest detrimental effect on a manufacturing company is machine downtime. Manufacturers are compelled to determine the root cause of the issue, the best course of action, and the estimated cost when production maintenance is reactive rather than proactive. The baseline and the related statistics provide companies with the information they have to spot issues before they occur.
5. **Minimize Costs:** Information insights on operations, manufacturing, advertising, and revenues can guide businesses in the direction of profitability. The aforementioned benefits of IIoT, including condition monitoring, reduced errors, enhanced quality assurance, and maximize throughput, will all boost a manufacturer's earnings. The leaders of a manufacturing organization can view the insights that are accessible at any time and from any location as the most crucial tool supplied by industrial IoT.

TECHNOLOGIES EXPLORED

Anyone working in IIoT environments and projects understands which technologies they should be watching, evaluating, and perhaps deploying. The IIoT's ideas and techniques will affect technical areas like architecture and network design in addition to business strategy and risk management.

Blockchain

As a result of connecting people, things, and places, the Internet of Things (IoT) creates and captures opportunities for value. Inanimate objects now contain sophisticated electronics, sensors, and actuators that all broadcast data to the Internet of Things.The mainstream adoption of IoT has been impeded by a severe security problem. Due to its various security flaws, Internet of Things (IoT)-connected devices are a prominent target for Distributed Denial of Service (DDoS) attacks. When several exploited computer systems flood a target—such as a central server—with a huge number of simultaneous data requests, DDoS assaults cause a denial of service for users of the targeted system. Cybercriminals can exploit the lack of security on unprotected IoT devices to hack them into initiating DDoS attacks, making them an easy target.

Another ground-breaking innovation that might help with some of the scalability and security challenges with the IoT is blockchain, commonly referred to as distributed ledger technology (DLT) Uddin et al. (2021). Blockchain is a "game changer" in the information age. A distributed digital ledger that is accessible over the Internet and shared by all system users is the essential building element of a blockchain system. An event or activity that has been verified and recorded in the ledger cannot be changed or deleted. It gives a user community a way to gather and disseminate information. High-level security (Alladi et al., 2019) is imposed through blockchain, which uses decentralized, distributed ledgers to authenticate and authorize encrypted device-generated data. Millions of devices are used in a distributed ledger to compute and store data. In contrast to the old approach, the failure of one device, server, or network won't have an impact on the entire Internet of Things ecosystem.

Blockchain Layers

The blockchain or distributed ledger (DLT) operates according to a predefined protocol, with the network's many computers (or nodes) reaching a "consensus" to validate transactional data (Panarello et al., 2018). Each node adds, checks, and modifies new entries as they come in. Blockchains have a layered design that enables this particular type of transaction authentication. Blockchain technology contains five layers:

1. **The Hardware Layer:** A data server has a secure backup copy of the blockchain data. Our computers request access to this data from the server whenever we use any blockchain applications or visit the internet. Through these networks, one node can easily and quickly connect with other nodes to share data. This data interchange is made possible by the client-server architecture.
2. **The Data Layer:** A chain of hashed blocks containing transactional records is known as a "blockchain." The blockchain's very first block is called the Genesis block. The next step is to link each new block that is added to the blockchain to the Genesis block using an iterative technique. And as a result, the blockchain keeps expanding. Every transaction is "digitally signed" using the sender's wallet private key. Because this key is only visible to the sender, it is impossible for anyone else to view or change the data. Additionally protecting the owner's identity is the digital signature, which is encrypted for the maximum level of protection. This concept is known as "finality" in the context of blockchain technology.
3. **The Network Layer:** To determine whether a transaction is legitimate, the P2P infrastructure enables many nodes to exchange transaction data. This indicates that in order to communicate fast, every node on the network must be able to find other nodes. The network layer enables this "inter-node

communication". Since it oversees node detection, block generation, and block adding, this layer is often referred to as the "Propagation Layer."

4. **The Consensus Layer:** Transaction authentication is handled by this layer, one of the most important for blockchain functionality. Because transactions won't be validated without this layer, the system won't operate as intended. The protocol is carried out by this layer and requires a specific number of nodes to validate a single transaction. Since multiple nodes process each transaction, they must all reach the same decision and approve it as valid. Many blocks may happen concurrently, leading to a branch in the blockchain, as a result of a large number of nodes processing transactions, bundling them, and adding them to the blockchain. However, the consensus layer ensures that this disagreement is settled and that a single chain block addition is always necessary.
5. **Application and Presentation Layer:** The applications that end users employ to create blockchain network connection make up the application layer. The application layer is made up of scripts, frameworks, UIs (user interfaces), APIs (application programming interfaces), chaincode, smart contracts, and Dapps (decentralized applications). The protocols of the application layer are divided into the application layer and the execution layer. With the use of APIs, the blockchain network functions as the back-end system for the applications. The execution layer is made up of underlying protocols, smart contracts, and chaincode.

Objectives of Blockchain in IIOT

The emergence of blockchain technology has had numerous positive effects in a range of businesses in trustless environments (Frizzo-Barker et al., 2020). Following listed are some of the objectives of blockchain technology:

1. **Decentralization:** By removing the need for a reliable third party in the IoT network, blockchain's decentralized system offers a viable method for overcoming bottleneck and one-point failure issues. The functionality of the blockchain and IoT network is unaffected by the failure of a blockchain node. Blockchain data is typically maintained across numerous P2P nodes, and the system is extremely resilient to hostile hacking attempts and technological failures.
2. **Improved Security:** A trusted party can check and approve transactions that are initiated by the blockchain technology, and data transit and storage are both encrypted. Blockchain technology keeps track of all transactions, makes it apparent who has access and who is dealing. By removing a single point of failure, encrypting data, and allowing for immediate identification of the weak spot in the entire network, blockchain adds a layer of protection.
3. **Lower Costs:** The entire ecosystem can be proactive at a lower cost by automating the blockchain's transaction validation and processing stages.
4. **Speed of Transactions:** This is particularly valid for supply chain transactions involving a number of suppliers, producers, distributors, and customers. Untrusted parties can communicate data directly with one another thanks to the blockchain, which functions in part as a shared ledger. This speeds up transactions by getting rid of manual procedures.

Challenges of Blockchain in IIoT

IIOT is a high risk, high reward endeavor. In this section, we'll look at some of the biggest industrial IIoT challenges, risks, and pitfalls that business leaders could experience. In this part, the following pitfalls will be covered. The following is a summary of the challenges in handling IoT data on the blockchain is listed below:

1. **Power Consumption, Performance, and Security Trade-Offs:** Due to the high processing requirements of blockchain algorithms, the creation of these technology-based apps for devices with constrained resources has been significantly slowed down.
2. **Data Concurrency and Throughput Issue:** Due to the constant data transmission from IoT devices, IoT systems have high concurrency. The blockchain's throughput is limited by its complex consensus processes and cryptographic security protocol. In a chain-structured ledger, faster synchronization of new blocks across blockchain nodes is required, which might boost throughput. To fulfill the demand for frequent transactions in IoT systems, the challenge is to increase blockchain's throughput.
3. **IoT Connectivity:** In order to exchange IoT data with potential stakeholders, it is predicted that IoT devices would be linked to advanced processing, storage, and networking capabilities. In order to provide new business prospects for the creation of new applications and services in a variety of fields, the Internet of Things' (IoT) restricted the ability to connect with blockchain technology.
4. **Blockchain for Massive Data Handling:** The whole distributed ledger is kept locally by each participant in the blockchain network. Each node adds the confirmed block to their local ledger when a new block is broadcast over the whole P2P network.
5. **Maintaining Privacy and Transparency:** Transparency of transactions can be ensured through blockchain, which is a must in some applications, like finance. However, when storing and accessing IoT, user confidentiality may suffer (Biswas et al., 2019).

Edge Computing

Data is the keystone of modern business, enabling real-time management over crucial corporate operations and procedures while also offering insightful business information. Businesses today are drowning in data, and vast volumes of data may be routinely gathered from sensors and IoT devices working in real time from far-flung locations and hostile operating environments practically anywhere in the globe. Moving continually expanding rivers of real-world data is not well suited for the traditional computer paradigm, which is based on a centralized data center and everyday internet. Such initiatives may be hampered by bandwidth restrictions, latency problems, and unpredictably occurring network interruptions. Edge computing architecture is being used by businesses to address these data concerns.

Numerous new smart devices (Premsankar et al., 2018) are being connected to the internet every day, which creates Big Data in turn. This data needs reliable and powerful cloud storage in order to be saved and protected from illegal access. Furthermore, this data requires a lot of processing power in order to yield any useful information. Many business operations are looking to technology to improve the performance of IoT devices and save running costs.

The proximity of processing power to physical devices or data sources is advantageous for the Internet of Things (IoT). In order to analyze IoT device data and respond faster or address issues, it must be analyzed at the edge rather than traveling back to a central location. It has both downstream and upstream streams. By downstream, we mean that the IoT devices are receiving data from the cloud. The cloud services are served by this stream. For the Edge and cloud paradigms to function, both of these streams are essential. The term "edge" refers to any network resource with processing power in the line of sight between a cloud and IoT devices (Shi & Dustdar, 2016); Kaur et al., 2018). Figure 2 indicates the detailed architecture for the Edge computing.

In the layer below, IoT end user devices are depicted. IoT/end user devices serve as both data consumers and providers (Goyal et al., 2020). These include all of the smart devices that are now on the market, such as smart watches, cars, homes, TVs, parking systems, smart lights, and smart speakers. Smart cameras, computers, and other devices. The edge can be made up of devices like gateways, routers, tiny servers, and micro data centers. They can perform computer tasks and access information and services over the cloud at the same time. Providing services, offloading computing, managing the Internet of Things, caching, storing, and analyzing data are all functions that are completed at the edge.

The Edge must be built with dependability, security, and privacy protection in mind in order to efficiently meet the criteria. All of the databases, servers, and supercomputers that store large amounts of data and carry out complicated computations are together referred to as the cloud.

Benefits of Edge Computing in IoT

The following are listed some of the benefits of edge computing in IIoT:

Figure 2. Edge computing architecture

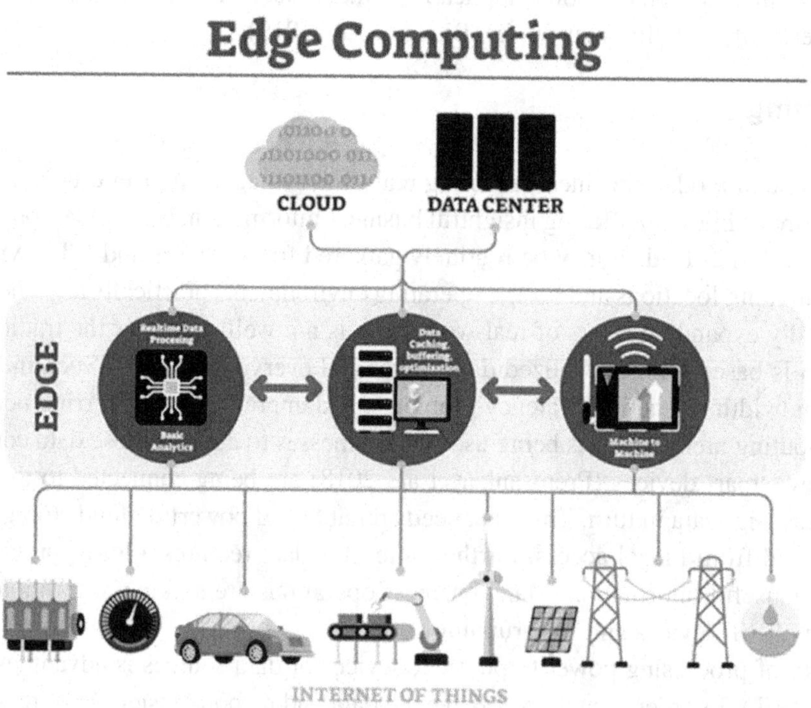

1. **Low Bandwidth Requirements**: The cloud's incredible computing capability made it incredibly effective to assign all of the computational work to it. But as more devices rely on the cloud for services, more bandwidth is needed, and as a result, things are almost at a standstill. The network's bottleneck is becoming bandwidth as the number of devices rises. Big data cannot be transmitted to the cloud in a timely manner.
2. **Low Energy Requirements:** All electronic devices will eventually be a part of the Internet of Things. The number of items on the edge will shortly surpass one billion. The previous typical cloud computing approach would be unable to manage the massive amounts of data produced by these devices. Consequently, the generated huge data won't ever be transported to the cloud; rather, the edge will utilize it (Kaur et al., 2018).
3. **Better Privacy:** IoT sensor data can produce sensitive and private information. There is a danger that the data generated by these IoT sensors will be hacked or incorrectly recorded if it is transmitted to the shared cloud. If the data is processed close to the source device and the user, it will be safer and less susceptible to hacking.
4. **Improves Efficiency and Reliability of Business:** Lower data traffic and cloud storage lead to more effective corporate operations. Additionally, connection problems won't be as severe as they are for other IoT services that rely on the cloud.
5. **Unlimited Scalability:** Edge computing, in contrast to cloud computing, enables scalability of the IoT network as needed, independent of the cost or availability of current storage.

Challenges of Edge Computing in IIoT

Here are some of the challenges of edge computing in IIoT:

1. **Management of Resources:** The Internet of Things and edge computing, which are both distributed systems, have issues with resource management (Gordon et al., 2012). The problem of managing resources can be solved using a variety of techniques, including auction-based systems and optimization.
2. **Confidentiality and Security:** IoT benefits from the dispersed architecture are numerous. However, a significant concern with this distributed structure is its privacy and security. When it comes to confidentiality, it has been noted that data gathered from IoT devices is kept at edge nodes, which might be riskier than cloud servers. The issue of data security and confidentiality must therefore be taken into consideration.
3. **Advanced Communication:** Cellular networks are evolving to lower network latency by increasing output, updating often, supporting interconnected groups (in a high number), and using technologies like Massive MIMO (Multiple-Input and Multiple Output), Millimeter-Wave, and UDNs (Ultra-Dense Networks). Edge computing technology will proliferate as a blend of the preceding technologies when communication technology progresses or evolves, which may prove to be inevitable.
4. **Data Abstraction:** This is still a complex problem in edge computing even if it has been overcome in cloud computing. Data abstraction is necessary yet challenging due to the availability of Internet of Things devices and data suppliers. The information from a security camera is continuously recorded throughout the day, but it is only stored in the database for a brief amount of time when no one is really using it, after which it is removed.

5. **Differentiation:** Since there will be a lot of IoT devices, the edge should be able to distinguish between those that are most important and those that are least important. Healthcare gadgets (fall detection, heart rate detection), trespass alarms, failure alerts, smoke detectors, fire detectors, and other entertainment services should be given the highest priority in a smart home environment.

IoT SECURITY

IoT security can be viewed as a cybersecurity approach and defense mechanism that guards against the threat of cyberattacks that explicitly target physically connected IoT devices. The integrity and protection provided by the least secure device eventually cause the network security posture to deteriorate as a whole (Abomhara & Køien, 2014).

Before going to IoT security challenges and principles lets look into IOT layered architecture (Zhao & Ge, 2013). The layered architecture is given in Figure 3.

1. **Hardware or Physical Layer:** The "Sensors" layer is a common name for the perception layer or physical layer in the Internet of Things. With the help of sensors and actuators, this layer's job is to collect environmental data. This layer detects, gathers, processes, and transmits data to the network layer.

Figure 3. IIoT layered architecture

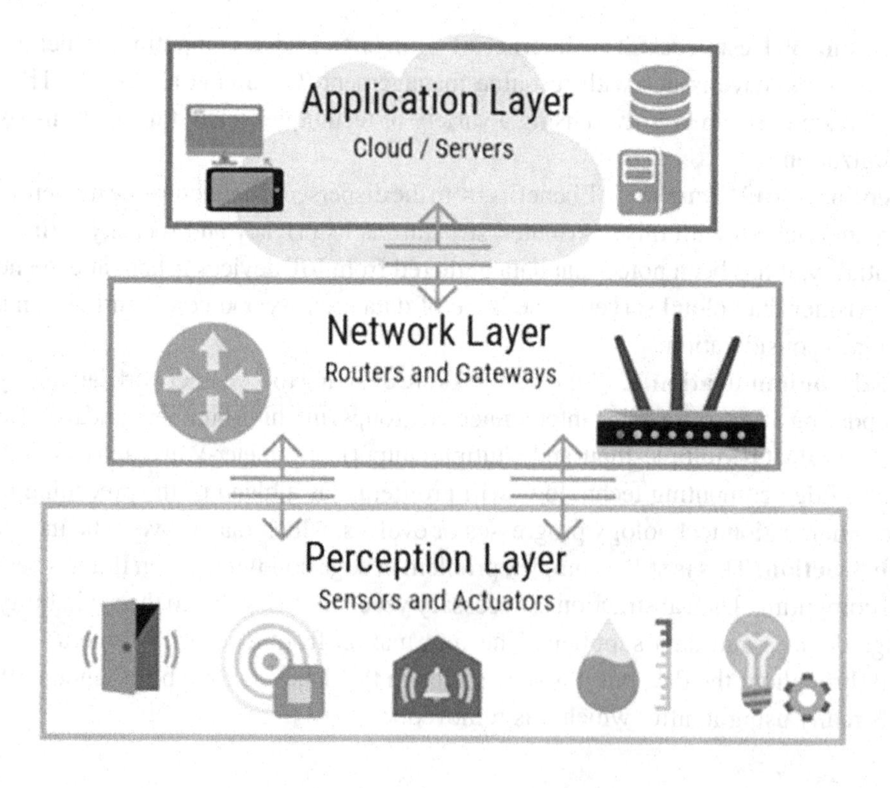

2. **Network Layer:** Data transmission and routing to various IoT hubs and devices are handled by the network layer of the Internet of Things. Cloud computing platforms, Internet gateways, switching and routing devices, among other things, are operated at this layer using gadgets like WiFi, LTE, Bluetooth, 3G, Zigbee, and other relatively progressive technologies.
3. **Application Layer:** Data confidentiality, integrity, and validity are ensured by the application layer. The goal of IoT or the development of a smart environment is accomplished at this layer.

IoT Security Issues

IoT security goals include availability, integrity, and confidentiality (CIA). Problems are made more difficult by the IoT's multiple restrictions and limitations relating to its components and devices, computing and power sources, and even its heterogeneous and widespread nature. The needs of security for the IoT as a whole and the security issues particular to each IoT layer are covered in this session (Atzori et al., 2012). Following are the some of the IOT security concerns discussed in below:

1. **Confidentiality:** Privacy and confidentiality are substantially identical concepts. Measures to maintain confidentiality are intended to guard against unauthorized access to sensitive information. Internal objects (devices connected to the network), external items, machines, services, and humans can all be users in the Internet of Things (devices not connected to the network) (Leo et al., 2014).
2. **Integrity:** Since the Internet of Things (IoT) depends on sharing data among many different devices, it is essential to ensure that the data is accurate, that it comes from the right sender, and that it is not tampered with during transmission due to malicious or unintentional interference. To enforce the integrity feature, end-to-end security in IoT communication can be maintained.
3. **Availability:** As many smart gadgets as is practical will be connected by IoT. IoT users ought always have access to the necessary data. However, the IoT needs more than simply data to function; in order to fulfill its promises, devices and services also need to be reachable and quick to respond to user requests.
4. **Authentication:** Every IoT object needs to be able to clearly identify and authenticate other IoT devices. Because there are so many distinct entities (devices, people, services, service providers, and processing units) engaged in the IoT, this procedure can be exceedingly challenging. In addition, there are times when objects may need to interact with others for the first time (objects they do not know), which can be very difficult (Farooq et al., 2015).
5. **Lightweight Solutions:** Since the computational and power capacities of IoT participating devices are constrained, lightweight security solutions have been proposed. When designing and implementing protocols for the encryption or authentication of data and devices in the Internet of Things, it should be remembered that:
6. **Policies:** Policies and standards must be in place to guarantee that data is managed, secured, and sent effectively. But more crucially, a system must be put in place to police these regulations and make sure that every organization complies with the rules. The Service Level Agreements must be specifically cited in each service that is used (SLAs). Current security guidelines for computers and networks might not be appropriate given the varied and dynamic nature of the Internet of Things.

Security Challenges in Each Layer of IoT

Every IoT layer is susceptible to security issues and attacks. As a result of an insider assault, these can be active or passive and originate from internal or external networks (Abomhara & Køien, 2014). A proactive attack kills the service, whereas a passive attack simply observes IoT network data without interfering with it.

1. **Hardware Layer:** Three security flaws exist in the Internet of Things perception layer. First and foremost is the wireless broadcasts' strength. Disruptive waves may have an effect on how effectively IoT sensor nodes communicate with one another using wireless technologies. As a result, the attackers as well as the owner can intercept the sensor node in IoT devices. Third, considering how frequently IoT nodes are moved, network topology is dynamic by nature.
2. **Network Layer:** DoS attacks are possible at the network layer of the Internet of Things, as was already mentioned. By using traffic analysis, eavesdropping, and passive monitoring in addition to DoS attacks, the adversary can target secrecy and privacy at the network layer (Abomhara & Køien, 2014). These attacks are highly likely to occur duc to remote access methods and device data exchange. After a Man-in-the-Middle assault, which is extremely vulnerable to the network layer, eavesdropping can happen. Communication over the internet is limited to machine-to-machine interactions, whereas IoT communication is not. For the machine-to-machine communication (Roman et al., 2013) feature introduced by the IoT, compatibility poses a security issue.
3. **Application Layer:** There are several problems with application security because there are still no international standards or laws that control communication and application development for the Internet of Things. The integration of all applications to secure data privacy and identity authentication is highly challenging because different applications have different authentication procedures.

IoT-Enabled AI Applications

In IoT applications and deployments, artificial intelligence is becoming more and more important. Over the past two years, investments and acquisitions in firms that combine AI and IoT have increased. IoT platform software from top suppliers now includes integrated AI features including machine learning-based analytics. This process we often called as "AIoT".

The goal of artificial intelligence (AI) is to enable computers to reason similarly to humans. This breakthrough will hasten the transformation of industries into digital ones. The planet can become autonomous by connecting things like people, animals, plants, machinery, appliances, dirt, stones, lakes, buildings, and anything else that comes to mind. We require both a data analysis (DA) (Witten & Frank, 2002) module and a machine learning (ML) (Michalski et al., 2013) module that simulates human learning in order to make the environment and its physical objects truly autonomous.

The Internet of Things (IoT) (Hassan et al., 2017), which envisions a society populated by intelligent devices put everywhere and interconnected by other (Fortino & Trunfio, 2014) communication channels like Bluetooth, infrared, etc., is one of the most brilliant concepts driving this trend. There will be links among individuals, between individuals and objects, and between objects themselves. The "Internet of Everything" (Sun et al., 2021) contends that every tangible, inanimate, or digital object is connected to others through a communication channel.

A broad concept, the Internet of Things (IoT) encompasses an abundance of sensors, actuators, data storage, and data processing capabilities. As a result, every IoT-enabled device can detect its surroundings, communicate, store, and analyze the collected data, and then take the required action.

The processing step completely determines the last step of behaving appropriately. The degree of processing or action that an IoT service is capable of doing determines its genuine level of intelligence. The usefulness and ability to adapt as the data changes will be constrained in a non-smart IoT system.

Here are a few examples of current IoT services that use AI to power their operations:

1. **Voice Assistants:** A user's personal assistant on a tabletop is provided via cloud-based voice services called voice assistants. They carry out a number of tasks via third-party applications and other nearby smart devices. They can perform a variety of tasks based on the user's voice commands, such as answering questions, dialing cabs, booking reservations at restaurants, playing music, turning on/off smart lights, and many more. Examples: Alexa, Siri, Google assistant.
2. **Robots:** In the past few years, this field of robotics has made advancements that have made it feasible to build robots that are more like humans and are able to interact with people while comprehending, recreating, and showing some human emotions. Since they have numerous sensors, actuators, and AI that enables constant self-learning and adaptation, robots are IoTs in and of themselves. Examples:
    a. A social humanoid robot named Sophia from Hanson Robotics is remarkably human-like and has more than 50 different facial expressions.
    b. Robotic kitchen that can prepare food of a professional standard using its recipe collection.
3. **Smart Devices:** Along with voice assistants and robots, the Internet of Things also includes intelligent objects and gadgets that aid human labour. Applications for computer vision, deep neural networks, transfer learning, object recognition, voice recognition, speech, and expression identification, and so on. Examples: smart watch, SkyBell, Smart Oven, etc.

IoT With 5G

5G Technology flourishing to the entire technology. When compared to previous generations, 5G technology is capable of significantly greater performance. Under optimum conditions, it produces the best results. Download speeds of up to 20 gigabits (GB) per second, which is 200 times faster than current 4G technology. In comparison to 4G technology, which has a latency of 50 milliseconds or more, 5G technology has a latency of less than one millisecond (the time it takes to send data from the source to the destination).

Additional advantages of 5G for IoT include extremely low latency (down to 1 ms), increased reliability, massive network capacity, increased availability, and a more consistent user experience due to 5G's more stable connections. IoT devices can be used for a variety of purposes, including real-time monitoring, healthcare, maritime surveys, and others.

The main 5G new use cases benefiting IoT are:

1. **Ultra-Reliable Low Latency:** The amount of time a signal spends traveling from its source to its receiver and vice versa is known as its latency. The ability to remotely control equipment in real-time is made feasible by 5G latency, which operates quicker than human visual processing. Among the many advantages that URL capabilities will provide are robotic procedures.

2. **Enhanced Mobile Broadband:** It typically provides faster data rates than traditional mobile broadband services, which improves the consumer experience. It will once again be advantageous for several IoT applications and offer blazing-fast speeds and excellent service quality (QoS).
3. **Massive Machine Type Communications:** Numerous devices or "things" can communicate with one another virtually or entirely without the need for human intervention thanks to the MTC communication module. New applications are being created for MTCs that service a lot of "things" in the 5G world, introducing the so-called enormous MTC (mMTC), IoT devices connected to 5G are made to use low power and data through mMTC. If devices consumed a lot of power and data, they would be very expensive and have short battery lives, making them unusable. In the ecosystems of the Smart City, Smart Home, Smart Farm, or Smart Factory, this is especially crucial for IoT devices.

The Benefits of 5G in IIoT

Following are some of the benefits of 5G in IIOT listed below:

1. **Telemetry Applications:** It is possible to use mobile telemetry with 5G capabilities to monitor temperature in railroad networks, which will help railroad station managers design safer routes for the trains to avoid numerous accidents. Data is also sent by telemetry to utilities. If people use more energy than is necessary, they may be charged more and held accountable.
2. **Smart Grid Automation:** The smart power grid can be automated according to 5G, one of the fastest-growing technologies that is considerably superior than the traditional electric system.IoT devices can offer users dependable services based on telemetry and AMI, and smart grid technology makes integration easier and is a self-sufficient network system used to control, analyse, and monitor problems that guarantee everyone will have access to reliable and safe electricity.
3. **Remote Surveillance:** With the use of ultra HD video, remote video surveillance over a 5G network using IoT sensors can be utilised to keep an eye on facilities, production lines, and high-security zones. 5G sensors and the low latency, high speed communications enable video analytics.
4. **Smart Traffic Management:** Intelligent transportation is made possible by 5G - IoT sensors gather real-time data from on-the-road infrastructure and vehicles and issue timely alerts as traffic cabinets integrate these systems at specific metropolitan street crossings. Regular application of these technologies can result in cost savings, system dependability, traffic safety, and flow.
5. **Medical Applications:** Low latency makes it possible to monitor patients while they are being transported and to set up hospitals for speedy treatment. Transmitting Ultra HD video between ambulances and hospitals through Internet of Things sensors. Smart wearables and sensors built into healthcare systems can aid in maintaining constant communication with the surgeon during operations and allow patients to receive treatment sooner than would be possible manually.

Challenges of 5G in IoT

Here are some challenges of 5G in IIoT is listed below:

1. **Huge Data Volume:** Technology developments cause each network's data volume to increase yearly, and this trend is accelerating. Each network needs to be able to handle a huge amount of data because many apps now allow high-resolution video calling, live streaming, downloading, etc.

2. **Device-to-Device Communication**: By leveraging one mobile device as a data hub for additional devices that are unable to connect to base station signals, a recent innovation known as "D2D communication" seeks to improve mobile connectivity. One of the most effective forms of communication in emergency situations (like natural disasters) where connectivity is poor or unavailable is device to device communication. To establish D2D communication, however, complicated data transmission protocols are necessary.
3. **Frequency Bands:** 5G requires frequencies up to 300GHz, in contrast to 4G LTE, which is currently operational on defined frequency bands below 6GHz. Some bands, most commonly referred to as mmWaves, can transport far more capacity and provide a 20-fold improvement above the fastest theoretical throughput for LTE.
4. **Deployment and Coverage:** Even though 5G promises a significant increase in capacity and speed, its less extensive coverage will require more infrastructure. Beamforming is a method for concentrating or aiming highly focussed radio waves at higher frequencies.

CHALLENGES IN IIoT

The physical world is gradually transforming into a digital world from the present life due to the smart technology and devices that allow users and devices to be in constant communication with each other. Because of machine learning, artificial intelligence and other advancements, digital world is now more efficient. Uneven levels of productivity, efficiency, and performance in IIoT systems increase when machine-to-machine communication and big data analytics are combined in one sector of the economy (Muller et al., 2018). It has been determined that 72% of those companies will lose market share in the future if they fail to implement a large data strategy (Rehman et al., 2019).

The following are a few of the challenges that IIoT faces.

1. **Expensive Investment:** One of the more obvious problems with industrial IoT is the high adoption cost. One of the main promises of industrial IoT is that it will boost productivity, provide better access to business intelligence, and increase manufacturing efficiency while lowering costs. Moreover, organizations struggle to justify expenditure when they are unsure of the expected return on investment and lack the experience in the implementation of connected systems (Sicari et al., 2016).
2. **Data Storage and Management in a Secure Environment:** According to an ABI Research survey, 68% of industrial specialists highlighted privacy and data security concerns as a major barrier to adopting technological innovations (Luo et al., 2020); Al-Otaibi, 2021). This is especially relevant in the context of IIoT, where protecting and controlling IoT networks, devices, applications, and data transfer is essential for success (Sisinni et al., 2018b).
3. **Collaborations with Heterogeneous IIoT Systems:** A technological revolution with the Internet of Things (IoT) link billions of objects and gadgets to the Internet. The communication protocols, network connectivity, and application models of various devices on the market differ (Noaman et al., 2022). Industry uses various standards and technologies that communicate via various protocols. Such diversity in IoT is not possible because developers frequently lack the resources required to understand the specifics of the constrained devices and networks (Huang et al., 2020). As a result of these heterogeneity challenges, large-scale IoT implementation is difficult (Noaman et al., 2022).

4. **Energy Efficiency:** Numerous IIoT applications must operate for years on batteries (Rault et al., 2014). As a result, low-power sensors that do not require battery replacement over their lifetime must be developed. As a result, demand for energy-efficient designs is increasing. Numerous energy-efficient wireless sensor network (WSN) schemes have been suggested in recent years, but these techniques are not fully relevant to IIoT (Rault et al., 2014).
5. **Trust in IIoT:** IoT devices come in a wide range of sizes and memory capacities. Because of this, such devices lack the sophisticated processing power required to support functionality on cryptography and their operating systems sometimes can't be updated to handle new threats. IoT devices are easy targets for tampering as a result of all of this and their tendency to be placed in risky areas, which causes numerous security issues in IIoT (Lee et al., 2022).
6. **Analytics Challenges:** While executing IoT architecture, data analyst must include data processing, cleansing, and representation (Astarloa et al., 2016). There is certainly enough space for the functionality factor, and this factor can easily add real-time or predictive analytics to an IoT solution (Jaidka et al., 2020).

FUTURE OF IIoT

The IIoT is rapidly expanding, with new developments occurring on a daily basis in this exciting new field (Sari et al., 2020). Manufacturing, transportation, and energy are being transformed by IoT and IIoT technologies. Through better data-driven insights, connectivity enables businesses to create new business models and improve performance (Chalapathi et al., 2021). The following are ten trends that can be expected in the future of the IIoT.

1. **Increased Connected Devices:** The quantity of IIoT-connected devices is predicted to exceed double by 2023 (Mili´c & Babi´c, 2020). The next generation of IIoT devices will support current emerging techniques like artificial intelligence, cloud computing, mobility, machine learning, augmented reality, and virtual reality (Hazra et al., 2021). Across the web of things (WoT), such "intelligent" devices are emphasizing increased interoperability to support telemetry data exchange.
2. **Manufacturing as a Service for the IoT:** A new manufacturing strategy known as "Industrial IoT Manufacturing as a Service" is based on a pay-as-you-go model and relies entirely on external suppliers for all components (Chalapathi et al., 2021). Manufacturers agree to uphold specific standards and specifications in this type of arrangement, but they are free to select the specific technologies that will be used to implement them (Hazra et al., 2021). According to many, by 2022, this trend will have shaped IIoT in some ways.
3. **Improvement in Edge and Cloud Computing:** With the capacity to gather data from remote devices on the edge of a network, the utilization of the edge computing and cloud computing will rapidly expand (Hong et al., 2019). The most recent IIoT devices have the ability to communicate and deliver crucial data and instructions through local gateways that link them directly to the cloud for almost instantaneous access by an application or service (Chen et al., 2019). Response time increases and latency is decreased with this kind of implementation.
4. **Predictive Maintenance:** Predictive maintenance is another IIOT trend expected in 2023 (Chen et al., 2019a). Manufacturers can anticipate operational failures while a system or device is still online and under warranty thanks to advancements in IIoT technologies (Chen et al., 2019b). By

using this kind of analysis, businesses can decrease unplanned downtime while enhancing overall equipment performance, cutting costs, and raising uptime.

5. **Data Management Process:** Manufacturers must implement advanced analytics capable of processing huge amounts of information quickly in order to gain valuable insights from IIoT data (Chen et al., 2019b). Process data management (PDM) will thus primarily target manufacturers by 2023 as they implement Industrial Internet of Things solutions to link remote devices and/or sensors via the IIoT (Kaur et al., 2018). By doing this, they will be able to safeguard it from cyberattacks while collecting and analyzing enormous amounts of data from their operations and individual processes.
6. **Digital Twins:** Digital twins are a more recent idea in the Internet of Things in Industry, where they give physical objects a digital representation and allow those objects to communicate with each other and outside users (Zhang et al., 2021). This technology enables online monitoring of IIoT devices with the aim of improving performance, foreseeing faults, and reducing downtime. Digital twins (virtual replicas) will be a significant IIoT trend by 2023 for managing and monitoring assets and equipment that are located remotely (Malakuti & Gr¨uner, 2018). In order to make effective decisions, manufacturers can also use digital twins to combine data from various devices and sensors into a composite view of the system.
7. **Location Tracking:** Using location tracking, an IIoT application, businesses can keep an eye on the whereabouts and health of their assets, machinery, and employees in real-time. Utilizing location tracking tools to make sure that materials are available when needed can increase manufacturing operations' efficiency (Platenius-Mohr et al., 2020). Additionally, it gives manufacturers a better understanding of where their goods are in the supply chain, which makes it simpler to spot potential issues and delays (Liu et al., 2020).
8. **IIOT Blockchain for Supply Chain:** Manufacturers must satisfy consumer demands for supply chain transparency, which may include certifications and audits. Supply chain solutions powered by blockchain will also aid in preserving the quality of the final product after it leaves the factory (Gandhewar et al., 2019). Manufacturers will use blockchain IIoT solutions to increase supply chain visibility, enabling them to better satisfy consumer demands for product quality and transparency (Zelbst et al., 2019).
9. **Smart Manufacturing:** A significant IIoT trend will be "smart manufacturing," which offers a unified view of all factory activities and aids businesses in increasing operational effectiveness. Manufacturers can use smart manufacturing to identify irregularities in their workflows and stop potential problems before they start (Malik et al., 2021). Manufacturers can now access real-time data about the state of their equipment and production procedures thanks to this technology. On the factory floor, this applies to information about motors, pumps, and other machinery (Abuhasel & Khan, 2020).
10. **Machine Learning:** To stay competitive, manufacturers need to manage their assets well and develop quality products. By spotting patterns in processes and workflows that can increase efficiency, machine learning can benefit manufacturers (Bu et al., 2021). Additionally, machine learning can anticipate problems in a process before they arise to help decrease downtime and enhance the overall health of the factory (Chen et al., 2019c). Additionally, it can assist manufacturers in deriving useful insights from their data that alter business outcomes (Ambika, 2020).

CONCLUSION

Global industrial systems are being deeply integrated with cutting-edge computing, analysis, and sensing technologies through the Industrial Internet. The proposed chapter presented the evolution of IIOT from IOT, benefits of IIoT, various IIoT software and hardware technologies which are utilized in industries. Further, the chapter discusses the key techniques, system architecture, and open issues and challenges along with future of IIoT in industrial development. With the development of the industrial internet of things (IIoT), businesses have never-before-seen opportunities to optimise operations, enhance customer experiences, and generate substantial revenue increases. Predictive maintenance will receive more attention in the future of IIoT, along with improved device communication and more accessible pricing for businesses of all sizes to benefit from connected facilities' commercial advantages. Among these advantages are increased productivity, cost savings, immediate control, and the sensing of opportunities and problems.

REFERENCES

Aazam, M., Zeadally, S., & Harras, K. (2018). Deploying fog computing in industrial internet of things and industry 4.0. IEEE Transactions on Industrial Informatics.

Abomhara, M., & Køien, G. M. (2014). Security and privacy in the internet of things: Current status and open issues. *2014 International Conference on Privacy and Security in Mobile Systems (PRISMS)*, 1–8. 10.1109/PRISMS.2014.6970594

Abuhasel, K. A., & Khan, M. A. (2020). A secure industrial internet of things (iiot) framework for resource management in smart manufacturing. *IEEE Access : Practical Innovations, Open Solutions*, *8*, 117354–117364. doi:10.1109/ACCESS.2020.3004711

Al-Otaibi, Y. D. (2021). Distributed multi-party security computation framework for heterogeneous internet of things (iot) devices. *Soft Computing*, *25*(18), 12131–12144. doi:10.100700500-021-05864-5

Alladi, T., Chamola, V., Parizi, R., & Choo, K.-K. R. (2019). Blockchain applications for industry 4.0 and industrial iot: A review. IEEE Access.

Ambika, P. (2020). Machine learning and deep learning algorithms on the industrial internet of things (iiot). *Advances in Computers*, *117*(1), 321–338. doi:10.1016/bs.adcom.2019.10.007

Astarloa, A., Bidarte, U., Jim'enez, J., Zuloaga, A., & L'azaro, J. (2016). Intelligent gateway for industry 4.0-compliant production. In *IECON 2016-42nd Annual Conference of the IEEE Industrial Electronics Society* (pp. 4902–4907). IEEE.

Atzori, L., Iera, A., Morabito, G., & Nitti, M. (2012). The social internet of things (siot) – when social networks meet the internet of things: Concept, architecture and network characterization. *Computer Networks*, *56*(16), 56. doi:10.1016/j.comnet.2012.07.010

Behrendt, F. (2019). Cycling the smart and sustainable city: Analyzing ec policy documents on internet of things, mobility and transport, and smart cities. *Sustainability*, *11*(3), 763. doi:10.3390u11030763

Biswas, S., Li, F., Maharjan, S., Mohanty, S., Wang, Y., & Sharif, K. (2019). Pobt: A lightweight consensus algorithm for scalable iot business blockchain. *IEEE Internet of Things Journal*, *7*, 2327–4662.

Bu, L., Zhang, Y., Liu, H., Yuan, X., Guo, J., & Han, S. (2021). An iiot-driven and ai-enabled framework for smart manufacturing system based on three-terminal collaborative platform. *Advanced Engineering Informatics*, *50*, 101370. doi:10.1016/j.aei.2021.101370

Chalapathi, G. S. S., Chamola, V., Vaish, A., & Buyya, R. (2021). Industrial internet of things (iiot) applications of edge and fog computing: A review and future directions. Fog/edge computing for security, privacy, and applications, 293–325.

Chen, B., Wan, J., Lan, Y., Imran, M., Li, D., & Guizani, N. (2019). Improving cognitive ability of edge intelligent iiot through machine learning. *IEEE Network*, *33*(5), 61–67. doi:10.1109/MNET.001.1800505

Farooq, M., Waseem, M., Khairi, A., & Mazhar, S. (2015). A critical analysis on the security concerns of internet of things (iot). *International Journal of Computers and Applications*, *111*, 1–6. doi:10.5120/19547-1280

Fortino, G., & Trunfio, P. (2014). *Internet of things based on smart objects: Technology, middleware and applications*. Springer. doi:10.1007/978-3-319-00491-4

Frizzo-Barker, J., Chow-White, P. A., Adams, P. R., Mentanko, J., Ha, D., & Green, S. (2020). Blockchain as a disruptive technology for business: A systematic review. *International Journal of Information Management*, *51*, 102029. doi:10.1016/j.ijinfomgt.2019.10.014

Gandhewar, R., Gaurav, A., Kokate, K., Khetan, H., & Kamat, H. (2019). Cloud based framework for iiot application with asset management. In *2019 3rd International conference on Electronics, Communication and Aerospace Technology (ICECA)* (pp. 920–925). IEEE. 10.1109/ICECA.2019.8821897

Gordon, M., Jamshidi, D., Mahlke, S., Mao, Z., & Chen, X. (2012). Comet: Code offload by migrating execution transparently. Academic Press.

Goyal, S., Sharma, N., Kaushik, I., Bhushan, B., & Kumar, A. (2020). Precedence & issues of iot based on edge computing. *2020 IEEE 9th International Conference on Communication Systems and Network Technologies (CSNT)*, 72–77. 10.1109/CSNT48778.2020.9115789

Hassan, Q. F., & Madani, S. A. (2017). *Internet of things: Challenges, advances, and applications*. Academic Press.

Hazra, A., Adhikari, M., Amgoth, T., & Srirama, S. N. (2021). A comprehensive survey on interoperability for iiot: taxonomy, standards, and future directions. *ACM Computing Surveys*, *55*(1), 1–35.

Hong, Z., Chen, W., Huang, H., Guo, S., & Zheng, Z. (2019). Multi-hop cooperative computation offloading for industrial iot–edge–cloud computing environments. *IEEE Transactions on Parallel and Distributed Systems*, *30*(12), 2759–2774. doi:10.1109/TPDS.2019.2926979

Huang, J., Zhang, C., & Zhang, J. (2020). A multi-queue approach of energy efficient task scheduling for sensor hubs. *Chinese Journal of Electronics*, *29*(2), 242–247. doi:10.1049/cje.2020.02.001

Jaidka, H., Sharma, N., & Singh, R. (2020). Evolution of iot to iiot: Applications & challenges. *Proceedings of the International Conference on Innovative Computing & Communications (ICICC).*

Karmakar, A., Dey, N., Baral, T., Chowdhury, M., & Rehan, M. (2019). Industrial internet of things. *RE:view*, 1–6.

Kaur, K., Garg, S., Aujla, G. S., Kumar, N., Rodrigues, J. J. P. C., & Guizani, M. (2018). Edge computing in the industrial internet of things environment: Software-defined-networks-based edge-cloud interplay. *IEEE Communications Magazine*, *56*(2), 44–51. doi:10.1109/MCOM.2018.1700622

Khajenasiri, I., Estebsari, A., Verhelst, M., & Gielen, G. (2017). A review on internet of things solutions for intelligent energy control in buildings for smart city applications. *Energy Procedia, 111*, 770–779. 10.1016/j.egypro.2017.03.239

Lee, K., Romzi, P., Hanaysha, J., Alzoubi, H., & Alshurideh, M. (2022). Investigating the impact of benefits and challenges of iot adoption on supply chain performance and organizational performance: An empirical study in malaysia. *Uncertain Supply Chain Management, 10*(2), 537–550. doi:10.5267/j.uscm.2021.11.009

Leo, M., Battisti, F., Carli, M., & Neri, A. (2014). A federated architecture approach for internet of things security. *2014 Euro Med Telco Conference (EMTC),* 1–5. 10.1109/EMTC.2014.6996632

Liao, Y., Rocha Loures, E., & Deschamps, F. (2018). *Industrial internet of things: A systematic literature review and insights.* Academic Press.

Liu, S., Guo, C., Al-Turjman, F., Muhammad, K., & de Albuquerque, V. H. C. (2020). Reliability of response region: A novel mechanism in visual tracking by edge computing for iiot environments. *Mechanical Systems and Signal Processing, 138*, 106537. doi:10.1016/j.ymssp.2019.106537

Luo, X., Yin, L., Li, C., Wang, C., Fang, F., Zhu, C., & Tian, Z. (2020). A lightweight privacy-preserving communication protocol for heterogeneous iot environment. *IEEE Access : Practical Innovations, Open Solutions, 8*, 67192–67204. doi:10.1109/ACCESS.2020.2978525

Malakuti, S., & Gr¨uner, S. (2018). Architectural aspects of digital twins in iiot systems. *Proceedings of the 12th European Conference on Software Architecture*, 1–2. 10.1145/3241403.3241417

Malik, N., Alkhatib, K., Sun, Y., Knight, E., & Jararweh, Y. (2021). A comprehensive review of blockchain applications in industrial internet of things and supply chain systems. *Applied Stochastic Models in Business and Industry, 37*(3), 391–412. doi:10.1002/asmb.2621

Meyer, O., Rauhoeft, G., Schel, D., & Stock, D. (2018). *Industrial internet of things: covering standardization gaps for the next generation of reconfigurable production systems*. Academic Press.

Michalski, R. S., Carbonell, J. G., & Mitchell, T. M. (2013). *Machine learning: An artificial intelligence approach*. Springer Science & Business Media.

Mili’c, S. D., & Babi’c, B. M. (2020). Toward the future—Upgrading existing remote monitoring concepts to iiot concepts. *IEEE Internet of Things Journal, 7*(12), 11693–11700. doi:10.1109/JIOT.2020.2999196

M¨uller, J. M., Kiel, D., & Voigt, K.-I. (2018). What drives the implementation of industry 4.0? the role of opportunities and challenges in the context of sustainability. *Sustainability*, *10*(1), 247. doi:10.3390u10010247

Noaman, M., Khan, M. S., Abrar, M. F., Ali, S., Alvi, A., & Saleem, M. A. (2022). Challenges in integration of heterogeneous internet of things. *Scientific Programming*, *2022*, 2022. doi:10.1155/2022/8626882

Panarello, A., Tapas, N., Merlino, G., Longo, F., & Puliafito, A. (2018). Blockchain and iot integration: A systematic survey. *Sensors (Basel)*, *18*(8), 18. doi:10.339018082575 PMID:30082633

Platenius-Mohr, M., Malakuti, S., Gr¨uner, S., Schmitt, J., & Goldschmidt, T. (2020). File-and api-based interoperability of digital twins by model transformation: An iiot case study using asset administration shell. *Future Generation Computer Systems*, *113*, 94–105. doi:10.1016/j.future.2020.07.004

Premsankar, G., Francesco, M., & Taleb, T. (2018). Edge computing for the internet of things: A case study. IEEE Internet of Things Journal.

Rault, T., Bouabdallah, A., & Challal, Y. (2014). Energy efficiency in wireless sensor networks: A top-down survey. *Computer Networks*, *67*, 104–122. doi:10.1016/j.comnet.2014.03.027

Riahi Sfar, A., Natalizio, E., Challal, Y., & Chtourou, Z. (2018). A roadmap for security challenges in the internet of things. *Digital Communications and Networks*, *4*(2), 118–137. doi:10.1016/j.dcan.2017.04.003

Roman, R., Zhou, J., & Lopez, J. (2013). On the features and challenges of security and privacy in distributed internet of things. *Computer Networks*, *57*(10), 2266–2279. doi:10.1016/j.comnet.2012.12.018

Sari, A., Lekidis, A., & Butun, I. (2020). Industrial networks and iiot: Now and future trends. In *Industrial IoT* (pp. 3–55). Springer. doi:10.1007/978-3-030-42500-5_1

Shi, W., & Dustdar, S. (2016). The promise of edge computing. *Computer*, *49*(5), 78–81. doi:10.1109/MC.2016.145

Sicari, S., Cappiello, C., De Pellegrini, F., Miorandi, D., & Coen-Porisini, A. (2016). A security-and quality-aware system architecture for internet of things. *Information Systems Frontiers*, *18*(4), 665–677. doi:10.100710796-014-9538-x

Sisinni, E., Saifullah, A., Han, S., Jennehag, U., & Gidlund, M. (2018a). Industrial internet of things: Challenges, opportunities, and directions. *IEEE Transactions on Industrial Informatics*, *14*(11), 4724–4734. doi:10.1109/TII.2018.2852491

Sun, N., Li, T., Song, G., & Xia, H. (2021). Network security technology of intelligent information terminal based on mobile internet of things. *Mobile Information Systems*.

Uddin, M. A., Stranieri, A., Gondal, I., & Balasubramanian, V. (2021). A survey on the adoption of blockchain in iot: Challenges and solutions. Blockchain. *Research and Applications*, *2*(2), 100006. doi:10.1016/j.bcra.2021.100006

ur Rehman, M. H., Yaqoob, I., Salah, K., Imran, M., Jayaraman, P. P., & Perera, C. (2019). The role of big data analytics in industrial internet of things. *Future Generation Computer Systems*, *99*, 247–259.

Witten, I. H., & Frank, E. (2002). Data mining: Practical machine learning tools and techniques with java implementations. *SIGMOD Record*, *31*(1), 76–77. doi:10.1145/507338.507355

Yu, Y., Li, Y., Tian, J., & Liu, J. (2018). Blockchain-based solutions to security and privacy issues in the internet of things. *IEEE Wireless Communications*, *25*(6), 12–18. doi:10.1109/MWC.2017.1800116

Zelbst, P. J., Green, K. W., Sower, V. E., & Bond, P. L. (2019). The impact of rfid, iiot, and blockchain technologies on supply chain transparency. *Journal of Manufacturing Technology Management*, *31*(3), 441–457. doi:10.1108/JMTM-03-2019-0118

Zhang, P., Wang, C., Jiang, C., & Han, Z. (2021). Deep reinforcement learning assisted federated learning algorithm for data management of iiot. *IEEE Transactions on Industrial Informatics*, *17*(12), 8475–8484. doi:10.1109/TII.2021.3064351

Zhao, K., & Ge, L. (2013). A survey on the internet of things security. *2013 Ninth International Conference on Computational Intelligence and Security*, 663–667. 10.1109/CIS.2013.145

Zhu, C., Rodrigues, J., Shu, L., & Yang, L. (2018). Trust-based communication for the industrial internet of things. *IEEE Communications Magazine*, *56*(2), 16–22. doi:10.1109/MCOM.2018.1700592

KEY TERMS AND DEFINITIONS

Artificial Intelligence: Artificial intelligence is the simulation of human intelligence processes by machines, especially computer systems.

Blockchain: A blockchain is "a distributed database that maintains a continuously growing list of ordered records, called blocks." These blocks are linked using cryptography. Each block contains a cryptographic hash of the previous block, a timestamp, and transaction data.

Cloud Computing: The practice of using a network of remote servers hosted on the internet to store, manage, and process data, rather than a local server or a personal computer.

Data Analytics: Data analytics converts raw data into actionable insights. It includes a range of tools, technologies, and processes used to find trends and solve problems by using data.

Decentralization: The term decentralization can be used to denote security where data is spread out over multiple nodes that reduce the likelihood of any one point or aspect having a negative impact on the system.

Edge Computing: Edge computing is a distributed information technology (IT) architecture in which client data is processed at the periphery of the network, as close to the originating source as possible.

Internet: Internet is a global network that connects billions of computers across the world with each other and to the World Wide Web.

Machine Learning: Machine learning is a growing technology which enables computers to learn automatically from past data. Machine learning uses various algorithms for building mathematical models and making predictions using historical data or information.

Network Security: Network security is a set of rules and configurations designed to protect the integrity, confidentiality, and accessibility of computer.

Things: A thing, in the context of the internet of things (IoT), refers to any entity such as a device that forms a network and can transfer data with other devices.

Chapter 4
Making Cities Smarter:
IoT and SDN Applications, Challenges, and Future Trends

Wasswa Shafik
https://orcid.org/0000-0002-9320-3186
School of Digital Science, Universiti Brunei Darussalam, Bandar Seri Begawan, Brunei Darussalam & Digital Connectivity Research Laboratory, Kampala, Uganda

ABSTRACT

The internet of things (IoT) entails all devices that can get onto the internet. This is mainly because of the technological advancement. This exponential growth of IoT increases on the dense nodes with a huge data volume on the network that affect the collision and network congestion probabilities. This chapter presents a comprehensive description of the central and supporting innovations that are used to make cities smarter, focusing on the fifth generation (5G) IoT paradigm from a software-based network viewpoint. Furthermore, the main initiatives of international significance are discussed. Also, the chapter presents software-defined networking (SDN), IoT, and network function virtualization (NFV) challenges as it relates to the user privacy and security, IoT security, energy consumption, integration of IoT with subsystems, and architecture design. A segment of the top five future trends that are made and will make cities smarter is conclusively included.

INTRODUCTION

The IoT is a concept where electronic devices and components communicate with other through the internet (Al-Turjman et al., 2022). IoT systems employ sensors to perceive and gather data then sent via a specific gateway to the cloud or even a command centers for additional archiving, processing, analytics, and decision-making. The system's actuator then receives a corresponding command in accordance with the decision in reaction to the identified data volumes. IoT addresses some of these issues in energy plant management. Internet-connected devices can detect any operational malfunction or irregular decrease in energy efficiency using IoT sensors, signaling the need for maintenance (Bahalul Haque et al., 2022).

DOI: 10.4018/978-1-7998-9266-3.ch004

The research shows that the latest IoT-based energy plant saves approximately 237 million euros over its lifespan, while an existing plant of the same capacity will save fifty million dollars.

Data processing in the factory's structure is a critical component since it enables data in the cloud networks to be evaluated to assist managers in making more effective decisions in real time. The depreciation of machinery and mechanical devices is a major problem in factories about monitoring and sustaining manufacturing properties. The proper system size can be chosen using a suitable IoT platform and tools to minimize corrosion and related protection costs (Parameswaran et al., 2022). The autonomous system certainly not exceeds the threshold limits due to the IoT-based conditional control. In short, this means the system will last longer with fewer failures. Furthermore, errors that result in energy loss are expected to be addressed. In the IoT-based net, energy is always needed for proper execution, like in databases. All assets linked to the grid in such systems can communicate with one another. Furthermore, data on any asset's energy demand is available. This relationship will ensure that energy delivery is perfectly always controlled in all areas. In terms of smart grids' collaborative effect, different parts of the city can be linked together.

A fog node in fog computing is any device capable of computing, storing data, and using the internet. Examples of these machines are personal computers and factory controllers, among others. In this computing model, fog processes and stores IoT data locally on IoT devices rather than sending it to the cloud. Enhanced secure services, including latency and network traffic reduction, are among one of the identifiable benefits of this approach (Zhao et al., 2022). As a result, distinct cloud and fog computing provide processing and computing services that are quicker and more secure. This allows for quicker decision-making and appropriate action. In the IoT domain, among the most urgent issues are security and energy usage (William et al., 2022). IoT devices, for instance, have restrictions owing to heterogeneity in energy resources and computation, which could result in communication bottlenecks and the adoption of security solutions. Figure 1 shows the graphical structure of the chapter.

Temperature-supported sensors are utilized to track cooling and heating fluctuations within the system. Without doubt, temperatures are crucial and widely used various environmental metrics. The approach of converting mechanical to electrical energy is the basic concept of energy. Thermal or temperature conversion are used to achieve these energy conversions, allowing energy to be handled appropriately to conserve energy. The IoT technology application is broad, entailing demand side, services, regulation, and markets, utilities and generation, transmission, and distribution are shown in Figure 2. All these are used to regulate a variety of processes in a smart city.

Humidity sensors differentiate between the volume of moisture and the humidity of the air. The relative humidity is described as the ratio of air moisture to the maximum moisture amount at a specific air temperature. The sensors (for instance, humidity sensors) have a broad range of uses in the energy industry. They're commonly used in the wind energy industry, for example. When wind turbines are located, offshore, humidity sensors are even more important (Yun et al., 2021). For continuous moisture monitoring, these sensors found nacelle mounted and at the bottom of wind turbines. Consequently, operations are more stable, more efficient, and have lower energy costs since operators are better equipped to respond promptly to changes or variations in the operating circumstances of the turbine.

Light sensors: To measure a light's luminance or brightness, light sensors are utilized. They are employed in a variety of commercial and residential appliances, depending on the energy usage. Lighting is the main source of this energy, which we use in houses, which is 15% approximately of the overall electricity use. On a global scale, lighting consumes approximately 20% of all energy (Pandey et al., 2022). This can reduce the amount of energy used for interior illumination. Motion sensors are applied

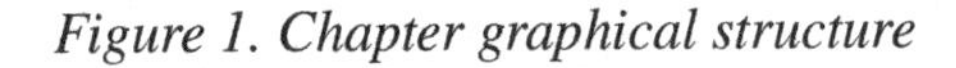

Figure 1. Chapter graphical structure

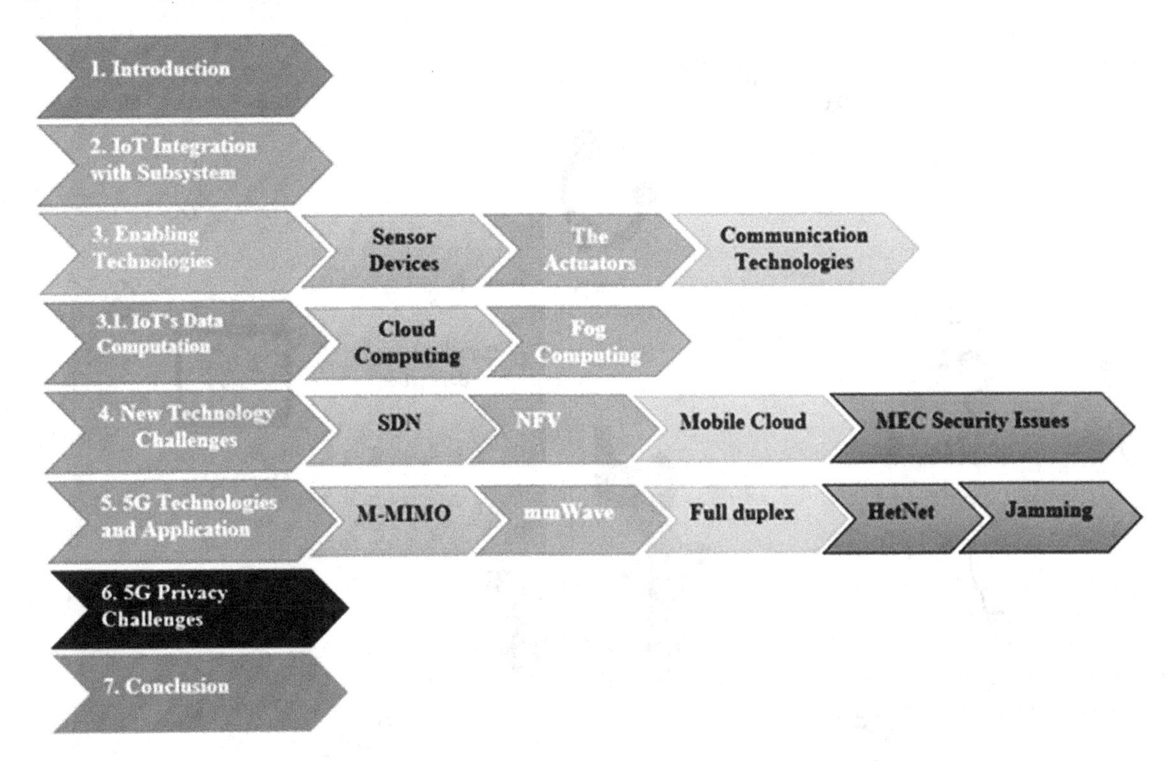

to detect infrared light radiated from an object(s) in its environment. The human presence in space, for instance, can be identified using motion sensors. When no movements are noticed in the room, light control switches off the lights. Buildings' energy use is reduced. This is also practically true for air conditioning setups, accounting for approximately 40% of the total energy consumption in the building.

Proximity sensors are more used in detecting the presence of the surrounding object(s) with no real contact requirement. Wind energy generation is an example of how proximity sensors can be used. In wind turbines, these sensors provide long-term output and accurate position sensing. The best illustrations of this type of sensor are rotor speed monitoring and brake wear tracking which are just a few of the applications of proximity sensors in wind turbines.

In actor pneumatics for the power of pneumatic actuators, compressed air is used. Pneumatic actuators consist of a piston to provide motive force. Since this kind of process does not require motive power, then these actuator(s) control processes, which require fast and precise reactions. Hydraulic actuators use liquid to create motion. Hydraulic actuators are made up of a cylinder that operates mechanically using hydraulic force. The mechanical motion produces a rotatory, oscillatory production or linear. The actuator(s) are applied in industry in processing application control that involves large and high-level speed forces.

Thermal actuators generate physical motion using a heat source. Thermal actuators are devices that transform thermal into kinetic movement energies. These thermostatic actuators typically consist of a substance that senses temperature and is coated through the diaphragm that rubs up next to a plug to move the piston. Electric actuators produce motion by using external energy sources, such as batteries.

Figure 2. IoT applications in a smart integrated energy system

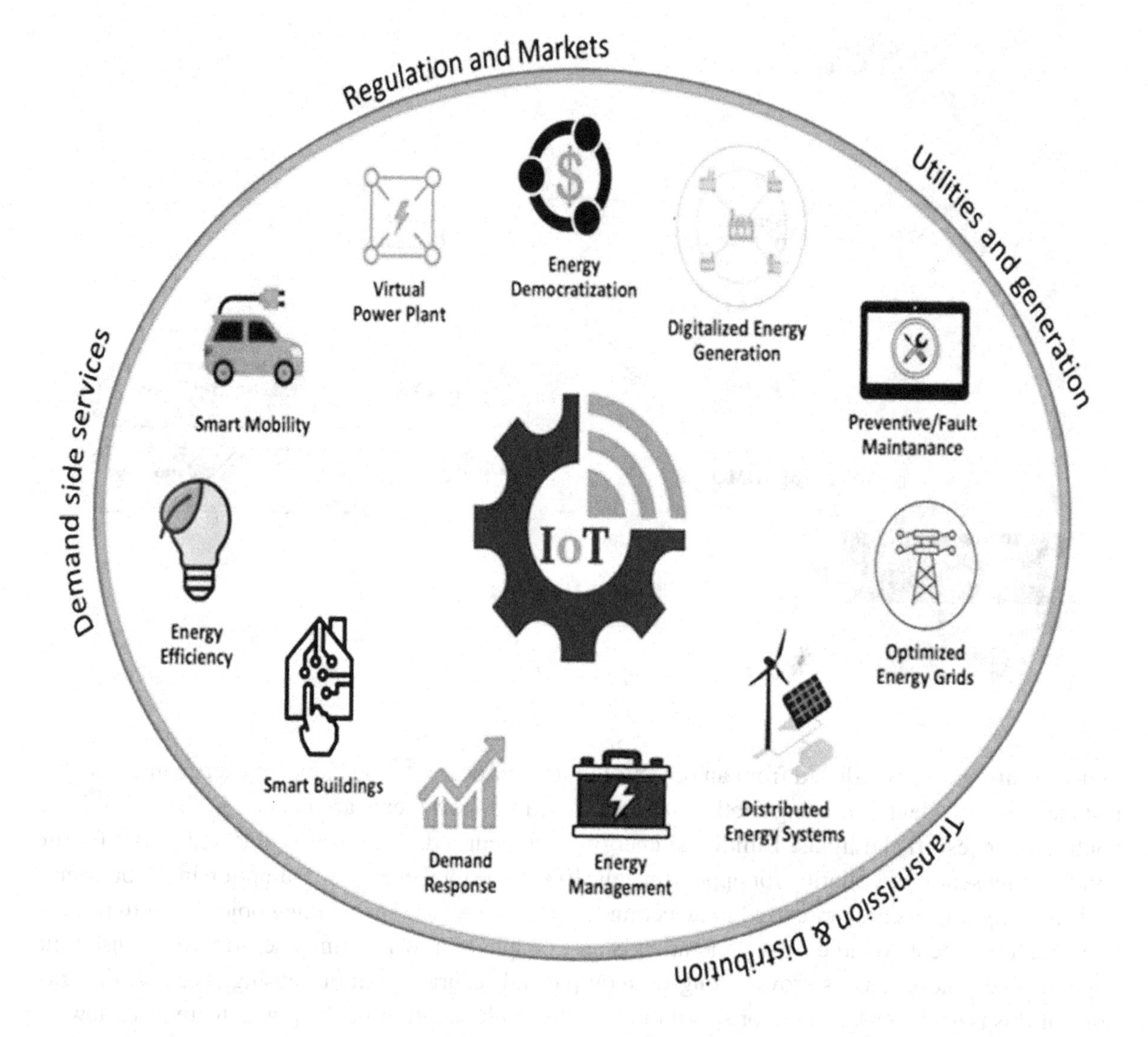

Electric actuators are mechanical devices that can transform in a rotary motion or in a single linear, the electrical energy is converted to kinetic energy.

Intelligent home interfaces require smart meters, ZigBee Things, and electrical equipment to be interoperable so that energy can be consumed efficiently. A ZigBee-based control device that measures and transfers the energy of home appliances attempting lower energy utilization. Experimental field tests with ZigBee in monitoring wind and solar energy systems were done with an acceptable justification. The study justified that ZigBee Things have the capacity to operate in smart metering systems and distributed renewable generation.

The Bluetooth (BLE) is frequently utilized in the energy industry for energy consumption in commercial and residential buildings. An intelligent home energy control system that controls BLE for transmission between household appliances. Equally, a fuzzy centered method for intelligent energy control in home mechanization is sustainable. Effectiveness of an application for home energy conservation by creating a WSN-based Zigbee network to reduce energy costs for customers. A satellite is a very large coverage area that can sustain machine-to-machine (M2M) low data rate function modes. For the support of IoT devices and computers in remote areas, satellite technology is appropriate. Solar and wind power facilities should use satellite established IoT connectivity to exchange data.

Batteries were recharged in conventional grids because of adapters using electric cables and ACϟDC inverters. Using inductive charging technology, these kinds of batteries have the capacity to be wirelessly charged in the smart grid. Also, end-user energy trends are scrutinized by compiling data across an IoT based platform, for instance, a period when electric car or cell phone is charging. The devices can then be charged at the closest tuner battery charging position in the a given time slot when available. Another benefit is that the use of IoT can develop regulation and tracking of battery-powered cars, allowing for the adjustment of energy distribution and the assurance of electricity supply to these vehicles. This would significantly reduce excessive energy usage.

IoT enabling technologies critically help in tracking processes during production, monitoring equipment may include gateway systems. Wireless networking like ZigBee, Wi-Fi, Z-wave, Bluetooth, for instance, and the Local Zone Network (LAN) can be connected to the two equipment bits. Moreover, sensors mounted on each segment of an industrial plant can detect components with higher energy consumption than their nominal levels, making IoT more effective. This makes it easy to monitor any part, repair its failures, and maximize its energy consumption. In smart factories, this eliminates major energy losses.

By selecting a more practical option with a shorter distance and the shortest route, passengers can save both time and money. People's ability to prepare for and control their people will be better equipped to manage their schedules and plans. Trips within the city will therefore be quicker and use less energy. This might significantly cut down on carbon dioxide emission(s) and other air pollutants from transportation. Numerous studies have been done on the creation of effective communication protocols that allow distributed computing systems to provide energy-efficient communication (Yao et al., 2021). As a solution, the employment of radio optimization techniques including cooperative communication and modulation optimization has been suggested. In addition, the usage of multitrack routing techniques and cluster design are energy-efficient routing solutions that have been recognized as another alternative.

Green M2M communication is another example, that allows for low-power transmission and more effective networking protocols by algorithmic and distributed computing techniques. WSNs' sensor nodes have the option of sleeping and waking up only when necessary. The nodes' power consumption can also be decreased by using radio optimization techniques like modulation optimization or collaborative networking. Additionally, efficient solutions will be offered by energy-efficient routing techniques, for example, cluster topology or multipath routing. After all, the approaches and demonstrates presented and assessed above will help IoT systems use less energy resources.

Techniques used to help IoT, and related services are well emphasized in the forthcoming sixth-generation (6G) networks. Most current IoT methods, however, are established on terrestrial connections, for instance, as long ranges, and narrowband IoT, among others. The satellite is a crucial element for 5G and beyond networks or 6G to offer the desired seamless exposure and constant services. Satellites can support vast geo-distributed IoT devices and similarly provide highly efficient backhaul links for IoT-based terrestrial networks (Shokoor et al., 2022). IoT, however, faces difficulties like those linked to the absence of a general controller, system collection, and various attacks. For the IoT in network layer, IoT-functioning apps, and physical layer, there is generally essential to offer an architecture with the ability to balance the requirement for security and decrease energy consumption.

Absolute control of the network, intelligent management, connectivity, high usability, and remote control, among others are done by the SDN control. Moreover, the capacity to install centralized and secure network services including routing, security, energy usage, and bandwidth management is provided by the SDN controller, which can also prevent unwanted access to network resources. The Sustainable Development goals as presented by the UN (https://sdgs.un.org/goals/) discovered that energy conservation is one of the essential aspects of sustainable evolution. Besides, long-term economic advantages of energy production can be accomplished by decreasing fuel prices, producing energy, and reducing energy-related emissions to improve energy quality and optimize energy storage.

Environmental surveillance, healthcare systems and facilities, effective energy storage within the buildings, and drone centered services are all examples of IoT applications. The collection of IoT modules, including communication protocols, data storage, and sensor device computation, is suitably needed. In theory, the data amount obtained from the devices is immense. This necessitates the creation of data storage executed either in the cloud fog or edge of network (Shafik et al., 2020). The data in storage, that is applied for analysis, is the sixth feature of IoT platform. After the data has been stored, analytics can be performed offline or in real-time. Decisions on the operation of the application are made using data analytics. The contributions of the chapter reported are as summarized below:

1. The study presents a detailed analysis related to the IoTs integration in the smart cities.
2. Emphases the discovery of major IoTs that are applicable in city development.
3. Presents the IoT' data communications and some identified enabling technologies (for instance, Satellite, Zigbee, BLE, Weightless, Sigfox, LTE-M) with their technological traits.
4. A comprehensive and analytical study on the new trending improved technologies like Multi-Access Edge Computing (MEC), mobile cloud, NFV, and SDN.
5. The study further highlighted and discussed some identified main 5G technologies like millimeter wave (mmWave), Full duplex, heterogeneous networks (HetNet), Jamming, and Multiple Input Multiple Output (M-MIMO).
6. Some future research trends to support the researchers in developing a model that cut across the main gaps are briefly presented.

The remaining sections of this chapter are structured into seven sections. Section 2 depicts IoT integration with subsystems presenting user privacy, security issues, IoT standards, and architectural design. Section 3 presents enabling technologies in three ways, sensor devices, actuators, and communications (like satellite, Zigbee, Bluetooth, weightless, Sigfox, LTE-M, NB-IoT, and Lora. Further present IoT's data computing focusing on cloud and fog computing. Section 4 demonstrates new technology challenges like security in SDN and NFV, mobile clouds, and Multi-Access Edge Computing (MEC) security issues. Section 5 presents 5G technologies as used in smart cities, for instance, M-MIMO, mmWave, Full duplex (FD), HetNets, and Jamming. Some 5G privacy challenges are illustrated in Section 6 and future motivations. The conclusion is presented lastly in Section 7.

INTERNET OF THINGS INTEGRATION WITH SUBSYSTEMS

The major problem is the incorporation of the IoT based system into the power system subsystems. Since power sector subsystems are distinctive use of different sensor and data communication high-tech. The approaches to seek solutions to the integration are difficult, considering the IoT specifications of the subsystem, concerning the simulation of the integrated structure of the energy system.

User Privacy

Confidentiality applies to the right to protect the privacy of human or cooperative electrical users when shared with an entity. Access to the right details, for instance, energy users and energy-efficient "Things" referring to the devices that connect and share resources over the internet is still challenging (Yang et al., 2021). These types of data collected through IoT enable better decision-making which can impact energy efficiency, delivery, and consumption.

Security Issues

The usage of connectivity and IoT technologies in energy systems raises the risk of cyberattacks on user data and energy systems from production to distribution and consumption. These strings are described as the energy sector's protection challenge. Furthermore, IoT-established energy systems are commonly used in the energy sector to provide services across broad geographical areas (Singh et al., 2022). Because of their widespread use, IoT devices are more vulnerable to cyberattacks, an encryption-based approach is suggested to protect energy data from cyber-related attacks.

IoT Standards

A variety of hardware of varying specifications to relate a single computer to a broad number of computers is still used. A new challenge arises in the incoherent usage of IoT devices with various standards. IoT-enabled networks provide two categories of standard: network protocols, communications protocols, data security, data aggregation standards, and privacy regulatory standards. The IoT standards are one of the barriers to the introduction of standards around non-structuring data storage, safety, and security, as well as legal standards for data markets (Javed et al., 2022). To overcome the challenge of standardization, IoT-related systems are proposed. Another option is for the cooperating parties to establish open knowledge templates and protocols for the standards.

Architecture Design

IoT-enabling systems consist of a range of technologies with an increase in intelligent things and sensors get linked to each other. IoT should be able to communicate in an autonomous and ad hoc way at any time with any relevant services. In other words, IoT systems are built with dynamic, decentralized, and mobile properties based on their application. Because of the features and needs of an IoT application, no single solution can be found for all these applications in a reference architecture. Heterogeneous architectures that are open and standard-compliant are therefore required for IoT systems (Wang et al., 2022). The architectures do not restrict users to use IoT communications that are fixed and terminal.

IoT devices play a key role in decreasing energy deficiencies in heating, ventilation, and air conditioning (HVAC) systems with advances in field technology. HVAC systems, for example, will minimize service in unoccupied areas, resulting in substantial reductions in energy usage and losses. The IoT also manages energy shortages in the lighting scheme. Moreover, an effective revision of the real data will translate the load(s) from high speed(s) to low speeds. This promotes substantially cost-effective usage of electricity and the reduction of green technology gas emissions. IoT can result in a flexible and versatile market response and more effective monitoring and demand control.

INTERNET OF THINGS ENABLING TECHNOLOGIES

The IoT system uses various enabling technologies that aid its proper application, for instance, sensors to perceive and gather data, which is then sent via gateways to command centers extra archiving, processing, analytics, and decision-making (Szatkowski et al., 2022). The system's actuator then receives a corresponding command in accordance with the decision in response to the sensed data. Description of the current technologies that allow IoT in this section is presented, as there are several different devices and data computing approaches. Then we show how these innovations are applied in the energy sector using examples from the literature.

The Sensing Devices

Sensors essentially, are components of IoT and store and relay real-time data. The Sensor improves the effectiveness and efficiency of IoT and are crucial to its performance. Sensors of various types have been designed for various applications. For example, in healthcare systems and facilities, environmental monitoring, public protection, and agriculture are all examples of these applications. Many of these sensors are utilized in the energy field as well. Sensor availability in the energy sector saves both capital and energy (Sumathi et al., 2022). Sensor device research and future trends are also aimed at developing sensor applications and market awareness, as well as developing unique facilities to boost renewable energy production.

The Actuators

The actuators are mechanical "Things" that convert a particular kind of energies into motion(s). Actuators convert electrical input(s) after robotic systems into motion and then operate on the IoT systems' devices and machines. Actuators may create a variety of patterns of motion, incorporating linear, oscillatory, and rotating motions. Pneumatic actuators are often used to operate valves in energy field. Higher energy efficiency is possible thanks to the usage of electric control-valve actuator technology. Several actuators are intended exclusively for the energy industry, as detailed (https://www.link.com/businessareas/energy/), which include resolutions like reducing energy loss while opening hatches together with brake locking in wind turbines, as well as generating solar motion in tracking panels. Many published studies within the literature show how actuators can be applied in IoT applications.

COMMUNICATION TECHNOLOGIES

Wireless communications are critical to the IoT's activation. Wireless systems link sensor-supported "Things" to IoT gateway as well as allowing data communications end-to-end amongst IoT components. Different wireless standards are used to build wireless networks, and the one used to depend on the application's specifications. Energy from renewable sources like solar and wind power plants, for example, is often found in remote locations. As a result (Rivera et al., 2022), maintaining dependable Internet of Things communications in inaccessible locations is quite difficult. The use of IoT Things in these locations requires the choice of the best communications, which can provide a stable connection and enable real-moment data transmission while minimizing resource consumption.

Wireless Local Area Network (WLAN): The energy sector was the focus of the study, which evaluates NB-IoT as a prospective smart grid communication result by comparing its latency and data rates to existing communication technologies. Weightless is an open WLAN standard built to interact with several Internet of Things devices (http://www.weightless.org/). There are feasible in nature and support smart energy metering. Weightless can be employed in intelligent home IoT applications for intelligent grid networks and intelligent metering.

Narrowband IoT (NB-IoT) technically makes efficient networking mechanisms (LPWAN) that allow vast numbers of IoT devices and services to connect at a very latent rate (https://www.3gpp.org/1733-niot/) at a high data rate. Narrowband IoTs (NB-IoT) becomes an enhanced coverage alternative and low-cost battery life span. Because of the latency and physical characteristics, NB-IoTs suggests a potential resolution for smart energy supply nets through low-cost connections for intelligent meters. Sigfox is an ultra-narrowband wide-ranging network technology (https://www.sigfox.com/). For IoT applications, Sigfox enables devices to transmit with low powers. Regarding the suitability of this technology in the energy business, the technical improvements mention Sigfox, low power for intelligent computation for real-time household energies. Moreover, the analysis of a variety of low-energy wide-area network tools determines that Sigfox is a good option for use with electrical connectors and sensors in smart buildings.

Long-Range (LoRa) focuses on IoT-focused wireless networking technologies (https://loraalliance.org/). Lora is simply defined as communication technology at low cost with the capacity's battery life extension of the IoT Things for some years. Lora can also be used to create long-distance broadcasts with very low power consumption. Implementing a building energy management system employing LoRa technology attempts to optimize energy utilization.

ZigBee is a system of communication in design with small-scale applications in mind to create a personal area network (https://zigbee.org/). The Zigbee is an implementable system with low data exchange, and an extremely dependable network solution is designed. The Zigbee in addition uses a network mesh specification in which devices have multiple interconnections. Zigbee also uses a mesh network nature, which connects things through multiple interconnections. The overall contact range of Zigbee is increased when mesh networking is used, which is up to 100 meters. Much of Zigbee's energy market IoT solutions provide more effective methods for energy use.

Bluetooth Low Energy (BLE), this technology is a bit related to Wi-Fi although a small range. This technology permits data to be switched using rapid wavelengths (https://www.bluetooth.com/). The construction of an immediate personal area network is made possible by the BLE's lower installation cost and typical range of 0 to 30 m. The BLE concentrates on small-scale Internet of Things applications that allow for low-power data transmission between devices. Using this technology, the energy sector will create new channels for machine-to-machine and machine-to-human communication.

Wireless fidelity (Wi-Fi) together with other short-range technologies has been thoroughly examined for Internet of Things applications in the energy industry (https://www.wi-fi.org/). Wi-Fi is not the best option in the energy industries due to its high-power needs. Moreover, Wi-Fi's power consumption makes it a poor choice for the energy industry. Better options for applications in the energy sector include NB-IoT technology, ZigBee technology, BLE, and new LPWAN technologies like LoRa, and LTE-M functioning in the unlicensed band. Because of the development of a trustworthy, feasible by these new LPWAN technologies. Table 1 assesses each of the technologies when utilized with IoT to exhibit the variations between them.

INTERNET OF THINGS DATA AND COMPUTATION

Analysis and computation produced by the IoT permit a more in-depth understanding of the system, as well as make appropriate decisions about the system's energy consumption. Big data (BD) is processed and analyzed efficiently due to its traits, like high variety and velocity. Traditional methods for processing BD, such as collecting it for computation local hard drives for analysis. To handle BD, sophisticated analytic and computing techniques are required. We will illustrate the cloud and fog computing that are frequently used for computing and processing BD in the following sections.

Cloud Computing

Data processing technique that enables data streaming computation from IoT "Things" by providing services, programs, storage, and computing over the internet. The terms "cloud" and "Internet" are interchangeable, but "computing" specifically refers to computational services. Cloud computation incorporates all Internet-based applications, software, and data center-based hardware systems. This computing permits the execution of large data amounts while also providing sophisticated computation capabilities (Ji & Chen, 2022).

The key merits of using cloud system (i) lowering hardware expenditures; (ii) processing energy increment and storage capacities; and (iii) using multiple core-based architectures to make data processing simpler. Besides, it is also considered a secure framework for providing essential services and storage from a remote location. Besides, the expense of purchasing hardware and applications and the

Table 1. Different wireless technology comparisons (H = High; L = Low; M = Moderate)

Technologies	Data Range	Range	Security	Battery Life	Installation	Application
Satellite	100 kbps	Extremely long >1500 km	H	H	Costly	Solar and Wind power plant
Zigbee	250 kbps	≤100 m	L	L	L	Intelligent metering
BLE	1Mbps	≤50 m	H	L (Very low)	L	Smart home appliances
Weightless	100kbps	<5 km	H	Low (Very low)	L	Smart meter
Sigfox	100bps	≤50 m	H	Low (7 to 8 years)	M	Smart buildings
LTE-M	.2-1 Mbps	≤200 km	H	Low (7 to 8 years)	M	Smart metering
NB-IoT	≤100 kbps	≤50 km	H	High (1 to 2 years)	L	Smart grid communications
LoRA	.3-38.4 kbps	≤50 km	H	Very L (8 to 10 years)	L	Intelligent lighting

cost of running IoT, data processing algorithms are reduced by the cloud computer, which significantly decreases the number of resources required for data computation.

Fog Computation

This type of computation is done between the network edge and/or cloud. More efficient methods are needed due to some latencies and limitations in the bandwidth of centralized data processing services (Prasanalakshmi et al., 2022). It is a distributed computing architecture and an expansion of cloud that connects computing and analytical capabilities to the web edge.

NEW TECHNOLOGY CHALLENGES

Challenges of Security in SDN

SDN facilitates the creativity of the interaction network and easier network operation by promoting fully programmable and consolidating the control plans of the network theoretically. For potential networks, both aspects are critically significant; they also expose the network to security threats, however. For starter, in the data forwarding components, the flow rules are revised or changed by the SDN controller (Ibrahim et al., 2022). It is possible to easily identify this data traffic management, rendering it a visible network entity and making it a favorite Denial-of-Service (DoS) attack alternative. Likewise, the central control of network resources in the case of saturation attacks also would cause the controller bottleneck for the whole network. By supporting programmability, many network tasks could be introduced as SDN applications (Rangsietti & Kodali, 2022). If unintended malware is exposed to access to harmful applications or critical application programming interfaces (APIs), the havoc could be extended throughout the network.

The new architecture of SDN (such as Open Flow) needs information about routing components before the controller modifies the flow forwarding instructions to hold traffic flow requirements. Therefore, because the forwarding components, like, Open Flow switches, have minimal capacity to buffer unsolicited (transmission control protocol (TCP) or user datagram protocol) flows, the data components may also be vulnerable. Also, this reliance on the controller allows the management control plane station to be resistant to safety threats, not like the present voluntary usage of safety protocols and the time-consuming wait to retrieve them on big networks (Jha, 2022). The problem of controller availability can be addressed by redundant or multiple controllers or improved resistance to security attacks (details are presented in Table 2).

CHALLENGES OF SECURITY IN NFV

Although NFV is of considerable significance to future communication networks, it faces fundamental safety problems like anonymity, credibility, legitimacy, and non-rejection. With respect to its use on mobile networks, this is also stated that for virtualized telecommunications services, the present NFV solutions will not even have adequate protection and separation (Nogales Dorado, 2022). The complicated VNF architecture contributes to configuration bugs and hence data breaches are one of the key problems persisting in the usage of NFV on mobile systems.

Table 2. A summarized Security attacks related to SDN (MitM = Man-in-the-Middle attacks, DPI = Deep Packet Inspection)

Challenges		Issues	Loophole	Possible Solution
Security	Targeted Points			
• Forged or faked traffic flows • MitM attacks • Replay Attacks	Data Channel	User Privacy	Maintaining User's personal information	Asking for users' permission
• DoS attack	Centralized control elements	The precursor to ransom attacks	Injection attack	Securing infrastructure
• Configuration attacks	SDN (virtual) switches, routers	Misconfiguration Attacks	Insecure configuration option	Proper file and directory permissions
• Saturation attacks	SDN controller and switches	Active side-channel attack	Continuous compromising security	Detect saturation threat/attack
• Penetration attacks	Virtual resources, clouds	Vulnerability in software	Security flaw	Security vulnerabilities Assessing
• TCP level attacks	SDN controller-switch communication	IP address spoofing	MAC discrepancies	Packet Filtering with PPI
• Unauthorized access via apps • Insecure storage of apps • Buggy software	Application Plane	The complexed root cause of the problem	Un prioritize test code	Establish a standardized process
• Compromised availability	Control plane	Packet loss or connectivity	Spotting	Application of AI

In addition, Virtual network functions (VNFs) are susceptible to conventional cyber security attacks like impersonating, DoS, and sniffing. A specific set of virtual machine risks are also prone to NFV, like hypervisor hijacking, side-channel attacks, flooding attacks, VM (virtual machine) migration-based attacks, malware injection, and Cloud-specific threats as well as. In contrast, only insider threats (similar malicious investigators) are vulnerable to the implementation of private NFVs, since remote system access is stopped (Hatim et al., 2022). Because of the infrastructure's mutual usability, by injecting malware or altering traffic on the network, a deceptive attacker or an infected VNF vendor may interfere with the infrastructure's operations.

Operative intrusion and abuse of common services are known as assaults on the NFV at the technology level. Since physical infrastructure services are readily available, an attacker can interact with infrastructure operations by installing malware or exploiting network traffic. The survivor will have no gain from shared or devoted services in these forms of resource abuse attacks. A major problem is also the preservation of trust in virtualized NFV applications. Typically, a trustworthy employee installs and configures physical network equipment and the device's trust is created. However, since VNFs are dynamically fetched from the cloud, to stop malicious VNFs, some form of trust mechanism is necessary.

CHALLENGES OF SECURITY IN NFV

Although NFV is of considerable significance to future communication networks, it faces fundamental safety problems like anonymity, credibility, legitimacy, and non-rejection. With respect to its use on mobile networks, this is also stated that for virtualized telecommunications services, the present NFV solutions will not even have adequate protection and separation (Nogales Dorado, 2022). The complicated VNF architecture contributes to configuration bugs and hence data breaches are one of the key problems persisting in the usage of NFV on mobile systems.

In addition, Virtual network functions (VNFs) are susceptible to conventional cyber security attacks like impersonating, DoS, and sniffing. A specific set of virtual machine risks are also prone to NFV, like hypervisor hijacking, side-channel attacks, flooding attacks, VM (virtual machine) migration-based attacks, malware injection, and Cloud-specific threats as well as. In contrast, only insider threats (similar malicious investigators) are vulnerable to the implementation of private NFVs, since remote system access is stopped (Hatim et al., 2022). Because of the infrastructure's mutual usability, by injecting malware or altering traffic on the network, a deceptive attacker or an infected VNF vendor may interfere with the infrastructure's operations.

Operative intrusion and abuse of common services are known as assaults on the NFV at the technology level. Since physical infrastructure services are readily available, an attacker can interact with infrastructure operations by installing malware or exploiting network traffic. The survivor will have no gain from shared or devoted services in these forms of resource abuse attacks. A major problem is also the preservation of trust in virtualized NFV applications. Typically, a trustworthy employee installs and configures physical network equipment and the device's trust is created. However, since VNFs are dynamically fetched from the cloud, to stop malicious VNFs, some form of trust mechanism is necessary as Table 3 demonstrates.

MOBILE CLOUDS AND MEC SECURITY ISSUES

Because cloud storage services include numerous properties which are shared between clients, it is possible for a user to take down the entire system's efficiency by spreading unwanted activities, accessing further services, or accessing other users' resources secretively (Tanwar et al., 2022). On the other side, MEC integrates various related technologies that interoperate in an open environment in which service providers use virtual machines and shared networks to provide and support the apps to the final customers (Alazab et al., 2022). Provided that MEC is comparatively infantile, along with the proliferation of MEC technology, cyber security attacks (summary is demonstrated in Table 4) and protection problems are expected.

Table 3. A summarized Security attack related to NFV

Challenges		Issues	Loophole	Possible Solution
Security	Targeted Points			
• Resource (slice) Theft	Hypervisor, shared cloud resources	The precursor to ransom attacks	Injection attack	Securing infrastructure
• Configuration attacks	SDN (virtual) switches, router	Misconfiguration Attacks	Insecure configuration option	Proper file and directory permissions
• Single point failure • Malicious insider • DDoS attack	NFVO layer	Deployment	Implementation flaws	Use of detection
• VM Migration attacks • Side-channel attacks • Scheduler Attacks	VM (Virtual Machines)	VM sprawl, network congestion, server hardware failures	Implementation flaws	VM threat detectors

Table 4. A summarized security attacks related to Cloud

Challenges		Issues	Loophole	Possible Solution
Security Attacks	Targeted Points			
• DoS attack	Centralized control elements	Deployment	Implementation flaws	Use of detection
• Signaling storms	5G core network elements	Mobile broadband network architecture evolution	Overload the bandwidth at the cell	Detection and mitigation of signaling storms
• User identity theft	User information databases	Identity Fraud	Phishing and spoofing	Add an authentication step and use strong passwords.
• Timing attacks	Subscriber location	Timing attacks against network servers	RSA encryption-breaking RSA.	RSA blinding
• Reset and IP spoofing	Control channels	Detect inconsistencies	Deploying packet filtering	Monitoring networks for atypical activity
• Semantic information attacks	Subscriber location	Phishing e-mail, multimedia masquerading	Exploitation of information	domain-appropriate methodologies Application

The front-end risk environment varies from physical threat to risk depending on the program. Risks are primarily geared at the mobile cloud servers on the back-end network. The nature of these threats will vary from DoS-HX (extensible markup language DoS and hypertext transfer protocol) attacks to data replication. DoS attacks, Wi-Fi sniffing, address spoofing, and account hijacking for network-based cybersecurity threats are possible vulnerabilities (Ren et al., 2022). The cloud supported IoT setting and the open APIs wherein programmers and developers assist MEC apps and end users with content, are the main security concerns on the MEC side.

In MEC, the requirement for exposed APIs is primarily to offer merged network backing and donations to various service providers and developers. The acceptance of open APIs, however, also creates vulnerabilities from which adversaries may initiate different attacks on the MEC ecosystem through the context of third parties. This has sparked research into related security technologies that are channeled into MEC node security, including the IoT nodes and MEC servers. In this case, MitM (man-in-the-middle) attacks, DoS attacks, privacy leaks, malicious mode problems, and VM misuse are typical challenges (Tian et al., 2022).

ISSUES AROUND 5G TECHNOLOGY

On the physical layer, there are significant numbers of scholarly works in comparison to the gentle growth made in 3GPP protection. PHY-SEC (Physical layer security) delivers information secrecy via developing the wireless transmitting medium's intrinsic random nature. In view of efficiency and performance, PHY-SEC provides several advantages compared to upper-layer mechanisms. Computational load and workload related to higher-layer safety protocols are not impacted.

Measures for PHY-SEC encryption secure the contact stage of the network, the higher layer authentication methods secure the stored data after the communication process. They also are complementing one another and could be well integrated to form security policies that efficiently protect the sharing of sensitive and confidential data from 5G (Kumar & Singh, 2022). Shannon's knowledge-theoretical confidentiality study can be traced back to the origins of similar works, and it stated that the protection functionality depends on the amount of knowledge eavesdroppers are aware of. The confidentiality power refers to the overall rate of transmission of data to legal users, with a confidentiality restriction on the access to data to an eavesdropper.

In addition, the main physical layer generation emerges as a promising path that takes advantage of random physical layer features like channel state information, RSS (signal strength), or else stage data to produce and exchange. The hidden keys for physical layer verification can be generalized (Senger & Malik, 2021), the PHY-SEC was established with the advent of the Gaussian, multiple antenna, and relay channels together with non-degraded signals, respectively. With the launch of modern 5G radio technology such as M-MIMO, mmWave, cooperative relaying, three-dimensional MIMO, among other are new research paradigms has become a theoretical and functional confidentiality potential in the light of new channel conditions.

M-MIMO

Because of the gain of the antenna equivalent to the total number of base station antennas and M-MIMO stiffening on signal impact, several instructions of magnitude larger than an eavesdropper are derived from a lawful person's signal power. Consequently, there is a substantial improvement in the ability to avoid passive eavesdropping attacks. However, by leveraging the vulnerability of the channel estimation mechanism, a successful eavesdropper will pretend to being authentic and offer his own pilot series (Pradhan et al., 2022). Not only does the pilot contamination attack dramatically reduce the achievable confidentiality capability, but it is also difficult to detect. Potential approaches have been studied to detect successful attacks to resist active attacks, further study is needed.

mmWave

There are many important features of mmWave communications, like short-range and extremely directional transmissions, a broad variety of antennas, various rules of transmission, and exposure to obstruction consequences (Hussain & Kim, 2022; Nakano et al., 2022). It was shown that in the light of numerous eavesdropping, P2P mmWave wireless connection with a great number of transmitting antennas would reach great confidentiality throughout. The discrepancy here between the end user's spatial orientation. However, has a big influence on the secrecy performance of mmWave communication. Enhancing the confidentiality efficiency of mmWave networks, it is useful for the directional beam shaping antenna, which contributes to more oriented array gain (Soto-Valle et al., 2022). Wretchedly, by utilizing mmWave technologies that are strongly directional, by eavesdropping, the replicated signals generated through small-size items may be used to reduce the possible confidentiality level.

Full Duplex (FD)

With both openings and hurdles, it brings 5G security. Compared to traditional half-duplex communications, ideally, it will double the spectral quality. The technology makes additional artificial noise (AN) to be produced by the receiver to interfere with eavesdroppers (Ngo Hoang & Lee, 2022; Nguyen et al., 2022). It is possible to group the current research on FD protection into three classifications: Receiver and FD Sender, FD Receiver, and FD eavesdropper. Any antenna may be used by an FD receiver to transmit the signal, by the usage of other antennas to transmit simulated noise to deceive the eavesdropping.

HetNet

The 5G is considered a wonderful HetNet (heterogeneous network) which can easily merge different nodes to shape a multilevel hierarchy framework, including macro cell level, compact small cell levels, and system levels (Ogbodo et al., 2022a). The multitier architecture offers fresh possibilities for the physical layer defense design to improve confidentiality capability and fresh challenges. A user's strategy of creating communication networks under security constraints between various types of nodes is more complex. To protect data transfer and balance the demand on the HetNet (Mahantal & Lele, 2022), it is important to use current security-oriented phone organization regulations. Furthermore, one or more levels can be available on smartphone devices, such as open access and closed access. It is necessary to provide novel consumer suggestion rules such that the quality of service and confidentiality are condensed by this (Kunsei et al., 2022). In addition, Super HetNet can cause significant cross-tier interference, which has a negative effect on data transmission protection.

Jamming

The availability of mobile communication networks is closely linked to jamming. Malicious users may misuse it to mess with ordinary connectivity among lawful users. Jamming has allowed attackers to explore paths to strengthen eavesdropping capabilities and jamming signals to reduce the potential of secrecy. According to (Khan et al., 2022; Ogbodo et al., 2022b), constructive attackers can find ways to develop wiretap channel capabilities instead of minimizing confidentiality capabilities. During the uplink channel approximation method, for instance, an attacker can claim to be an authorized person and submit a symbol to a BSS pilot that might transmit signal power to the eavesdropping rather than the valid user. In the meantime, to secure messages from eavesdropping, jamming may be used.

5G PRIVACY CHALLENGES

The key privacy issues could come from data, location, and identification from the viewpoint of the user. Many mobile apps need specifics of the personal information of users prior to downloading. How the data is stored and for what reasons it will be used are seldom discussed by device developers and businesses (Saeed et al., 2022). Risks like timing attacks, semantic data attacks, and boundary attacks threaten consumer location data. Place confidentiality in 5G mobile networks could be revealed by access point classification techniques at the physical layer stage.

By running a fake base station, it is also possible to trigger these attacks that UE is considered to become the selected base station that has blocked entry to Temporary Mobile Subscriber Identity (TMSI); the subscriber would then reply with its IMSI (Yeh et al., 2022). Besides that, 5G networks have several entities, like virtual mobile network operators (VMNOs), communication services providers (CSPs), and communications infrastructure providers. For protection and privacy, both actors have different goals. Coordination of mismatch in the 5G network and privacy policies between such entities would be an issue.

In earlier decades, telecom carriers had complete access and ownership over all aspects of the system. However, as they would depend on new players such as CSPs, 5G network providers are losing direct ownership of the networks. Thus, 5G operators would lose maximum protection and privacy governance. Consumer and data protection are severely questioned in shared networks in which the same infrastructure is shared by multiple entities (like VMNOs and other rivals). Furthermore (Jain et al., 2022), while NFV and cloud-based data storage features are used for 5G networks, there are no physical limitations. Therefore, in cloud systems, 5G providers have no clear knowledge of the location of data storage. Since various countries, based on their chosen setting, have different degrees of data protection mechanisms, protection is threatened in case consumer data is stored in a cloud in another region (Ahmed et al., 2022; Shamshuddin & Jayalaxmi, 2022). A detailed overview of the MEC threat environment is given, in which the writers discuss the MEC framework with a broad spectrum of potential security threats, and why defense is among the key problems of the MEC as summarized in Table 5.

FIVE TRENDS THAT MAKE CITIES SMARTER

From a wider exploration of smart cities comprehensively, there are five main trends that make smart cities smarter, including intelligent healthcare, intelligent safety, intelligent energy, intelligent infrastructure, and intelligent citizens as briefly explained.

Intelligent Healthcare

A better environment for residents' health is greatly influenced by the community, as the current pandemic (Coronavirus disease (COVID-19) has demonstrated. By promoting preventive self-care in addition to diagnosis and treatment, smart technologies can ease the burden on healthcare ecosystems. By doing this, the emphasis is shifted from individual-centered healthcare to a communal approach.

Table 5. Summarized major IoT, NFV, and SDN in the current advancement of technology

Challenges	Issues	Solutions	Benefits
User's privacy	Maintain users' data	Seeking for users' consent	Permits improved decision-making
IoT Security	Cyber-attacks and Threats	Distributed control system Encryption	Improved security
Energy consumption	Effective energy usage	Distributed computing Approaches	Energy saving
	Elevated data rate transmissions	effective communication protocol designs	Energy saving
Standardisation	Giant IoT device operation	Distributed systems	Various technologies
	unpredictability among IoTs	Public information models and protocol	Technologies
Integration of IoT with subsystems	IoT with active systems merging	Integrated energy system modeling	Reduction in cost of maintenance
	Data management of IoT	Designing models	Subsystems and real-time data among devices
Architecture Design	Diversified technologies	Open standard Application	Ascendable
	Authentic end-to-end connections	Heterogeneous reference designs	Interlinking devices with people

Intelligent Safety

With greater application by law enforcement agencies, biometrics, facial recognition, smart cameras, and video surveillance have all gained popularity. Cities can use this technology to speed up reaction times, analyze crime forecasts, and spot patterns and trends in crime data. McKinsey analysis (https://www.mckinsey.com/business-functions/operations/our-insights/smart-cities-digital-solutions-for-a-more-livable-future#part1/), employing these technologies might cut down on robberies and assaults by 30% and fatalities by 80%. Cities will need to exercise caution when deploying these technologies to prevent discriminating against communities or demographic groupings.

Smart Energy

These cities may employ technology to monitor and improve their energy use in addition to making sustainable energy investments. By employing tried-and-true technology and methods, cities may reduce emissions by nearly 90% by 2050, according to a 2019 estimate by the Coalition for Urban Transitions. This calls for the employment of environmentally friendly and morally sound materials, resource- and environment-wise designs, renewable energy-driven systems, and usage-adaptive digital technology. New York's Schenectady is replacing its incandescent streetlights with LED ones that can be dimmed or altered in response to real-time data (accessed at: https://www.cityofschenectady.com/245/Connected-City---Smart-Lighting/).

Intelligent Infrastructure

By 2028, according to Gartner (https://www2.deloitte.com/us/en/insights/industry/public-sector/future-of-cities.html/), there will be more than four billion connected IoT devices in commercial smart buildings, all of which will be supported by networks infrastructure, for example, 5G, high-effectiveness Wi-Fi, and smart utilities like power, garbage, and water. In-demand technologies include traffic and wastewater management, IoT sensors, lighting sensors, and fire warning systems. Future cities might become smaller and closer-knit because of this.

Intelligent Citizens

Lastly, smart cities amplify the voices of their citizens. Citizens can instantly report local problems using apps, and they can pool their resources and share them on community networking sites. Cities are becoming more participative, transparent collaborative ecologies. Open data and developing technologies like transforming cities and cryptocurrency Miami (https://identityreview.com/miamicoin-transforming-cities-and-cryptocurrency/) are paving the way for cities to become more human-centered and multidirectional for government, and companies, among others.

CONCLUSION

Utilization of the identified trends will help the public to enjoy smart technologies amidst the technological dynamics. The chapter further presents a comprehensive description of the central and supporting innovations that are used to make cities smarter. Focusing on the 5G IoT paradigm from a privacy and security viewpoint; protection of software-based networks, and questions about privacy, among others. Besides, the main initiatives of international significance are discussed, in line with 5G security issues and beyond. Moreover, challenges of SDN and NFV as IoT is concerned are identified, including user privacy, IoT security, energy consumption, integration of IoT with subsystems, and architecture design.

REFERENCES

Ahmed, U., Lin, J. C.-W., & Srivastava, G. (2022). Privacy-Preserving Active Learning on the Internet of 5G Connected Artificial Intelligence of Things. *IEEE Internet of Things Magazine*, *5*(1), 126–129. doi:10.1109/IOTM.001.2100205

Al-Turjman, F., Zahmatkesh, H., & Shahroze, R. (2022). An overview of security and privacy in smart cities' IoT communications. *Transactions on Emerging Telecommunications Technologies*, *33*(3), e3677. doi:10.1002/ett.3677

Alazab, M., Manogaran, G., & Montenegro-Marin, C. E. (2022). Trust management for internet of things using cloud computing and security in smart cities. *Cluster Computing*, *25*(3), 1765–1777. doi:10.100710586-021-03427-9

Bahalul Haque, A. K. M., Bhushan, B., Nawar, A., Talha, K. R., & Ayesha, S. J. (2022). Attacks and Countermeasures in IoT Based Smart Healthcare Applications. In *Recent Advances in Internet of Things and Machine Learning* (pp. 67–90). Springer. doi:10.1007/978-3-030-90119-6_6

Hatim, J., Chaimae, S., & Habiba, C. (2022). Improved IOT/SDN Architecture with the Concept of NFV. *International Conference on Digital Technologies and Applications*, 294–301. 10.1007/978-3-031-01942-5_29

Hussain, N., & Kim, N. (2022). Integrated Microwave and mm-Wave MIMO Antenna Module with 360∘ Pattern Diversity For 5G Internet-of-Things. *IEEE Internet of Things Journal*, *9*(24), 24777–24789. doi:10.1109/JIOT.2022.3194676

Ibrahim, M., Hanif, M., Ahmad, S., Jamil, F., Sehar, T., Lee, Y., & Kim, D. (2022). *SDN Based DDos Mitigating Approach Using Traffic Entropy for IoT Network*. Academic Press.

Jain, S., Gupta, S., Sreelakshmi, K. K., & Rodrigues, J. J. (2022). Fog Computing In Enabling 5G-Driven Emerging Technologies for Development of Sustainable Smart City Infrastructures. *Cluster Computing*, *25*(2), 1–44. doi:10.100710586-021-03496-w

Javed, A. R., Hassan, M. A., Shahzad, F., Ahmed, W., Singh, S., Baker, T., & Gadekallu, T. R. (2022). Integration of blockchain technology and federated learning in vehicular (iot) networks: A comprehensive survey. *Sensors (Basel)*, *22*(12), 4394. doi:10.339022124394 PMID:35746176

Jha, M. (2022). Secure SDN Based IoT Network Through Blockchain for Smart Architectures. *2022 IEEE Region 10 Symposium (TENSYMP),* 1–6.

Ji, Y., & Chen, L. (2022). FedQNN: A Computation-Communication Efficient Federated Learning Framework for IoT with Low-bitwidth Neural Network Quantization. *IEEE Internet of Things Journal.*

Khan, A., Jhanjhi, N. Z., & Humayun, M. (2022). The Role of Cybersecurity in Smart Cities. In *Cyber Security Applications for Industry 4.0*. Chapman and Hall/CRC. doi:10.1201/9781003203087-9

Kumar, S., & Singh, H. (2022). A Comprehensive Review of Metamaterials/Metasurface-Based MIMO Antenna Array for 5G Millimeter-Wave Applications. *Journal of Superconductivity and Novel Magnetism, 35*(11), 1–25. doi:10.100710948-022-06408-0

Kunsei, H., Hoole, P. R., Pirapaharan, K., & Hoole, S. R. H. (2022). Tracking Everyone and Everything in Smart Cities with an ANN Driven Smart Antenna. In *Machine Learning Techniques for Smart City Applications: Trends and Solutions* (pp. 75–94). Springer. doi:10.1007/978-3-031-08859-9_7

Mahanta, N. R., & Lele, S. (2022). Evolving Trends of Artificial Intelligence and Robotics in Smart City Applications: Crafting Humane Built Environment. *Trust-Based Communication Systems for Internet of Things Applications,* 195–241.

Nakano, K., Motozuka, H., Wee, G. Y. H., Irie, M., Egami, A., Sakamoto, T., Takinami, K., & Takahashi, K. (2022). High-Capacity Data Collection Platform for Smart Cities Using IEEE 802.11 ad-Based Millimeter-Wave V2X Communication. *Wireless Communications and Mobile Computing*, 2022.

Ngo Hoang, T., & Lee, K. (2022). *Short-Packet URLLCs for Multihop MIMO Full-Duplex Relay Networks: Analytical and Deep-Learning-Based Real-Time Evaluation*. Academic Press.

Nguyen, M.-S. V., Sur, S. N., & Do, D.-T. (2022). A Comprehensive Review on Physical Layer Design for Smart Cities. *IoT and IoE Driven Smart Cities*, 1–19.

Nogales Dorado, B. (2022). *An NFV system to support adaptable multi-UAV service deployments.* Academic Press.

Ogbodo, E. U., Abu-Mahfouz, A. M., & Kurien, A. M. (2022). A Survey on 5G and LPWAN-IoT for Improved Smart Cities and Remote Area Applications: From the Aspect of Architecture and Security. *Sensors (Basel), 22*(16), 6313. doi:10.339022166313 PMID:36016078

Pandey, G., Kumar, C. R., Kumar, M., Gupta, K., Jha, S. S., & Jha, J. (n.d.). IOT-BDA Architecture for Smart Cities. In *IoT and Big Data Analytics for Smart Cities* (pp. 57–74). Chapman and Hall/CRC.

Parameswaran, T., Reddy, Y. P., Nagaveni, V., & Sathiyaraj, R. (n.d.). Era of Computational Big Data Analytics and IoT Techniques in Smart City Applications. *IoT and Big Data Analytics for Smart Cities*, 1–22.

Pradhan, D., Dash, A., Tun, H. M., & Wah, N. K. S. (2022). Improvement of Capacity and QoE: Distributed Massive MIMO (DM-MIMO). *Technology.*

Prasanalakshmi, B., Murugan, K., Srinivasan, K., Shridevi, S., Shamsudheen, S., & Hu, Y.-C. (2022). Improved authentication and computation of medical data transmission in the secure IoT using hyperelliptic curve cryptography. *The Journal of Supercomputing*, *78*(1), 361–378. doi:10.100711227-021-03861-x

Rangsietti, A. K., & Kodali, S. S. P. (2022). SDN-Enabled Network Virtualization and Its Applications. *Software Defined Networks: Architecture and Applications*, 231–277.

Ren, D., Gui, X., & Zhang, K. (2022). Adaptive Request Scheduling and Service Caching for MEC-Assisted IoT Networks: An Online Learning Approach. *IEEE Internet of Things Journal*, *9*(18), 17372–17386. doi:10.1109/JIOT.2022.3157677

Rivera, A. O. G., White, E. M., Acosta, J. C., & Tosh, D. (2022). Enabling Device Trustworthiness for SDN-Enabled Internet-of-Battlefield Things. *2022 IEEE Conference on Dependable and Secure Computing (DSC)*, 1–7. 10.1109/DSC54232.2022.9888903

Saeed, M. M., Hasan, M. K., Obaid, A. J., Saeed, R. A., Mokhtar, R. A., Ali, E. S., Akhtaruzzaman, M., Amanlou, S., & Hossain, A. Z. (2022). A comprehensive review on the users' identity privacy for 5G networks. *IET Communications*, *16*(5), 384–399. doi:10.1049/cmu2.12327

Senger, S., & Malik, P. K. (2021). A Comprehensive Survey of Massive-MIMO Based on 5G Antennas. *International Journal of RF and Microwave Computer-Aided Engineering*, e23496.

Shafik, W., Matinkhah, M., & Sanda, M. N. (2020). Network resource management drives machine learning: A survey and future research direction. Journal of Communications Technology. *Electronics and Computer Science*, *2020*, 1–15.

Shamshuddin, K., & Jayalaxmi, G. N. (2022). Privacy Preserving Scheme for Smart Transportation in 5G Integrated IoT. In *ICT with Intelligent Applications* (pp. 59–67). Springer. doi:10.1007/978-981-16-4177-0_8

Singh, P., Dixit, S., Sammanit, D., & Krishnan, P. (2022). The automated farmlands of tomorrow: An IoT integration with farmlands. *IOP Conference Series. Materials Science and Engineering*, *1218*(1), 012048. doi:10.1088/1757-899X/1218/1/012048

Soto-Valle, G., Hu, K., Holda, M., Cui, Y., & Tentzeris, M. (2022). Novel Additive Manufacturing-Enabled RF Devices for 5G/mmWave, IoT, Smart Skins, and Wireless Sensing Applications. *International Journal of High Speed Electronics and Systems, 31*(1-4), 2240017.

Sumathi, A. C., Ahalawat, A., & Rameshkumar, A. (2022). Early detection of DDoS attack using integrated SDN-Blockchain architecture for IoT. *2022 International Conference on Innovative Computing, Intelligent Communication and Smart Electrical Systems (ICSES),* 1–5. 10.1109/ICSES55317.2022.9914202

Szatkowski, J. M., Li, Y., & Du, L. (2022). Enabling Reconfigurable Naval SCADA Network through Software-Defined Networking. *2022 IEEE Transportation Electrification Conference & Expo (ITEC),* 214–218. 10.1109/ITEC53557.2022.9813872

Tanwar, S., Popat, A., Bhattacharya, P., Gupta, R., & Kumar, N. (2022). A taxonomy of energy optimization techniques for smart cities: Architecture and future directions. *Expert Systems: International Journal of Knowledge Engineering and Neural Networks*, *39*(5), e12703. doi:10.1111/exsy.12703

Tian, K., Chai, H., Liu, Y., & Liu, B. (2022). Edge Intelligence Empowered Dynamic Offloading and Resource Management of MEC for Smart City Internet of Things. *Electronics (Basel), 11*(6), 879. doi:10.3390/electronics11060879

Wang, B., Liu, X., & Zhang, Y. (2022). Future Prospects of BDS and IOT Integration Application. In *Internet of Things and BDS Application* (pp. 287–291). Springer. doi:10.1007/978-981-16-9194-2_7

William, P., Shrivastava, A., Chauhan, H., Nagpal, P., & Singh, P. (2022). Framework for Intelligent Smart City Deployment via Artificial Intelligence Software Networking. *2022 3rd International Conference on Intelligent Engineering and Management (ICIEM),* 455–460.

Yeh, L.-Y., Shen, N.-X., & Hwang, R.-H. (2022). Blockchain-Based Privacy-Preserving and Sustainable Data Query Service Over 5G-VANETs. *IEEE Transactions on Intelligent Transportation Systems.*

Zhao, L., Zhu, D., Shafik, W., Matinkhah, S. M., Ahmad, Z., Sharif, L., & Craig, A. (2022). Artificial intelligence analysis in cyber domain: A review. *International Journal of Distributed Sensor Networks, 18*(4). doi:10.1177/15501329221084882

KEY TERMS AND DEFINITIONS

Could Computing: On-demand computer system resources, primarily data storage and computing power, without user administration. Big clouds have functions in numerous data centers.

Enabling Technologies: Enabling technologies are innovations that can transform a user, state, culture or technology for general-purpose operations.

Fifth Generation (5G): A technology standard for broadband cellular networks, which cellular phone operators began deploying worldwide in 2019. It will replace 4G networks, which most cellphones use.

Internet of Things: Describes physical things with sensors, processing power, software, and other technologies that communicate over the Internet or other communications networks.

Massive MIMO: Multiple-input and multiple-output (MIMO) radio links use multipath propagation to increase capacity. HSPA+, IEEE 802.11n, WiMAX, 802.11ac, and long-term evolution all use MIMO.

Network Function Virtualization: A network architecture approach that uses information technology virtualization to virtualize entire classes of network node functions into building blocks that may be chained to design and deliver communication services.

Privacy: Ability to hide data, information and express oneself selectively depending on the situation and the location.

Sensor Devices: Are devices that detect and react to physical input, for instance, motion, light, pressure, heat, moisture, or many other natural phenomena could constitute the input.

Smart Cities: Technically edge-cutting urban area that uses several kinds of electronic sensors and approaches to gather precise data. The gained information from that data is unitized to manage resources, services, and assets efficiently; then, that data is employed to advance operations within the entire city.

Software-Defined Networking: A method to network administration that uses dynamic, programmatically efficient network design to increase performance and monitoring, similar to cloud computing.

Chapter 5
Innovative Model of Internet of Things for Industrial Applications

Jay Kumar Jain
https://orcid.org/0000-0002-9590-0006
Sagar Institute of Research and Technology, India

Dipti Chauhan
Prestige Institute of Engineering Management and Research, India

ABSTRACT

The internet of things is one of the most significant and promising innovations today. In this chapter, the authors proposed the dual probability-based energy estimation model in the wireless sensor network. The dual probability-based function measures the expected value of energy for the transmission of data. This function creates a subgroup of networks based on energy function and carries out the operation of energy management in the context sensor node data processing. This function also integrates cloud-based services with the sensor networks. The benefit of this function is that it increases the throughput of network and quality of service. The proposed model was simulated in MATLAB R-2014a environment, and the results were obtained using different scenarios of network density. Finally, the authors analyzed the performance of our proposed work with respect to the following metrics: data utility, energy consumptions, and data reconstruction error.

INTRODUCTION

The concept of the internet of things integrates all devices with the internet. It includes sensors, actuators, and edge devices. The life of network devices depends on the consumption and utility of energy in the sensor node or network. The wireless sensor network in cooperative technology is a challenge in IoTs. The application of IoTs is linked with the edge network (Nguyen et al., 2021). The edge network supports the concept of the dynamic nature of cloud-based services. The edge-based component has a problem with bandwidth and energy. Energy is a major factor in the sensor node for data transmission

DOI: 10.4018/978-1-7998-9266-3.ch005

and data receiving. Due to the mobile nature of the sensors node, the consumption of energy is very high and the life of IoT devices (Ullah et al., 2021). In the current decade, the minimization of energy in a wireless sensors network is a big issue. For the minimization of energy, various low-cost based energy protocols are designed. The success story of wireless sensor network deals with the success of IoT based services over cloud environments. The minimization of energy in wireless sensor networks is possible to use various routing and MAC layer based protocols. The duty cycle based routing protocol also reduces energy consumption during the transmission of data over the sink node. Some authors also suggested the cluster-based routing protocol for the minimization of energy in wireless sensors networks (Kaur et al., 2017).

The internet of things (IoT) is a group of internet enables things. The internet of things provides services to all societal areas. The things basically deal with electronic communication objects connected through the internet. The acceptability of IoTs is increasing every day due to easy installation and low-cost maintenance. The IoTs change the scenario of remote area data accessing, for accessing the remote area data, such as temperature, pressure, weather and fire event in the forest sensor networks are used. The sensors collect the information and transmit it to the base station with IoTs devices. Nowadays the IoTs application is integrated with cloud-based services. The cloud-based services are deployed over smart devices (Huang et al., 2014). The cloud services basically support the static infrastructure. The IoTs integrate these services with dynamic infrastructure. The dynamic nature of the cloud enhances the reachability of IoTs services to the most distant the universe. Along these lines, IoTs can bring forth colossal valuable applications and administrations that we never envisioned. With the progression in innovation, the device's processing power and storage capacity significantly expanded while their sizes diminished. These smart devices are normally outfitted with various kinds of sensors and actuators. In addition, the physical objects are progressively outfitted with RFID labels or other electronic standardized identifications that can be scanned by smart devices, e.g., smart mobile phones or small installed RFID scanner. IoTs is the Internet's stretching out and growing to the physical world and its related properties incorporate center, content, gathering, figuring, correspondences and network situations. These properties demonstrate the consistent association among individuals and objects or between the items and objects.

The current trend of research focuses on minimization of energy in terms of data transmission and data receiving in wireless sensor network for IoTs. The researchers proposed the techniques based on clustering, low route cost, heuristic based optimization to improve energy efficiency of sensor nodes. The sensor nodes usually work for long time in idle because it causes the waste of energy in WSNs. Now the concept of active and sleep mode, based on duty cycle, enhanced the efficiency of energy. The active and sleep mode increase the use of energy but degrade the performance of network in terms of quality of service, delay and loss of data packet. In this paper, dual probability based energy estimation function for the reduction of energy during the transmission of data to the sink node has been proposed (Sharma et al., 2021). The dual probability based energy function works on the concept of dynamic optimization of route cost and energy. The function of energy works on different layers of energy such as high energy, middle energy and low energy.

The grouping of energy layers deals with subgroups of sensor nodes. The reduction of energy cost increases the life of sensor nodes and networks (Jain & Sharma, 2017). The life of network enhances the reliability of IoTs based services over the cloud computing. Instead of dual probability based energy function, some compressed sensing technique for the minimization of energy in IoTs based services over the cloud networks have also been used. The contribution of this paper is summarized as follows:

1. Reducing the utilization of energy in wireless sensor networks for integration of IoTs. The dual probability energy function estimates the both end's energy requirements and transmits the data from source to thing elements.
2. Dual probability based energy function increase the lifetime of network and quality of service of connected devices.
3. Design of integration of energy function with wireless sensor network protocol.
4. Experimental simulation of three network scenarios of 50 nodes, 75 nodes and 100 nodes in MATLAB software and performance measurement of proposed dual probability-based energy function.

IoTs AND BACKGROUND OF CS

Although automation is a key driver, the technology discussions to a very large extent is addressing computer science problems with little or no reference to the automation requirements.

Automation in turn is driven by industrial requirements on sustainability, flexibility, efficiency and competitiveness. These in turn are driven by big societal questions like environmental sustainability, availability of energy and other raw material, as well as rapidly changing market trends. These drivers are pushing for both much larger and greater number of automation systems compared to the situation we have nowadays.

The current state of the art of automation appears to have an upper bound of about 105 I/O's. A reason for this is probably high engineering costs for large and complex. This have lead to work on moving the ISA-95 automation pyramid to an Internet cloud paradigm based on Service Oriented Architectures, SOA. Here cheaper hardware and expectations on reduced engineering costs have been strong motivators for the developments.

To have Internet technology support these developments, there are a number of automation technology gaps identified.

These gaps can seemingly not be addressed by current state of the art Internet technology. The identified automation technology gaps are:

- Interoperability of a wide range of IoT and legacy devices.
- Automation requirement on latency guarantee/prediction for communication and control computations.
- Scalability of automation systems enabling very large integrated automation systems.
- Multi stakeholder integration and operations agility.
- Security and related safety of automation systems.
- Ease of application engineering.

Furthermore, it is noted that most automation systems are physically and geographically local. To control some physical behavior measurement of that specific physics is required to facilitate its control and automation.

Based on these requirements and locality of automation, this current paper proposes the concept of local automation clouds. Local clouds should be able to perform the desired automation and control functionalities locally, while meeting the above requirements. To provide scalability, this implies that multiple local clouds should be able to interact with each other, though with relaxed real time requirements.

LOCAL CLOUDS FOR AUTOMATION

Extending previous work on automation clouds using a SOA approach, let us provide a basic definition of a local automation cloud. The technology gaps of Internet regarding automation key requirements are given above.

The objective of a local cloud is to provide a communication and computation environment suitable for automation. Based on the key requirements, it is argued that the cloud boundary shall be the "protective fence" to the communication and computations necessary to full fill the desired automation tasks.

Thus, protecting the automation functionalities like their time critical communication and computations from "external" impact. Thus, a basic local automation cloud is a protected network with no in-bound or out-bound communication. Which is in direct contrast to current main stream definition of cloud which a metaphor for the Internet.

The here proposed key properties of a local cloud are:

1. **Self-Contained:** No external resources needed to establish the local cloud
2. Provide a security fence to external networks
3. Interoperability between systems within a local cloud is established through services of information exchange_ Support for protocol and semantics transparency
4. **Automation Support:** Both design- and run-time
    - Support for automation system design, configuration, deployment, operation, and maintenance
    - Enabling event-based information exchange
    - Enabling information exchange audit
    - Support for service and communication QoS
5. Security in relation to bootstrapping, software update, and communication in general
6. Inter-cloud service exchanges secure service discovery, authentication and authorization, orchestration and data exchange with other local automation clouds

In the following above properties are detailed. For the rest of the paper the following definitions are used. A device is a hardware capable of hosting one or several software systems, a system is software capable of producing and/or consuming one or several service and, a service is an exchange of data between a service producing system, and a service consuming system.

Self-Contained

The self-contained property allows local operation that is independent of external resources. This feature allows for a closed cloud boundary. For this purpose, a local automation cloud is proposed to have:

- Device, System and Service registries keeping track and allowing for the discovery of Devices, software Systems and Services deployed within a local cloud. The Service registry provides data on Service interfaces, methods, data types and associated meta data. The System registry provides data on which Systems are registered with a local cloud, meta data of these registered System and the services these systems are designed to produce or consume. The Device registry provides unique device identity and device meta data.

- **Service Orchestration:** SoS (System of Systems) run time Configuration Provides orchestration rules defining which Service produced by which System is to be consumed by which other System.
- Service authentication and authorization provides authentication of service consumers and authorization for service consumption.

Security Fence

A secure fence to external networks since the local cloud is self-contained there is no need for interaction with any external resources. Thus, the physical communication layer will be separated from any other network. If inter cloud service interaction is needed this should be handled through what here defined as gatekeeper gateways.

Interoperability Between Systems

The service interoperability property provides a structure for information on interfaces, service protocols, methods, data types, semantics, encoding, and compression provided by a service producing system within the local cloud. Based on these, service consumers can be properly matched with service providers. Furthermore, through protocol and semantics translation devices/systems featuring different service protocols, semantics and encoding can be made interoperable.

Automation Support

The automation support properties shall provide support for at least:

- Engineering of automation systems based on IoT and System of Systems.
- Support for run-time configuration of service, system, and device
- Quality of Service monitoring and management
- Event handling support
- Operational audit and storage of historical data

Security

Provision for authentication of a service consumer, authorization of service consumption and protection of payload data. In addition methodology for secure deployment of device, system and service is necessary. It's also clear that a methodology for secure software update is required.

Inter-Cloud Service Exchange

To build large automation systems it's necessary that service exchanges can be made between local clouds. Here service exchange administration as service discovery, authorization, authentication and orchestration shall be possible between local clouds. If such service exchange between local clouds is orchestrated and authorized the service pay load should be possible to protect end to end.

The advancement of internet technology derived the concept of smart internet is called internet of things. The internet of things provides services in all areas of society. The things basically deal with electronic communication object, connected through the internet. The acceptability of IoTs is increasing day to day due to easy installation and low-cost maintenance. IoTs change the scenario of remote area data accessing; for the accessing of remote area data, such as temperature, pressure, weather and fire event in forest, sensors networks are being used. The sensors collect the information and transmit it to base station with IoT devices. Now a days, the IoT's application is integrated with cloud based services. The cloud based services are being deployed over the smart devices. The cloud services basically support the static infrastructure; IoTs integrates these services with dynamic infrastructure. The dynamic nature of cloud enhances the reachability of IoT services towards last person in the universe. The IoTs connects real world objects and embeds the intelligence in the system to smartly process the object specific information and take useful autonomous decisions. Thus, IoTs can give birth to enormous useful applications and services that we never imagined before. With the advancement in technology, the device's processing power and storage capabilities have been increased significantly while their sizes have been reduced. These smart devices are usually equipped with different types of sensors and actuators. Also, these devices are able to connect and communicate over the Internet that can enable a new range of opportunities. Moreover, the physical objects are increasingly equipped with RFID tags or other electronic bar codes that can be scanned by the smart devices, e.g., smart phones or small embedded RFID scanner. IoTs is the Internet's extension and expansion towards the physical world and its related properties include focus, content, collection, computing, communications and connectivity scenarios. These properties show the seamless connection between people and objects or between objects and objects.

Compressive Sensing (CS) was an idea starting from the signal processing era. The quality of compressive Sensing is its capacity to remake meager or compressible signal from a modest number of estimations without requiring any of the earlier learning about the signal structure. Compressive sensing is beneficial wherever the signals are meager in a known premise, estimations are costly, and calculations at the collector end are shoddy (Yao et al., 2014). These qualities totally coordinate WSNs. Contrasted and information pressure, applying compressive sensing in WSNs offers promising enhancements as low power sensor hubs are not for the most part appropriate for actualizing encoding of information pressure systems (Jain et al., 2012). Past survey articles in CS constrain their base recuperation calculations to linear programming and Greedy calculation (Chauhan et al., 2016). These strategies experience the ill effects of multifaceted nature, precision and speed issues. Bayesian CS (BCS) is a strategy, which uses a measurable portrayal of the signal to supplement the customary strategies. It can give better execution as far as exact information remaking or a diminished number of estimations. Be that as it may, there are a couple of works in the WSNs field, which advantage from the Bayesian system to beat their execution. TC-CSBP (Nguyen et al., 2017) is a conviction spread (BP) based procedure, which utilizes just a transient connection among sensor readings to remake the signal. Compressive sensing states that meager or compressible signs can be precisely or around recouped from various straight projections. An inadequate signal is a signal, which normally displays sparsity while compressible signal can be all around approximated with meager portrayal through changing to another space, where few the coefficients speak to the vast majority of the energy of the signs (Jain & Chauhan, 2021).

The rest of the chapter is organized as follows. Section II describes the related work in the field of energy minimization in wireless sensor networks. Section III describes the proposed dual probability-based energy function and the integration model of energy function with IoTs. Section IV presents the methodology and experimental results to the proposed algorithm and Section V concludes the chapter.

RELATED WORK

The diversity and applicability of IoTs are in every field of engineering and social service based applications (Li et al., 2012). The growing data transfer over the internet with smart devices uses the sensor network based technology (Jain & Sharma, 2013). The life and efficiency of sensor networks depend on the utilization of energy stored in tiny small batteries (Jain & Jain, 2022). If the battery will drain the life of network will be ended.

Wu et al. (2016) used the process of energy minimization for sink node message broadcasting for the communication. They proposed flooding algorithm for synchronous duty cycle.

Sun et al. (2010) proposed Dynamic Routing Algorithm for need Guaranteein low obligation cycled sensor systems. The two plans of dynamic sending basic leadership and need based calendar are utilized as a part of DRAG to accomplish need ensure in low obligation cycled sensor systems. Theyassess DRAG by means of broad reproductions and the outcomes demonstrate that DRAG can accomplish great execution in conveyance proportion and system delay.

Doudou et al. (2012) describes the management of energy utilization in sensors based network. They also used concept of duty cycle for the processing of MAC data flow in different working mode of sensors node in terms of active mode, sleep mode and some other derived mode of alive sensors nodes.

Langendoen et al. (2008) typical energy-efficient approaches trade off performance in terms of throughput and latency in return for network lifetimes in the order of years with nodes assembled out of commodity components and powered by a set of penlight batteries. Since even a low-power radio is consuming two to three orders of magnitude more energy when switched on than when in sleep mode, the focus of attention is on reducing the so-called idle listening overhead.

Jain et al. (2022) suggested that concept of cyber security involves guarding against hostile attack on digital systems such computers, servers, mobile devices, networks, and the data they are connected to. Accounting for cyber security where machine learning is used and using machine learning to enable cyber security are the two main components of combining cyber security and ML. We may benefit from this union in a number of ways, including by giving machine learning models better security, enhancing the effectiveness of cyber security techniques, and supporting the efficient detection of zero day threats with minimal human involvement. In this review paper, we combine ML and cyber security to talk about two distinct notions. We also talk about the benefits, problems, and difficulties of combining ML and cyber security. In addition, we explore several attacks and present a thorough analysis of various tactics in two different categories. Finally, we offer a few suggestions for future research.

Chakchouk et al. (2015) provide taxonomy for opportunistic routing proposals, based on their routing objectives as well as the optimization tools and approaches used in the routing design. Hence, five opportunistic routing classes are defined and studied in this paper, namely, geographic opportunistic routing, link-state-aware opportunistic routing, probabilistic opportunistic routing, optimization-based opportunistic routing, and cross-layer opportunistic routing.

Zhau et al. (2006) demonstrate the efficacy of LBKs in counteracting several notorious attacks against sensor networks. Finally, authors propose a location-based threshold-endorsement scheme, called LTE, to thwart the infamous bogus data injection attack, in which adversaries inject lots of bogus data into the network.

Yick et al. (2008) have summarized and compared different proposed designs, algorithms, protocols, and services. Moreover, we have highlighted possible improvements and research in each area.

Chen et al. (2013) developed optimization approaches concerning sensor, network, and energy. An individual sensor uses MicroRF-201 series for wireless communication empowered by Zigbee protocol, and its design supports low energy consumption and has a memory of the maxi-mum size. Reliability and real-time transmission have been enabled at the network level.

Jain et al. (2020) suggested that Mobile Ad-hoc networks are the most favored novel Multipath Progressive Routing protocol with low vitality utilization is proposed to take care of these issues. This protocol has three novel perspectives contrasted with the other on-request multipath protocols: it diminishes routing overhead altogether and accomplishes different hub disjoint directing ways. The protocol has a heap balance component. By utilizing the heap balance component, it is conceivable to decrease the traffic blockage which is likewise a significant contributing variable to improved system execution.

Jumira et al. (2013) discussed a routing scheme called energy-efficient beaconless geographic routing with energy supply (EBGRES) for wireless sensor networks. EBGRES provides loop-free, fully stateless, energy-efficient source-to-sink routing with minimal communication overhead without the help of prior neighborhood knowledge. It locally determines the duty-cycle of each node, based on an estimated energy budget for each period, which includes the currently available energy, the predicted energy consumption and the energy expected from the harvesting device.

Le et al. (2009) demonstrate an energy-efficient access control scheme based on ECC to overcome these problems and more importantly to provide dominant energy-efficiency. Through analysis and simulation based evaluations, authors show that the discussed scheme overcomes the security problems and has far better energy-efficiency compared to current scheme.

Chauhan et al. (2021) defined a novel IPv6 header compression mechanism (IP6HC) to compress the IPv6 header of the tunneled packets for IPv6 tunneling mechanism. The primary objective behind designing IPv6 header compression protocol is to provide a solution for reducing high header overhead due to multiple headers present in case of IPv6 tunneling mechanism.

Li et al. (2014) improved algorithm effectively reduces the mortality rate of nodes and decreases energy consumption in the network, balances the network energy load, and prolongs the lifetime of network.

Meng et al. (2012) improve Cluster-Tree algorithm by introducing a neighbor table. The improved routing protocol has better performance in less energy consumption and a longer life cycle of entire ZigBee network.

Uikey et al. (2013) discuss method to investigate their promises and their performance over different performance parameters, and discussed the required changes. To justify modification for the model, implementation of modifications made in the network is performed and comparing performance of the discussed technique with existing Zigbee cluster network tree.

El Rachkidy et al. (2012) discussed to the PiRAT has better performance than previous protocols in terms of packet loss, end-to-end delay, congestion and node overload.

DUAL PROBABILITY FUNCTION (DPF)

The dual probability function measures the same level of energy for the grouping of sensor node to transmit the sensor to sink node and IoTs get away. The dual probability measures the energy level of same node with similar probability for the measuring the level of energy for communication.

$P-DP(n)$: **measure** the probability of sensor nodes of same energy level:

- $sink\ probablity(level-Dp(n))$ of a sensors node n with respect to another sensors node

$$nlevel-prop_k(p,o)=\max\{n-probablity(o),n(p,o)\} \tag{1}$$

Where $n(p,o)$ is the similar probability between p and o.

- *sub level probablity* (slp) of a sensors node n

$$slp_k(p)=\left(\frac{1}{k}\sum_{o\in N_{(p,k)}} level-prob_k(p,o)\right)^{-1}, \tag{2}$$

Where $N_{(p,k)}$ is the set of n node of similar probability of energy of n.

- *sink node energy* of a sensors node n

$$DP\_OT_k(p)=\frac{1}{k}\sum_{0\in N_{(p,k)}}\frac{slp_k(o)}{slp_k(p)} \tag{3}$$

Here k is sub set group of sensors node for the same level of energy factor and N is total nodes in environments.

SYSTEM MODEL

Given $n_t \in R^{GP}$ collected at level energy $e\in E$, the goal of DP and DP_E is to assign an DP_{OT} value to n_t, for the value of energy $E<P$ of the n nodes that have been measured up to level energy E. All n sensor level and their corresponding DP_{OT} values in sink connection of subgroups. Hence measure the energy value of extended energy DP_{OT} values of new nodes can be calculated. the goal of DP and DP_E is to detect the same level for the whole network communication and not just for the n last sensor nodes where the available sink connection of subgroups is limited to P. DP_E is an extension to DP.

DP: Dual Probability
DP_OT: Dual Probability Outer Sensor Node
DP_E: Extension to Dual Probability

Algorithm 1: DP_E Estimation

1. Input: a sensors node n_t at level energy e
2. Output: DP_{OT} value $DP_{OT(t)}$
3. Estimate $N_{(n_t,k)}$ and $p-probab(n_t)$
4. for all $n \in N_{(n_t,k)}$ do
5. Estimate $level-prob(p_t,o)$ using Equation (1)
6. end for
7. $S_{node} \leftarrow Pn_{(n_t,k)}\{thesetofsinknoden_t\}$
8. for all $n \in S_{node}$ and $q \in P_{(o,k)}$ do
9. Node $k-prob(o)$ and $level-prob(q,o)$
10. if $oN_{(q,k)}$ then
11. $S_{node} \leftarrow S_{node} \cup \{q\}$
12. end if
13. end for
14. for all $o \in S_{node}$ do
15. Node $slp(o)$ and $DP_{OT}(\{Pn_{o,k}\})$
16. end for
17. Estimate slp and DP_{OTn}
18. return DP_{OT}

Algorithm 2: DP (Dual Probability)

1. Input: a set of sensor nodes $N=\{n_1,\ldots\ldots,n_n\}$ sink connection of subgroups size limit of L
2. Output: set of $DP_{OT}=\{DP_{OT}(n_1),\ldots\ldots\ldots,DP_{OT}(n_n)\}$ nodes
3. $i \leftarrow 0;$ {energy level}
4. for all $n_t \in n$ do
5. $DP_{OT}(n_t) \leftarrow$ connection (n_t)
6. if Number of sensor nodes in sink connection of subgroups
7. $(sc^i, N^i) \leftarrow GP-Probal(n^i)$
8. for all $sc_j^i \in n^i$ do
9. Estimate $k-probab(sc_j^i), slp(v_j^i), DP_{OT}(v_j^i)$
10. end for
11. return DP_{OT}

The sensor node is a normal node and its unit of the data collector. The PD node is a selected node for the transmission of data to the sink node. Figure 2 describes the sensor node connection in terms of subgroups for the level of energy. The levels of energy are different in two decision parameters, one is DP and the other is DP_OT.

Integration of IoTs with dual probability-based function with an application with an internet gateway. The measure probability value DP and DP_OT create different levels of energy groups for the integration of the application.

1. If $DP_{OT} = 0$, the energy level of all sensors node is the same level and direct connect each node with sink nodes
2. If $DP_{OT} > 0$ the value of the energy level of DP is lower and creates a more similar probability-based node connection.
3. If $DP_{OT} < 0$, the value of energy level is average and some nodes are not alive so gateway only maintains a connection.

$n_s = n_s - 1;$

`if (` $DP_{OT} > 0$ **`then`**

$S \leftarrow S \cup \{n_i\}$

```
end
end
Step 5.        Set the level of energy of each group selected sensor nodes.
```

`Return` S

Figure 1 represents the real scenario of the communication of wireless sensor networks with IoTs and cloud-based services. The network has been deployed, which sends the information to the sink node (Base station) and base station, integrated with the waterway through IoTs elements. The mention DPF represents the value of energy level for the selection of sink node to communicate to things.

Algorithm 3: Integration of Sink Nodes

`Input:` `Sensor nodes set` $s = \{n_i, i = 1,........, n\}$

`Output:` S `- the selected sensor nodes subset`

`Begin`

`Step 1.        Define the initial value of request` $R = \varnothing$

`Step 2.        Calculate` GP `for each sensor nodes,` $i = 1,.........., n$

`Step 3.` $n_f = n;$ `Select the sensor nodes`

`Then, set` $S \leftarrow S\{\ \}; S \leftarrow S \cup \{\ \}; n_s = n_s - 1$

`Step 4.` **`while`** $S \neq \varnothing$ **`do`**

`Measure` DP_{OT} `in (2) to ðnd` n_i `where` $i \in \{1, 2, ..., ns\}$;

Figure 1. WSN with IoTs

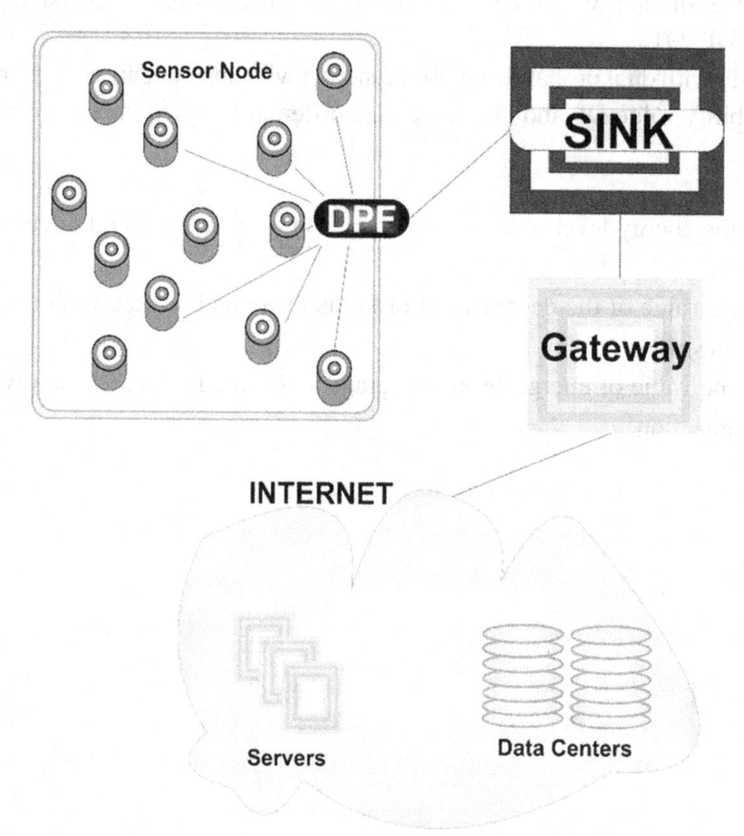

METHODOLOGY AND EXPERIMENTAL RESULT

Methodology

The minimization of the utilization of energy in data transmission and receiving derived novel energy estimation function is called dual probability (DP). The proposed DP approach is compared with the CP method. The dual probability-based function works in the mode of an energy level of the subgroup of sensors node. The subgroup of sensors node drives three-level energy factors high, low and average. The level of energy factor decides two decision parameters one is DP and the other is DP_OT. The DP decision parameter decides the subgroup of sensors node for the communication of sink nodes. The DP_OT takes the decision of data connection with sink nodes to IoTs. The dual probability-based function also measures the quality of data in terms of data utility of sensors node and IoTs elements. For the validation of the design model, we have used MATLAB R-2014a software and created four different network scenarios of a number of sensors nodes 50, 75, 100 and 500. For the transmission of data, it uses the CBR data packet and measures the performance in terms of small and combinations of large sensors network scenario and measures three parameters: data utility, energy efficiency, and data reconstruction error. These three parameters are taken for the evaluation result and compared with the compressive sensing technique. Instead of a compressive sensing technique, the design model increases the utilization of energy and enhanced the life of sensor networks. The efficiency of energy increased in terms of transmission of data to the sensors node to the sink node. The increased efficiency of energy depends on the proper integration of the probability factor and sink node energy level. The dual probability function measures the level of energy and groups the sensor node for communication. Despite the utilization of energy it also increases the data utility factor of sensors to the integration gateway for IoTs. The utility of data increases the performance of the network and life of sensors node. The data reconstruction of low energy node is also good in compression of CP model error reconstruction. The model goes in the dense network and increases the number of sensors node some little bit performance coincides with CP mode in terms of data utility, energy efficiency, and error reconstruction.

Simulation Environment

We evaluate the performance of our algorithm via simulations in MATLAB R-2014a. The simulation environment setup with the simulation parameters listed in Table 1. The simulated network is a We assume that all the sensor nodes and the sink node are randomly deployed in a circular sensing area. The initial energy of each sensor node is Joule.

EXPERIMENTAL RESULT

The performance metrics considered during simulation have their own significance for better network performance. The important definitions of parameters involving in this simulation are:

1. **Data Utility:** Network utilization is the amount of traffic on the network compared to the peak amount that the network can support. This is generally specified as a percentage.
2. **Energy Consumptions:** A useful measure of the efficiency of the network is the energy consumed per bit of data transferred.
3. **Data Reconstruction Error:** Data reconstruction is the rebuilding of a virtual environment such as the dynamic temperature, light, humidity, gas concentration and is based on the sensory data.

Table 1. Simulation parameter settings

Parameters	Value/Units
Size of the sensor network ($n_x m$)	400 * 400
Number of sensor nodes	50-500
Number of gateways	1
Number of the sink node	1
The initial energy of each sensor node	50 Joule
Deployment distribution	Random
Energy of gateway	50 Joule
Sensor communication radius (r)	200
Simulation Time	600 Seconds
Size of sensing data (bytes)	4000 Bytes
Location of the sink node	X= 200 Y=200

The design model is simulated in MATLAB R-2014a software. For the validation of methods, we have generated four network scenarios of 50, 75, 100 and 500 nodes. The data traffic used is a constant bit rate of 512. The process of energy function defines the value of energy as 12-14 joule. The energy distribution is uniform. The data is collected from the sink node in every 120 seconds. The transmission interval of data is 10 seconds. The total simulation time is 600 seconds. For the validation of models, we have used data utility, energy utilization parameters, and data error correction during the transmission. The results of DP based methods compare with the compressive sensing technique (CP) (Ullah et al., 2021).

Figure 2 shows the comparative result analysis between CP and DP method with the data utility and the number of nodes is 50 used in this simulation work. Due to enter the selection of path to data transfer to lower level interrogation to sink node data utility is increasing in comparison to CP. DP method gets the increasing percentage compared to CP method in data utility; when number of node is 10, 20, 30, 40, 50 then sequential increment in percentage 10%, 10%, 17.5%, 20% and 0% were found. Here, this proposed work as a DP method gives better performance compared to the CP method.

Figure 3 depicts that the comparative result analysis between CP and DP method with the energy consumption in transmission and the number of nodes is 50 in this simulation work. Due to the level selection of sensor node, it gives the fare path for the transmission of data now, the consumption of energy decreases in the method of DP. DP method gets the decreasing percentage compared to CP method in energy consumption, when the number of nodes is 10, 20, 30, 40, 50 sequentially decreased percentage 2%, 7%, 15%, 25%, and 33%. Here, this proposed work as a DP method gives better performance compared to the CP method.

Figure 4 shows the comparative result analysis between CP and DP method with the data reconstruction error and the number of nodes is 50 used in this simulation work. The value of data rejection is minimizing because rejection of a lower energy sensor node packet is not captured to the sink node so

Figure 2. Data utility with 50 nodes

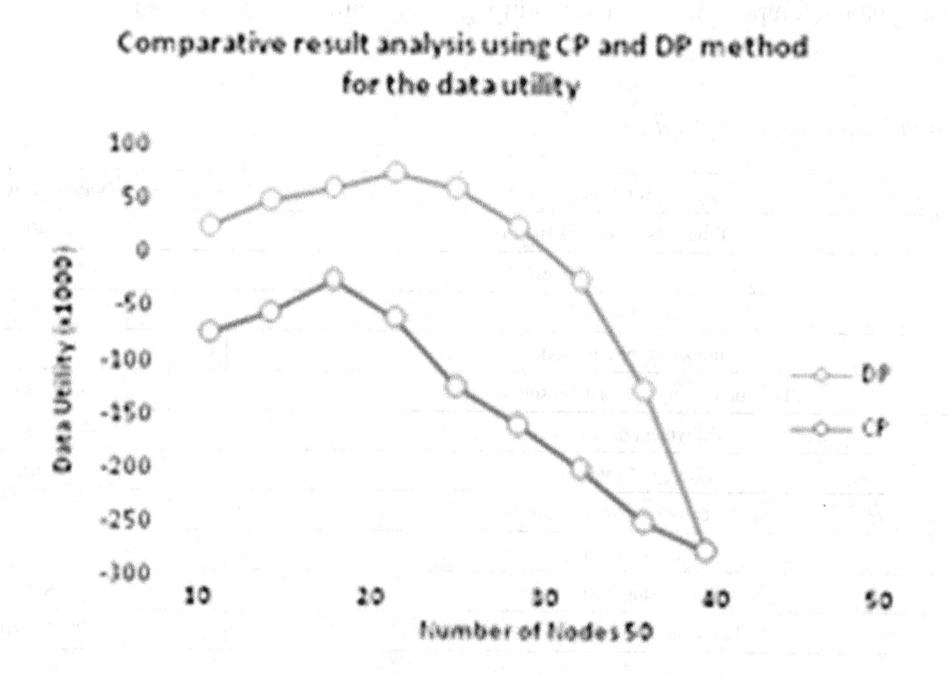

Figure 3. Energy consumption with 50 nodes

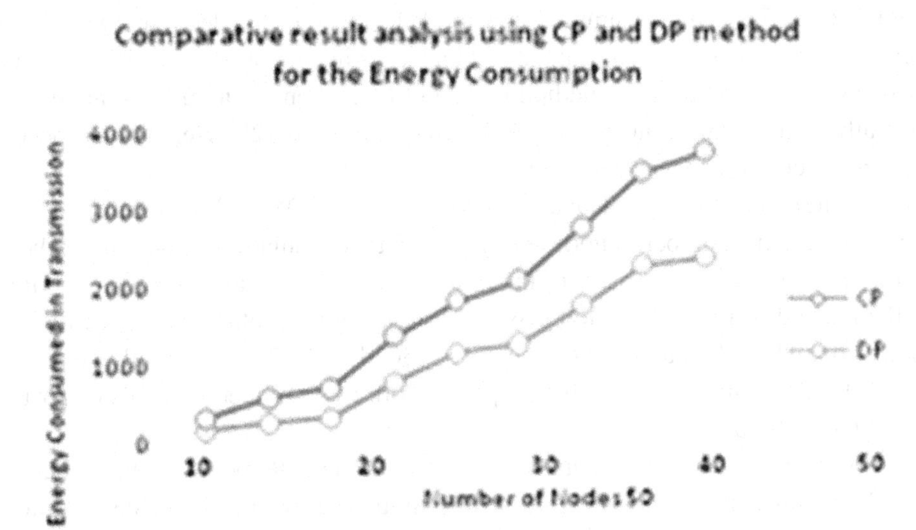

Figure 4. Data reconstruction error with 50 nodes

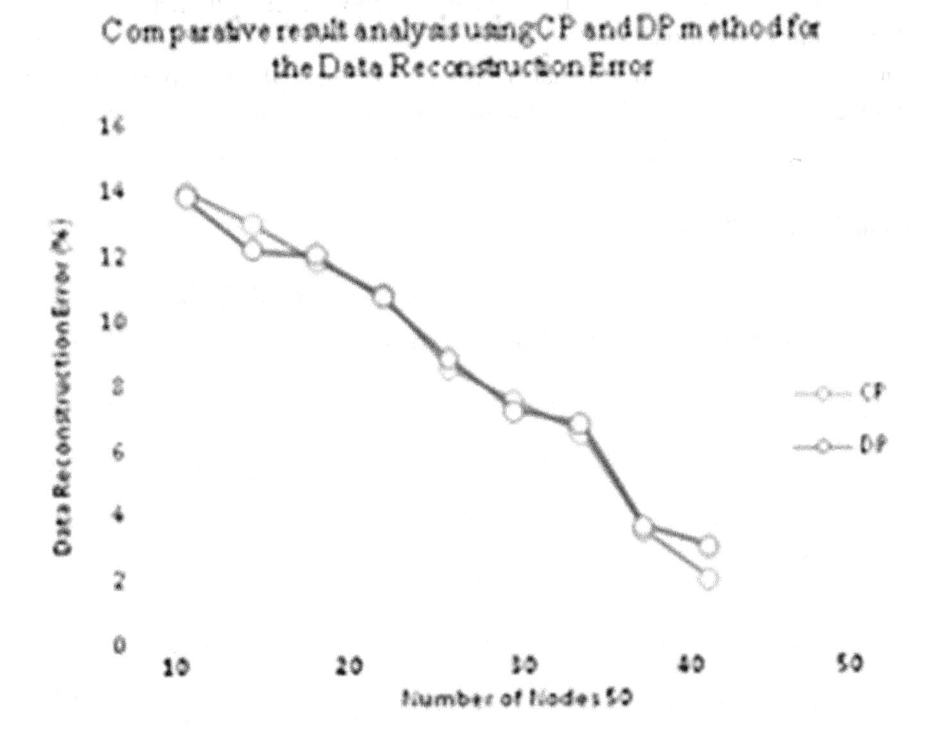

the value of reconstruction error is minimized. DP method gets the decreasing percentage compared to CP method in energy consumption, when number of node is 10, 20, 30, 40, 50 sequentially decreased percentage 0.2%, 0%, 0.5%, 0.8% and 1%. Here this proposed work as a DP method gives better performance compared to the CP method.

Figure 5 shows the comparative result analysis between CP and DP method with the data utility and the number of nodes is 75 used in my simulation work. Due to enter the selection of path to data transfer to lower level interrogation to sink node data utility increases in comparison of CP. DP method gets the increasing percentage compared to CP method in data utility, when the number of nodes is 15, 30, 45, 60, 75 sequentially increased percentage 15%, 10%, 20%, 20%, and 12%. Here this proposed work as a DP method gives better performance compared to the CP method.

Figure 6 shows the comparative result analysis between CP and DP method with the energy consumption in transmission and the number of nodes is 50 used in this simulation work. Due to the level selection of sensor node, it gives the fare path for the transmission of data now, the consumption of energy decreases in the method of DP. DP method gets the decreasing percentage compared to CP method in energy consumption, when the number of nodes is 15, 30, 45, 60, 75 sequentially decreased percentage 15%, 16%, 20%, 22%, and 5%. Here this proposed work as a DP method gives better performance compared to the CP method.

Figure 7 shows the comparative result analysis between CP and DP method with the data reconstruction error and the number of nodes is 75 used in this simulation work. The value of data rejection is minimizing due to rejection of lower energy sensor node packet are not captured to the sink node so the value of reconstruction error is minimized. DP method gets the decreasing percentage compared to CP method in energy consumption, when the number of nodes is 15, 30, 45, 60, 75 sequentially decreased percentage 12%, 11%, 10%, 13%, and 6%. Here this proposed work as a DP method gives better performance compared to the CP method.

Figure 8 shows the comparative result analysis between CP and DP method with the data utility and the number of nodes is 100 used in this simulation work. Due to enter the selection of path to data transfer

Figure 5. Data utility with 75 nodes

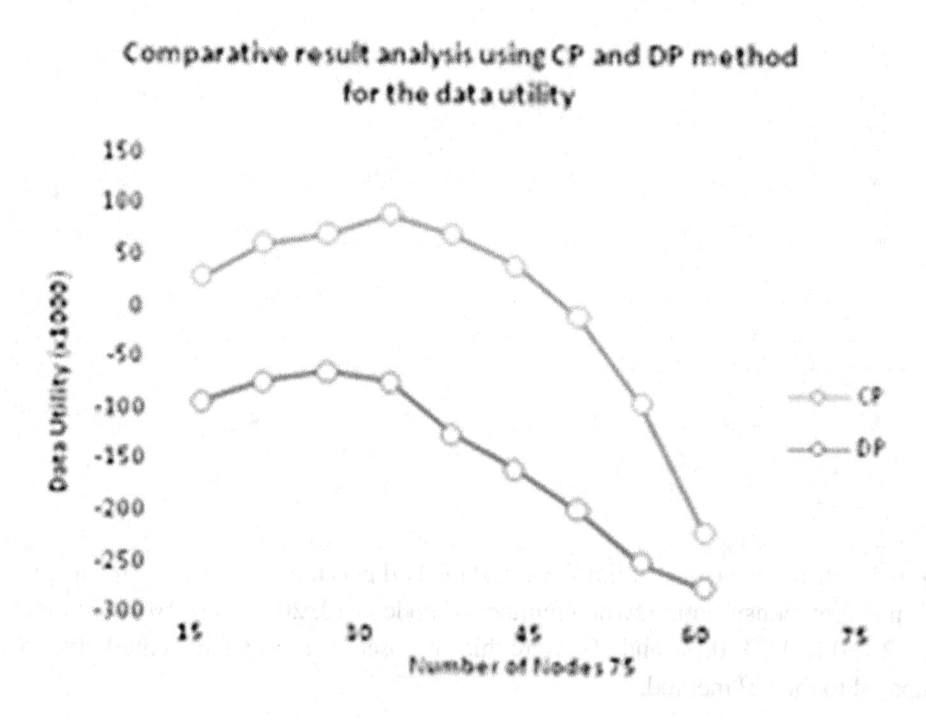

Figure 6. Energy consumption with 75 nodes

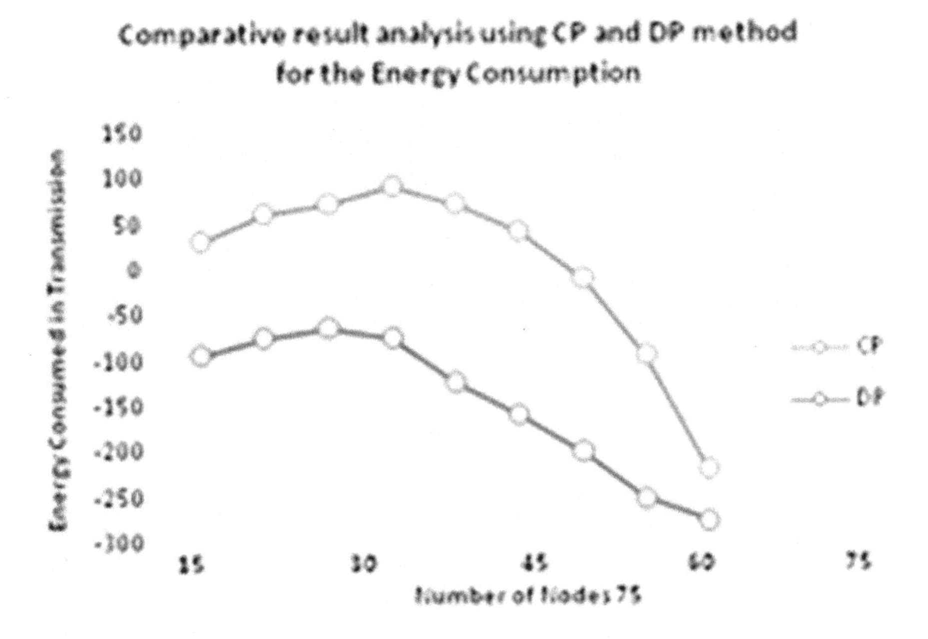

Figure 7. Data reconstruction error with 75 nodes

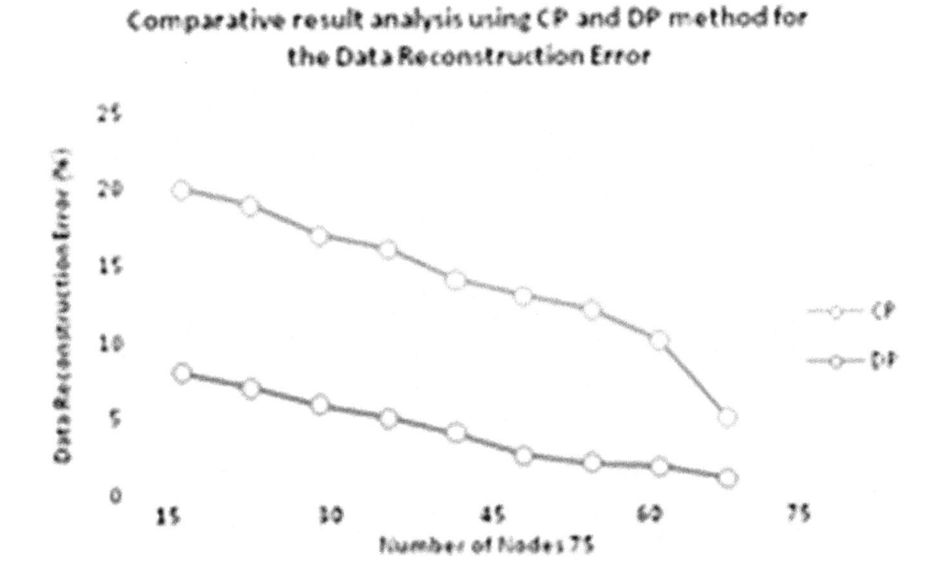

to lower level interrogation to sink node data utility is increasing in comparison to CP. DP method gets the increasing percentage compared to CP method in data utility, when number of node is 20, 40, 60, 80, 100 sequentially increased percentage 0.5%, 4%, 25%, 35% and 50%. Here this proposed work as a DP method gives better performance compared to the CP method.

Figure 9 shows the comparative result analysis between CP and DP method with the energy consumption in transmission and the number of nodes is 100 used in this simulation work. Due to the level selection of

Figure 8. Data utility with 100 nodes

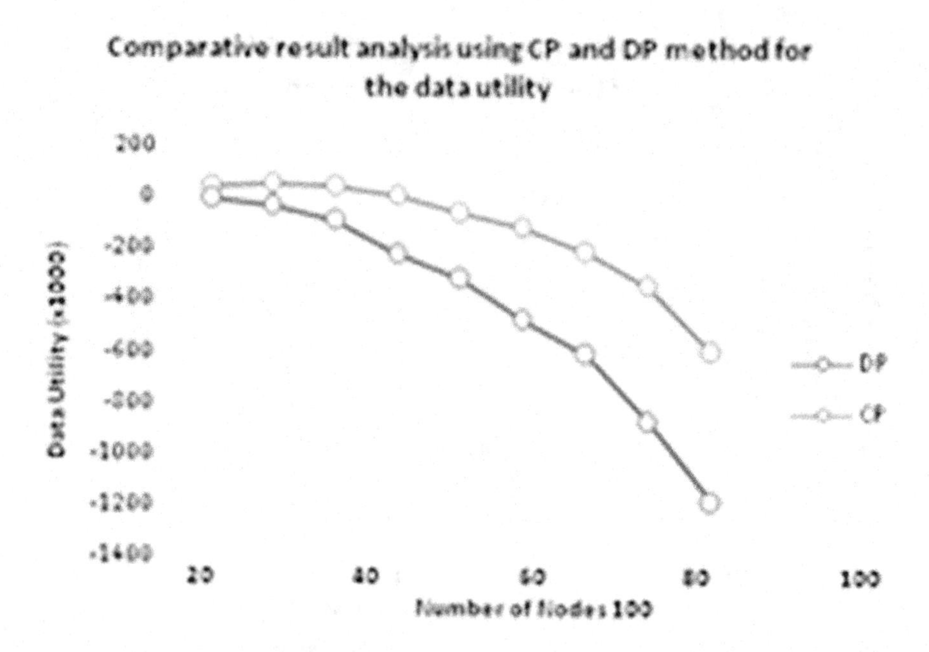

Figure 9. Energy consumption with 100 nodes

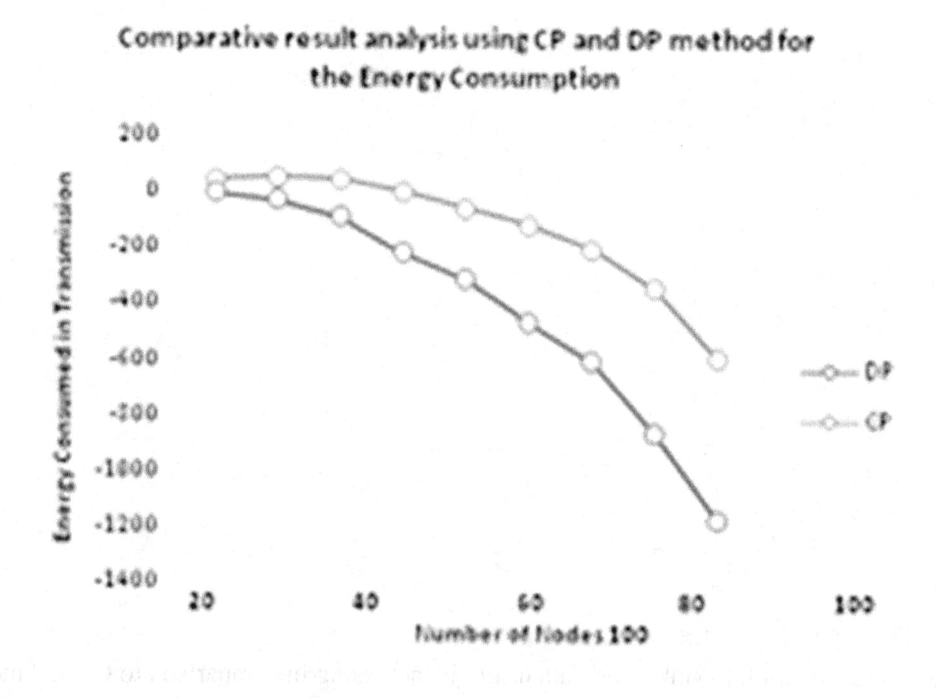

sensor node give the fare path for the transmission of data now, the consumption of energy decreases in the method of DP. DP method gets the decreasing percentage compared to CP method in energy consumption, when number of node is 20, 40, 60, 80, 100 sequentially decreased percentage 0.2%, 0.5%, 25%, 34% and 50%. Here this proposed work as a DP method gives better performance compared to the CP method.

Figure 10 shows the comparative result analysis between CP and DP method with the data reconstruction error and the number of nodes is 100 used in this simulation work. The value of data rejection is minimizing due to rejection of lower energy sensor nodes packet are not captured to the sink node so the value of reconstruction error is minimized. DP method gets the decreasing percentage compared method in data utility, when number of node is 20, 40, 60, 80, 100 sequentially increased percentage 0.5%, 4%, 25%, 35% and 50%. Here this proposed work as a DP method gives better performance compared to the CP method. to CP method in energy consumption, when number of node is 20, 40, 60, 80, 100 sequentially decreased percentage 0.3%, 0.1%, 0.6%, 0.8% and 1.5%. Here this proposed work as a DP method gives better performance compared to the CP method.

Figure 11 shows the comparative result analysis between CP and DP method with the data utility and the number of nodes is 500 used in this simulation work. Due to enter the selection of path to data transfer to lower level interrogation to sink node data utility is increasing in comparison to CP. DP method gets the increasing percentage compared to CP method in data utility, when the number of nodes is 100, 200, 300, 400, 500 sequentially increased percentage 30%, 27%, 0%, 36%, and 30%. Here this proposed work as a DP method gives better performance compared to the CP method.

Figure 12 shows the comparative result analysis between CP and DP method with the energy consumption in transmission and the number of nodes is 500 used in this simulation work. Due to the level selection of sensor node give the fare path for the transmission of data now, the consumption of energy decreases in the method of DP. DP method gets the decreasing percentage compared to the CP method in energy consumption, when the number of nodes is 100, 200, 300, 400, 500 sequentially decreased percentage 25%, 9%, 23%, 15%, and 20%. Here this proposed work as a DP method gives better performance compared to the CP method.

Figure 13 shows the comparative result analysis between CP and DP method with the data reconstruction error and the number of nodes is 500 used in this simulation work. The value of data rejection is

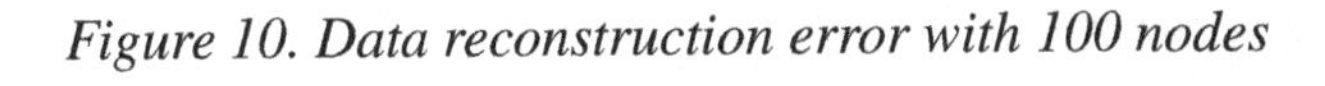
Figure 10. Data reconstruction error with 100 nodes

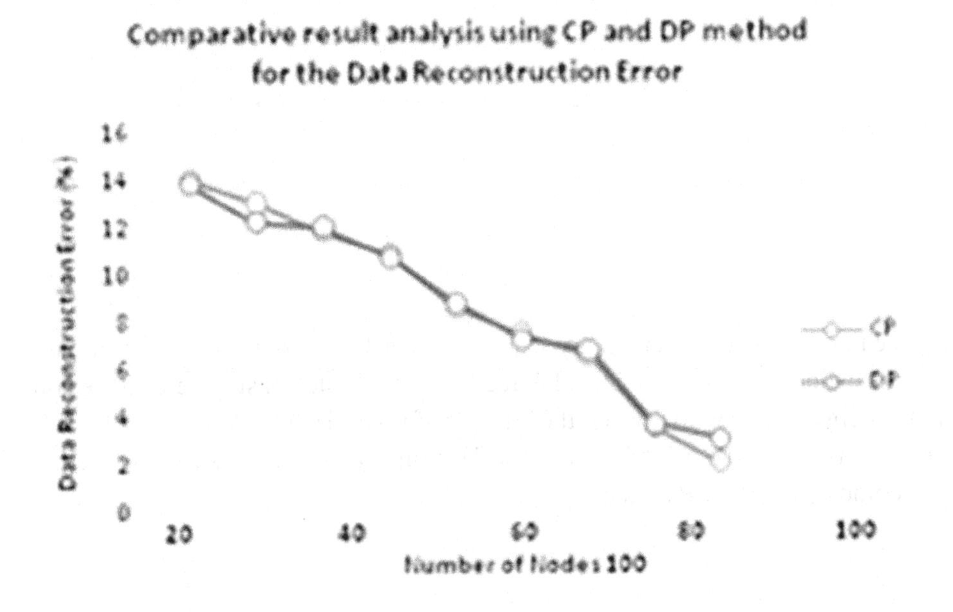

Figure 11. Data utility with 500 nodes

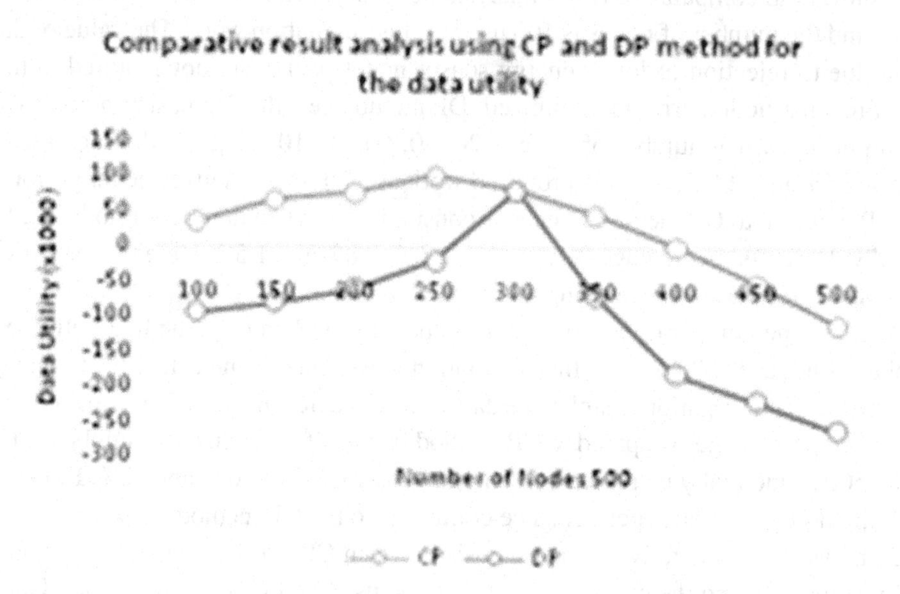

Figure 12. Energy consumption with 500 nodes

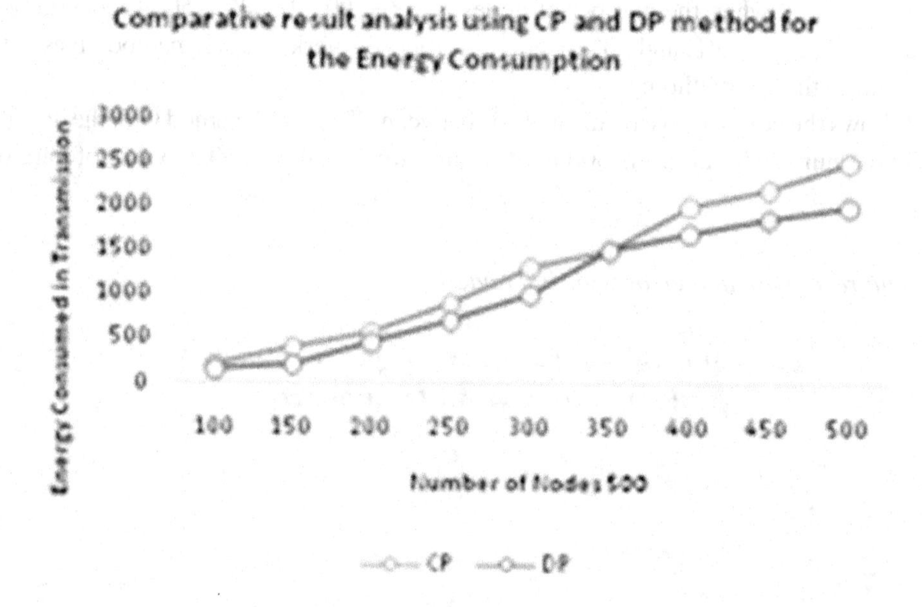

minimizing due to rejection of lower energy sensor node packet are not captured to the sink node so the value of reconstruction error is minimized. DP method gets the decreasing percentage compared to the CP method in energy consumption, when the number of nodes is 100, 200, 300, 400, 500 sequentially decreased percentage 12%, 7%, 5%, 54%, and 50%. Here this proposed work as a DP method gives better performance compared to the CP method.

Figure 13. Data reconstruction error with 500 nodes

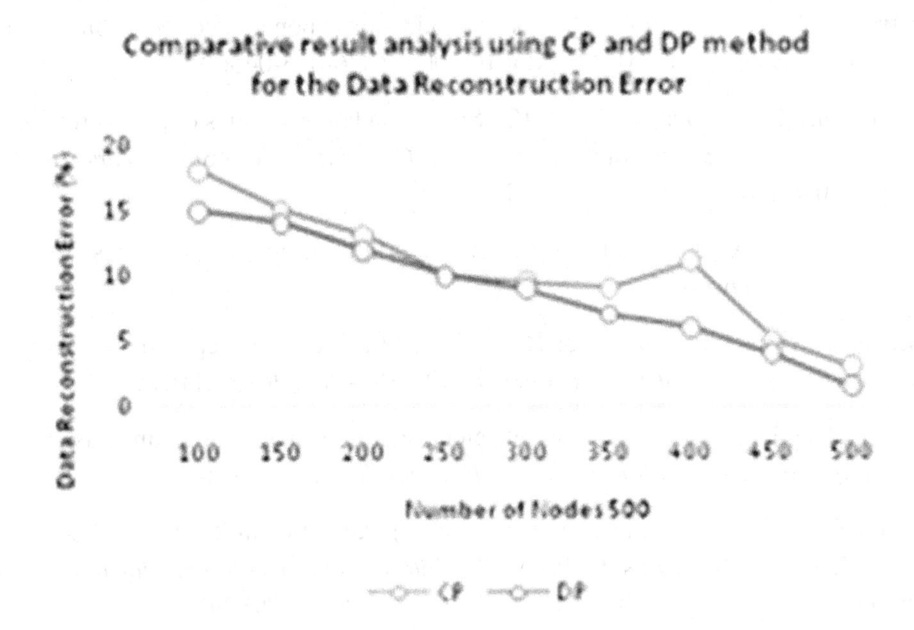

CONCLUSION

In this chapter we have designed a novel integration function of the sensor node with IoTs elements. The integration of sensor nodes with IoTs requires more energy and life of the sensor node expires. In dense network, the level of grouping of sensor nodes is very high due to the estimation of multiple functions of dual probability and the performance of network decreases. But the level of energy is minimized in terms of the level of impact of sensor nodes, the performance of the network again boosts. The dual probability-based function reduces the level value of sensors node grouping of data according to their design model. So the overall performance of the proposed model is better instead of the CP model. The design model works only for a single sink node communication patterns, in future work with multiple sink nodes and also integrates the dual probability function with a compressive sensing technique.

REFERENCES

Chakchouk, N. (2015). A survey on opportunistic routing in wireless communication networks. *IEEE Communications Surveys and Tutorials*, *17*(4), 2214–2241. doi:10.1109/COMST.2015.2411335

Chauhan, D., Jain, J. K., & Bahad, P. (2021). Performance evaluation of 802.11 A/G wireless networks with IP6HC. *Journal of Management Information and Decision Sciences*, *24*, 1–7.

Chauhan, D., Jain, J. K., & Sharma, S. (2016, December). An end-to-end header compression for multihop IPv6 tunnels with varying bandwidth. In *2016 Fifth international conference on eco-friendly computing and communication systems (ICECCS)* (pp. 84-88). IEEE. 10.1109/Eco-friendly.2016.7893247

Chen, D., Liu, Z., Wang, L., Dou, M., Chen, J., & Li, H. (2013). Natural disaster monitoring with wireless sensor networks: A case study of data-intensive applications upon low-cost scalable systems. *Mobile Networks and Applications*, *18*(5), 651–663. doi:10.100711036-013-0456-9

Doudou, M., Djenouri, D., & Badache, N. (2012). Survey on latency issues of asynchronous MAC protocols in delay-sensitive wireless sensor networks. *IEEE Communications Surveys and Tutorials*, *15*(2), 528–550. doi:10.1109/SURV.2012.040412.00075

El Rachkidy, N., Guitton, A., & Misson, M. (2012). Pivot routing improves wireless sensor networks performance. *Journal of Networks*.

Huang, J., Meng, Y., Gong, X., Liu, Y., & Duan, Q. (2014). A novel deployment scheme for green internet of things. *IEEE Internet of Things Journal*, *1*(2), 196–205. doi:10.1109/JIOT.2014.2301819

Jain, J. K., & Chauhan, D. (2021). An energy-efficient model for internet of things using compressive sensing. *Journal of Management Information and Decision Sciences*, *24*, 1–7.

Jain, J. K., Dangi, C. S., & Chauhan, D. (2020, November). An Efficient Multipath Productive Routing Protocol for Mobile Ad-hoc Networks. In *2020 IEEE International Conference for Innovation in Technology (INOCON)* (pp. 1-5). IEEE. 10.1109/INOCON50539.2020.9298291

Jain, J. K., Jain, D. K., & Gupta, A. (2012). Performance analysis of node-disjoint multipath routing for mobile ad-hoc networks based on QOS. *International Journal of Computer Science and Information Technologies*, *3*(5), 5000–5004.

Jain, J. K., & Sharma, S. (2013). Performance Evaluation of Hybrid Multipath Progressive Routing Protocol for MANETs. *International Journal of Computers and Applications*, *71*(18).

Jain, J. K., & Sharma, S. (2017). A Novel Approach of Routing Protocol using Hybrid Analysis for MANETs. *International Journal of Computer Science and Information Security*, *15*(1), 582.

Jain, J. K., Waoo, A. A., & Chauhan, D. (2022). *A Literature Review on Machine Learning for Cyber Security Issues*. Academic Press.

Jain, P., & Jain, J. K. (2022). First Order Control System Using Python Technology. In Innovations in Electrical and Electronic Engineering. *Proceedings of ICEEE*, *1*, 152–160.

Jumira, O., Wolhuter, R., & Zeadally, S. (2013). Energy-efficient beaconless geographic routing in energy harvested wireless sensor networks. *Concurrency and Computation*, *25*(1), 58–84. doi:10.1002/cpe.2838

Kaur, K., Dhand, T., Kumar, N., & Zeadally, S. (2017). Container-as-a-service at the edge: Trade-off between energy efficiency and service availability at fog nano data centers. *IEEE Wireless Communications*, *24*(3), 48–56. doi:10.1109/MWC.2017.1600427

Langendoen, K. (2008). Medium access control in wireless sensor networks. *Medium Access Control in Wireless Networks, 2*, 535-560.

Le, X. H., Lee, S., Butun, I., Khalid, M., Sankar, R., Kim, M., Han, M., Lee, Y.-K., & Lee, H. (2009). An energy-efficient access control scheme for wireless sensor networks based on elliptic curve cryptography. *Journal of Communications and Networks (Seoul)*, *11*(6), 599–606. doi:10.1109/JCN.2009.6388413

Li, S., Da Xu, L., & Wang, X. (2012). Compressed sensing signal and data acquisition in wireless sensor networks and internet of things. *IEEE Transactions on Industrial Informatics*, *9*(4), 2177–2186. doi:10.1109/TII.2012.2189222

Li, Y., Yuan, A. N., Liu, X., Du, Y. B., Huang, T. X., & Cui, H. X. (2014). A weight cluster-based hybrid routing algorithm of ZigBee network. *International Journal of Future Generation Communication and Networking*, *7*(2), 65–72. doi:10.14257/ijfgcn.2014.7.2.07

Meng, F. G., Jiang, B. C., & Wang, C. Y. (2012). An improvement of zigbee cluster-tree routing protocol. *Advanced Materials Research*, *588*, 1214–1217. doi:10.4028/www.scientific.net/AMR.588-589.1214

Nguyen, H. P., Le, P. Q. H., Pham, V. V., Nguyen, X. P., Balasubramaniam, D., & Hoang, A. T. (2021). Application of the Internet of Things in 3E (efficiency, economy, and environment) factor-based energy management as smart and sustainable strategy. *Energy Sources. Part A, Recovery, Utilization, and Environmental Effects*, 1–23. doi:10.1080/15567036.2021.1954110

Nguyen, T. D., Khan, J. Y., & Ngo, D. T. (2017). Energy harvested roadside IEEE 802.15. 4 wireless sensor networks for IoT applications. *Ad Hoc Networks*, *56*, 109–121. doi:10.1016/j.adhoc.2016.12.003

Sharma, H., Haque, A., & Blaabjerg, F. (2021). Machine learning in wireless sensor networks for smart cities: A survey. *Electronics (Basel)*, *10*(9), 1012. doi:10.3390/electronics10091012

Sun, G., & Xu, B. (2010). Dynamic routing algorithm for priority guarantee in low duty-cycled wireless sensor networks. *Wireless Algorithms, Systems, and Applications: 5th International Conference, WASA 2010, Beijing, China, August 15-17, 2010 Proceedings*, *5*, 146–156.

Uikey, R., & Sharma, S. (2013). Zigbee cluster tree performance improvement technique. *International Journal of Computers and Applications*, *62*(19).

Ullah, R., Abbas, A. W., Ullah, M., Khan, R. U., Khan, I. U., Aslam, N., & Aljameel, S. S. (2021). EEWMP: An IoT-based energy-efficient water management platform for smart irrigation. *Scientific Programming*, *2021*, 1–9. doi:10.1155/2021/5536884

Wu, S., Niu, J., Chou, W., & Guizani, M. (2016). Delay-aware energy optimization for flooding in duty-cycled wireless sensor networks. *IEEE Transactions on Wireless Communications*, *15*(12), 8449–8462. doi:10.1109/TWC.2016.2615296

Yao, Y., Cao, Q., & Vasilakos, A. V. (2014). EDAL: An energy-efficient, delay-aware, and lifetime-balancing data collection protocol for heterogeneous wireless sensor networks. *IEEE/ACM Transactions on Networking*, *23*(3), 810–823. doi:10.1109/TNET.2014.2306592

Yick, J., Mukherjee, B., & Ghosal, D. (2008). Wireless sensor network survey. *Computer Networks*, *52*(12), 2292–2330. doi:10.1016/j.comnet.2008.04.002

Zhang, Y., Liu, W., Lou, W., & Fang, Y. (2006). Location-based compromise-tolerant security mechanisms for wireless sensor networks. *IEEE Journal on Selected Areas in Communications*, *24*(2), 247–260. doi:10.1109/JSAC.2005.861382

KEY TERMS AND DEFINITIONS

Compressive Sensing: This is a signal processing technique that allows the reconstruction of a signal from only a small number of measurements or samples. It has applications in a variety of fields such as image processing, radar imaging, and medical imaging. The key idea behind compressive sensing is that a signal's sparsity or compressibility in a certain domain can be exploited such that fewer measurements are needed to accurately reconstruct the signal. This reduces data acquisition time and storage requirements, making it a useful tool in situations where data acquisition and storage resources are limited. Compressed sensing is a signal processing technique for efficiently acquiring and reconstructing a signal, by finding solutions to underdetermined linear systems. This is based on the principle that, through optimization, the sparsity of a signal can be exploited to recover it from far fewer samples than required by the Nyquist–Shannon sampling theorem.

Dual Probability: Dual probability sensor nodes are devices that use two different sensing mechanisms to detect the presence or absence of a particular event or condition. These sensors are frequently used in automation and control systems to detect and respond to changes in their environment or operating conditions.

Energy Consumptions: A useful measure of the efficiency of the network is the energy consumed per bit of data transferred. In Wireless Sensor Networks (WSNs), energy consumption is an important consideration since the network's longevity and performance depend on it.

Inter-Cloud Service Exchange: Inter-cloud service exchange refers to a marketplace that allows cloud service providers and consumers to exchange services and data across different clouds. It allows businesses to consume and provide services across multiple cloud providers.

Internet of Things: The Internet of Things (IoT) is a network of physical devices, vehicles, home appliances, and other items that are embedded with sensors, software, and network connectivity that enables these objects to collect and exchange data.

Local Cloud: This is a Cloud Computing model that allows companies and organizations to have their own private cloud infrastructure. This infrastructure is usually built on an on-premises server or data center, and it provides similar benefits to public cloud services, such as scalability, reliability, and cost-effectiveness.

RFID: RFID (Radio Frequency Identification) is a wireless technology used for identifying and tracking tagged objects. This technology is commonly used for inventory management, supply chain management, asset tracking, and access control. It is also used in contactless payment systems and in identifying pets and livestock.

Sensor Node: A sensor node is a small electronic device that is designed to collect and transmit data from the environment or surrounding area. It usually consists of a sensor, a microcontroller, a wireless communication module, and a power source.

Service-Oriented Architectures: Service-oriented architecture (SOA) is an architectural style that defines the use of services to support a wide range of business processes and activities. It is an approach to designing software systems that focuses on creating modular and loosely-coupled services that can be used independently and can communicate with each other using standardized protocols.

Wireless Sensor Network: It is the network of sensors that are connected wirelessly to each other and to a central computer or server. These networks are used to gather and transmit data from various environmental or industrial sources, such as temperature and humidity sensors, traffic monitoring sensors, or security cameras.

Chapter 6
5G Network Programmability Enabling Industry 4.0 Transformation

George Makropoulos
NCSR Demokritos, University of Athens, Greece

Dimitrios Fragkos
NCSR Demokritos, University of Peloponnese, Greece

Harilaos Koumaras
Institute of Informatics and Telecommunications, Greece

Nancy Alonistioti
N.K. University of Athens, Greece

Alexandros Kaloxylos
University of Peloponnese, Greece

Vaios Koumaras
Infolysis P.C., Greece

Theoni Dounia
Infolysis P.C., Greece

Christos Sakkas
Infolysis P.C., Greece

Dimitris Tsolkas
Fogus Innovations and Services P.C., Greece

ABSTRACT

The 5G system aims at enabling innovative services of vertical industries by utilizing network programmability to its full extension. The 3rd Generation Partnership Project (3GPP) has already established the foundations to provide third parties with 5G Core's capabilities by introducing the Common API Framework (CAPIF) and Service Enabler Layer Architecture (SEAL) on top of the services offered by the Network Exposure Function (NEF). However, a scalable, robust, and secure ecosystem should be set for the verticals that exploit those capabilities. In this context, cloud native approach is becoming a key enabler for that ecosystem, taking advantage of the inherent cloud-native characteristics of the 5G Service-Based Architecture (SBA). The chapter presents the capabilities that 5G provides to verticals as well as the ecosystem that is built around the exploitation of those capabilities. As a matter of better justification and exemplification, Industry 4.0 vertical is targeted while developments related to Vertical Application Enablers (VAEs) for the factory of the future are provided.

DOI: 10.4018/978-1-7998-9266-3.ch006

INTRODUCTION

We're at the dawn of the next industrial revolution, commonly known as Industry 4.0, which will deliver greater operational efficiencies and flexibility at lower costs. The transition towards Industry 4.0 will offer advances in every aspect, from remote monitoring to advanced analytics and maintenance. However, the key factor for Industry 4.0 is connectivity, so manufacturers are able to use data to gain insight about their assets, be informed and make decisions on how to optimise their processes.

The intense research work on 5G experimentation globally (Díaz-Zayas et al., 2020), has reached the point where the 5G capabilities and evolvements, are appealing to be the ideal enablers aiming to shape a new and dynamic ecosystem in mobile networks from both the technology and marketing perspectives (Kostakis et al., 2021). 5G networks are envisioned to achieve a wider variety of objectives in terms of higher multi Gbps data speeds, ultra-low latency, advanced reliability, increased network capacity and availability, as well as greater bandwidth and throughput (Koumaras et al., 2021). These characteristics are essential for effectively leveraging the various services that are taking place across the entire lifecycle of the operations and processes within the verticals related to Industry 4.0. Among these unique characteristics another crucial functionality that 5G networks provide and offers high business potential, is the network exposure, which can in turn enable new levels of programmability within core networks. The programmability provided by 5G networks will unveil a wide list of network capabilities and services to third-party developers allowing them to enhance existing use cases or even create new ones.

In the light of the above, the goal of this chapter is to advocate the optimal exploitation of the 5G technology towards the enhancement of the Industry 4.0 applications, and to present the new network architecture including the "open" core capabilities, as the new frontier for business innovation in industrial applications. Towards this direction, the first three sections of the chapter describe the capabilities that the 5G system will provide to the verticals along with the key concepts and services that are related to the 5G openness, whereas Section 4 introduces the Vertical Application Enabler concept related to the deployment of vertical applications. Section 5 introduces the realisation of the Non-Public Network (NPN) Infrastructure in the Industry 4.0, and Section 6 presents a Factory of the Future (FoF) use case in order to showcase the utilisation of 5G programmability and network exposure.

1. 5G CORE NETWORK CAPABILITIES TO VERTICAL INDUSTRIES

The enormous growth in connectivity, the high volume of traffic data and the broad range of business models nowadays, impose the need to move towards highly flexible infrastructures that are characterized by consistency in terms of performance and Quality of Service (QoS) provision. Moreover, given the fact that 5G networks will be the infrastructure leveraging a variety of verticals, the set of services per vertical industry, is mandatory to be able to meet a broad range of requirements. These requirements are related to the provisioning of enhanced capabilities in terms of programmability as well as efficient management of infrastructure resources (5GPPP, 2016).

In 2007, the telecom industry was upended by the launch of a new smartphone Operating System (OS) platform that disrupted the market by its unique openness features of programmability via the offered Application Programmable Interfaces (APIs). Exploiting the opportunity for innovation, third-party developers embraced the open Mobile phone OS and provided easy-to-use, programmable Software Development Kits (SDKs) and APIs to develop new apps and novel services (Ericsson, 2019). Due to the plurality of the developed apps, the App Store (Marketplace) launched very soon and the mobile app ecosystem provided a platform for programmers to reach potentially millions of customers thus becoming a successful business. Building on this paradigm, as depicted in Figure 1, the concept of network programmability through 5GCore creates opportunities that open up new avenues for growth and innovation beyond simply accelerating connectivity and data transfer. In particular, it offers even more disruption in applications' programmability, by combining the untapped capacity of multiple simultaneous network features and promising a new generation of applications that deliver an unprecedented user experience via the openness of the network. The business potential that 5G openness provides is high, taking into account that exposing the OS of a mobile phone to external developers impacted the mobile market, then the potential by opening up a whole mobile network is enormous and is expected to disrupt Industry 4.0.

1.1. NEF and API Exposure

The above-mentioned capabilities related to the programmability of the network, are materialised through the Service Based Architecture (SBA), adopted by the 5GC network, as illustrated in Figure 2. This new shifted approach towards the SBA, allows for the elimination of resource inefficiency and performance degradation associated with virtual machines and hypervisors, thereby improving the network in terms of flexibility, speed, and automation. The 5GC control plane Network Functions (NFs) communicate through API-calls that define the related Service Based Interfaces (SBIs). In this context, the Network Repository Function (NRF) allows other NFs to register their services, which can then be discovered by other NFs. This allows for a versatile implementation, in which each NF allows other approved NFs to access the available resources of the infrastructure.

In addition, the Network Exposure Function (NEF), provides a set of northbound APIs for exposing network data and receiving management commands. More precisely, NEF provides adaptors for connect-

Figure 1. Mobile OS opened up vs. mobile network opened up
Source: Ericsson (2019)

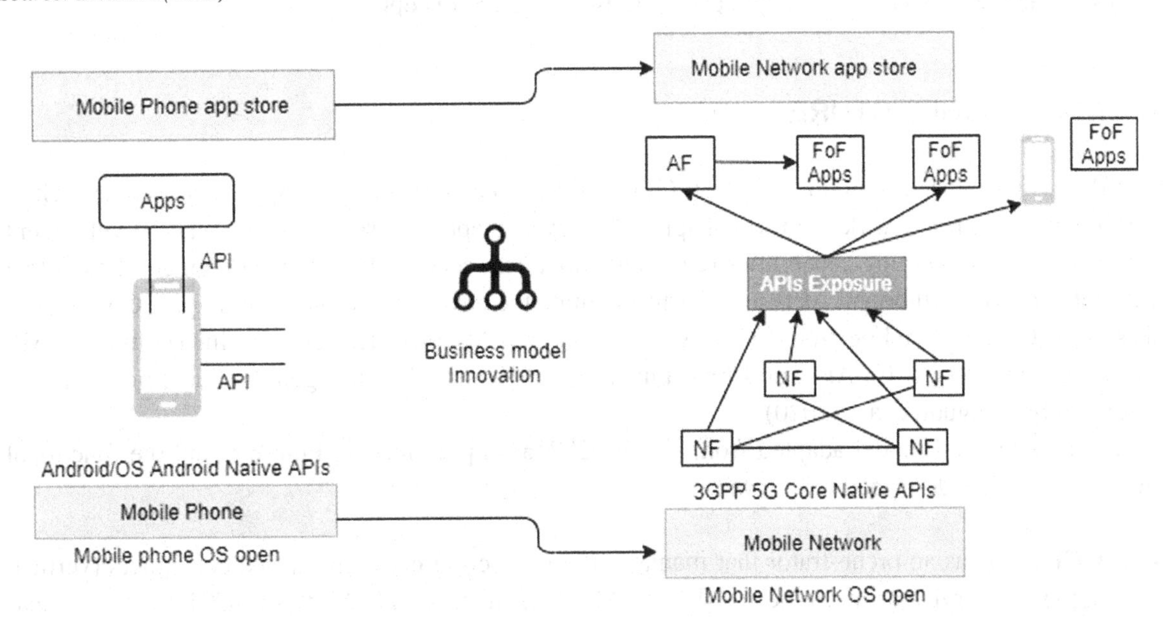

Figure 2. 5G system architecture

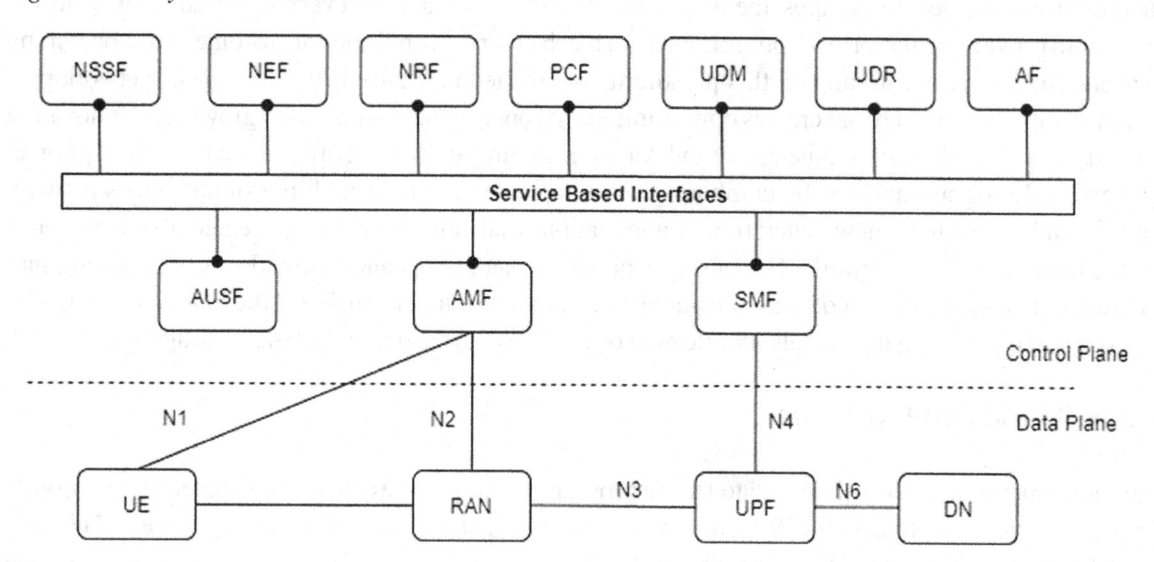

ing the southbound interfaces with the SBA to an exposure layer with northbound interfaces offered to third-party developers (Fragkos et al., 2021). The overall approach is illustrated in Figure 3. In this way, NEF facilitates the secure disclosure of network resources to 3rd parties, such as network slicing, edge computing, and machine learning utilizing the 5G system, fully compliant with the innovative paradigms that underpin a wide range of services (Tsolkas & Koumaras, 2022). The overall approach follows the concept of service producer-service consumer paradigm which is already established for cloud native services and now is adopted by the cellular network infrastructure.

3GPP has already established the foundations, and progressively performing work in order to provide 5GC Network capabilities to vertical industries. The key concepts that have emerged are the Common API Framework (CAPIF) and the Service Enabler Architecture Layer (SEAL) together with NEF. The following sections provide a brief synopsis of these two core concepts.

2. CAPIF ARCHITECTURE

CAPIF was introduced in 3GPP Rel. 15 (3GPP, 2021a), to enable a unified approach between 5GC's northbound APIs framework and vertical apps. The key concept is the standardization and development of the common supporting capabilities (e.g., authentication, service discovery, charging policies) that are applicable to northbound APIs in order to facilitate the development of vertical apps. CAPIF consists of the CAPIF Core Function (CCF), API Invokers and API provider domain which comprises API Exposing Function (AEF), API Publishing Function (APF) and API Management Function (AMF) as described by Tangudu et al., (2020).

The architectural model adapted from (3GPP, 2021a) is presented in Figure 4 and the functional entities are briefly described as follows:

- **CCF:** Acts as an orchestrator that manages the interaction between service consumers (vertical apps) and service providers (e.g., NEF, SEAL). The main responsibilities of CCF are authentica-

Figure 3. RESTful APIs for the service-based interfaces and northbound communication

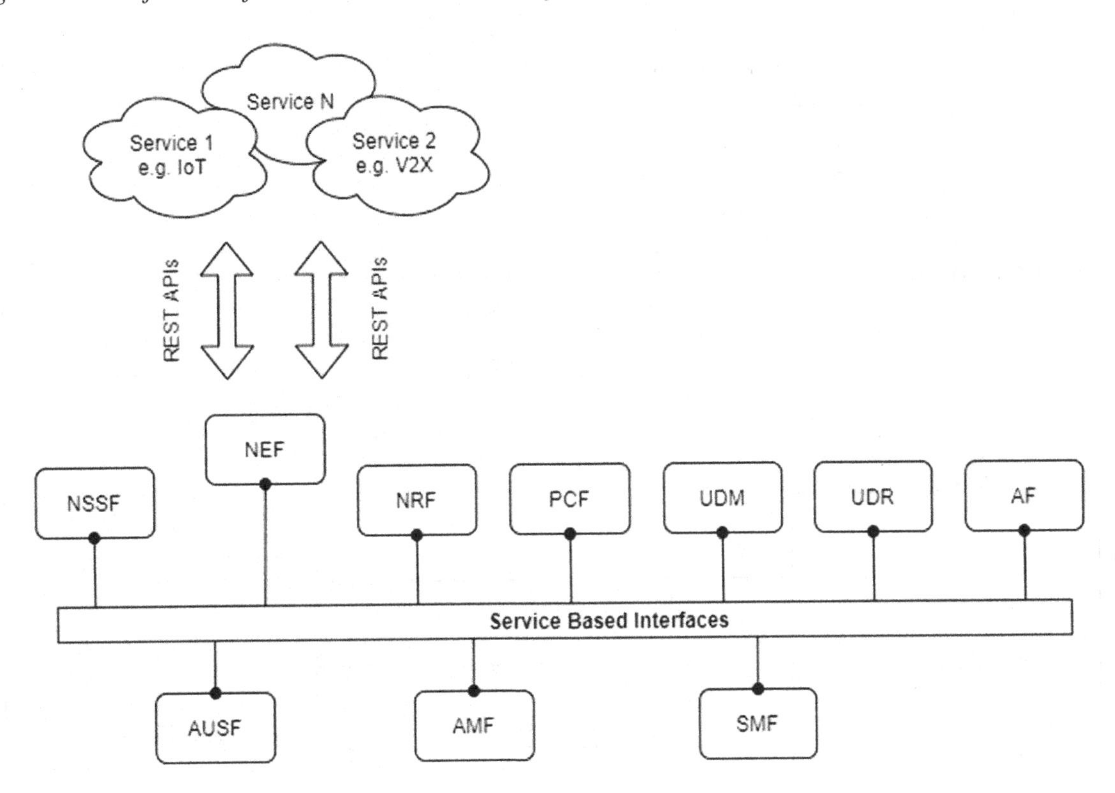

tion of the API invoker, authorization of the API invoker to access the available service APIs, monitoring the service API invocations.

- **API Invoker:** Represents the vertical app which consumes the service APIs utilizing CAPIF. API Invoker provides to the CCF the required information for authentication, discovers and then invokes the available service APIs.
- **AEF:** Is responsible for the exposure of the service APIs. Assuming that API Invokers are authorized by the CCF, AEF validates the authorization and subsequently provides the direct communication entry points to the service APIs. AEF may also authorize API invokers and record the invocations in log files.
- **APF:** Is responsible for the publication of the service APIs to CCF in order to enable the discovery capability to the API Invokers.
- **AMF:** supplies the API provider domain with administrative capabilities. Some of these capabilities include auditing the service API invocation logs received from the CCF, on-boarding/off-boarding new API invokers and monitoring the status of the service APIs.

3GPP considers two main architectural deployment models, centralized, when the CCF and API Provider domain functions are co-located, and distributed, when CCF and API Provider domain functions are not co-located and they are interacting through CAPIF-3/4/5 interfaces. Therefore, multiple CCFs can be deployed in the same PLMN trust domain (3GPP, 2021a). CAPIF is located within the PLMN operator network. Thus, there are two functional options for API Invokers; usually 3rd party applications, which have service agreement with PLMN operator, represent API invokers (i.e., API In-

Figure 4. Simplified CAPIF architecture
Source: 3GPP

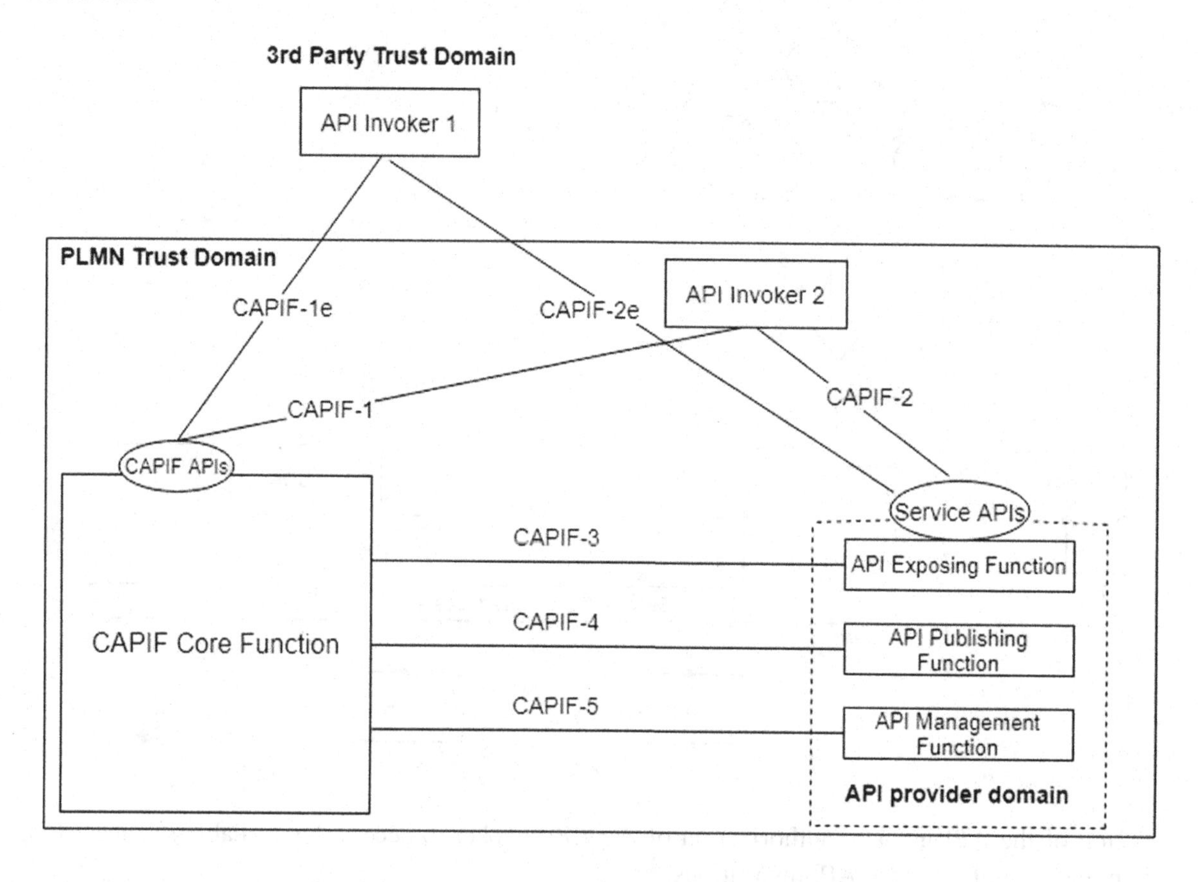

voker 1) but they may be co-located within the same PLMN trust domain (i.e., API Invoker 2). Whether third parties have business relationships with PLMN, they can provide their own service APIs to CCF through CAPIF-3e/4e/5e interfaces, but they need to act in accordance with the functionalities of the API provider domain. In order to be compliant with the overall architecture NEF and SEAL (i.e., SEAL server) supports the CAPIF's API provider domain capabilities, as specified in 3GPP (2021b, 2021c).

2.1. CAPIF Services

The available CAPIF services and their respective APIs according to 3GPP (2021a) are listed hereby. Services are divided into four categories, common, security, management, and internal connectivity services:

2.1.1. Common Services

Common services are fundamental to bridge the communication between services consumers (i.e., vertical app) and service providers (i.e., NEF, VAE, SEAL etc.). In the first place, service providers have to publish their services to the CCF. After the CCF becomes aware of the available services, a service provider initiates the procedure to discover the services. The common services are described as:

- **Discover (CAPIF_Discover_Service_API):** This service enables API Invokers to retrieve the available services that have been registered in CCF.
- **Publish (CAPIF_Publish_Service_API):** APF consumes this service to publish/unpublish a service API to the CCF. The publication includes details about the specific service API. APF can also update already published services.
- **Retrieve (CAPIF_Publish_Service_API):** APF requests from CCF information related with previous published services. When a publication occurs CAPIF registers all the related information in a repository (i.e., API registry).

2.1.2. Management Services

Considering that Invokers are onboarded and use the common services, management services assess the communication between service consumers and service providers, thus management services include:

- **Logging (CAPIF_Logging_API_Invocation):** Upon invocations (i.e., from API Invokers), CCF may store valuable information such as API invoker's ID, IP address, service API name etc. AEF utilizes this service to log API Invoker's service invocations onto CCF and potentially to access log files that have been stored previously in CCF.
- **Auditing (CAPIF_Auditing):** This service can be used to control CAPIF interactions with API Invokers (e.g., invocation events, onboarding events, authentication), which are stored in CCF. AMF initiates a request to fetch the respective log files.
- **Charging:** AEF can use this service to retrieve charging related information flows from the CCF.
- **Monitoring Events (CAPIF_Monitoring):** Monitoring event service is used by AMF in order to get notified whether an event occurs in the CCF. Some of the events are the availability of service APIs (e.g., active, inactive), changes in service APIs (e.g., after an update), service API invocations, API invoker status (e.g., onboarded, offboarded) and performance related events (e.g., load conditions).

2.1.3. Internal Connectivity

The services described below accommodate the various deployment scenarios of the CAPIF Architecture. As mentioned, CAPIF supports distributed deployments where multiple CCF can be realized. Moreover, Invokers can be part of the PLMN trust domain or third-party trust domains, and they may access the available services through different CCFs. The internal connectivity services that fulfill the described aspects of the CAPIF are listed below:

- **CCF Interconnection (CAPIF_Discover_Service_API/CAPIF_Publish_Service_API):** This service enables the interconnection between multiple CAPIF providers. Each CAPIF provider has a CCF which utilizes publish and discover services in order to interchange its APIs.
- **Topology Hiding (CAPIF_Routing_Info):** This service enables hiding the topology in the functional scenario where CAPIF includes PLMN trust domains, third party domains and API invokers access the service APIs from outside both the PLMN and third-party trust domains. In this case, API invokers access an AEF which acts as an entry point. Thus, the information for the entry

AEF is shared with API Invoker in the discovery service. Then, subsequently, AEF resolves the actual destination address of the requested service API and forwards the initial request.

The aforementioned services need to fulfill the authentication and authorization prerequisites. The capabilities of the services are presented under the assumption that API provider domain functions (i.e., AEF, APF, AMF) and API Invokers are already authorized by the CCF and they are active. The details on the security aspects are out of scope. However, more information can be found in 3GPP (2020).

3. SEAL ARCHITECTURE

SEAL was introduced in Rel. 16 to support easier and faster development and deployment of vertical apps (3GPP, 2019). While the demand to develop vertical app standards for different types of industries was continuously increasing, it became obvious that many auxiliary services, such as location management, are needed across multiple vertical apps. As a result, capturing these commonly used auxiliary services and offering them to verticals as a common service layer, will benefit both verticals, allowing them to focus only on the core features and functionality of the vertical app, and operators, saving them from enormous efforts and time to develop the corresponding services for each vertical. The afore described concept became reality with the standardization of SEAL architecture (3GPP, 2021c). SEAL architecture enables these common services to be consumed by vertical apps over 3GPP, CAPIF compliant, northbound APIs. SEAL architecture supports two functional models: on-network (i.e., SEAL-Uu), when the UE connects to the 3GPP network system to consume the service, and off-network (i.e., SEAL-PC5), when UEs connect to each other directly. The functional architecture is depicted in Figure 4. For simplification, we consider only the on-network model.

The main functional entities of SEAL architecture are the following:

- **Vertical Application Layer Client (VAL Client):** This entity provides the client-side functionalities of the corresponding vertical app (e.g., Vehicle to Everything (V2X) client).
- **Vertical Application Layer Server (VAL Server):** This entity provides the server-side functionalities of the corresponding vertical app (e.g., V2X application server). If CAPIF is supported, VAL server acts as an AEF to provide the service APIs to the Vertical Application Server (VAS) or another VAE server. It can also act like an API Invoker to consume the service APIs, whether they are provided by another VAL server.
- **SEAL Client:** This entity provides the client-side functionalities corresponding to a specific SEAL service (e.g., Location Management client)
- **SEAL Server:** This entity provides the server-side functionalities corresponding to a specific SEAL service (e.g., Location Management server). It can act as CAPIF's API exposing function.

Various deployment scenarios have been proposed in SEAL architecture, concerning the domain in which SEAL servers are deployed. The SEAL servers can be deployed: 1) in a single PLMN operator domain (centralized deployment), 2) in multiple PLMN operator domains, as distributed function, with or without interconnection between the SEAL servers, 3) in the VAL service provider domain or d) in a separate SEAL provider domain (3GPP, 2021c).

3.1. SEAL Services

The following section describes the common set of SEAL services designed to be used by vertical apps:

- **Location Management:** Enables the vertical app to have access to network location information of its corresponding UEs. More specifically, this service can send reports on-demand to a VAS about the location of its UEs, subscribe the VAS so as to receive notification when location information of UEs changes, share UE location information, etc.
- **Group Management:** Allows vertical apps to group UEs, thus enabling group management operations, such as enforcing group policies, edit group configurations, etc. The service also allows the vertical app to subscribe for and receive notifications when group information or status is modified.
- **Configuration Management:** Enables the vertical app to create and manage configuration on its UEs (provide initial configuration, edit configuration, notify server when configuration changes, etc.)
- **Identity Management:** This service is responsible for the authentication and authorization procedures of a vertical app user.
- **Key Management:** Enables a vertical app to support secure transfer of data by providing and storing encryption keys.
- **Network Resource Management:** Allows a vertical app to manage network resources by managing (create, modify, delete) unicast and/or multicast bearers.

4. VERTICAL APPLICATION ENABLERS

Within this frame of reference, 3GPP has introduced the concept of Vertical Application Enablers (VAEs) in Rel. 16 (3GPP, 2021), enabling the efficient use and deployment of vertical apps over 3GPP systems (3GPP, 2019). The specifications and the architecture are based on the notion of the VAE layer that interfaces with one or more Vertical applications. VAEs communicate via network-based interfaces that are well defined and version-controlled. The main focus of VAEs is to provide key capabilities, such as message distribution, service continuity, application resource management, dynamic group management and vertical application server APIs utilising the 5G system (5GPPP, 2021).

VAE layer acts as a support layer between SEAL layer and a specific vertical application layer (e.g., V2X application client and server). This intermediate layer enables the deployment of the actual vertical app, by utilizing SEAL/NEF APIs and translating all the underlying network data to vertical application specific. In the case of Industry 4.0 applications the VAE layer can be described as Factory of Future Application Enabler layer (FAE). The functional model of the VAE/FAE layer is depicted in Figure 5. Similarly, to SEAL architecture, VAE supports both on-network and off-network models. Note that, both VAE Client and VAE Server are mutually-exclusive with VAL Client (SEAL) and VAL Server (SEAL), respectively.

The main entities that comprise the VAE architecture are the following:

- **Vertical Application Specific Client:** Provides client-side functionalities corresponding to a specific vertical app (e.g., a platooning client in V2X use case).

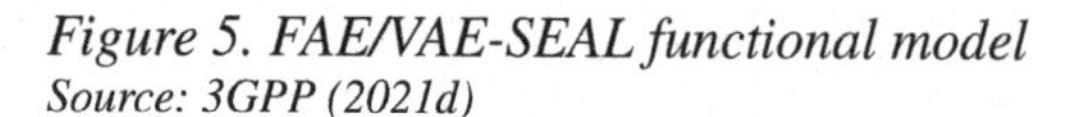

Figure 5. FAE/VAE-SEAL functional model
Source: 3GPP (2021d)

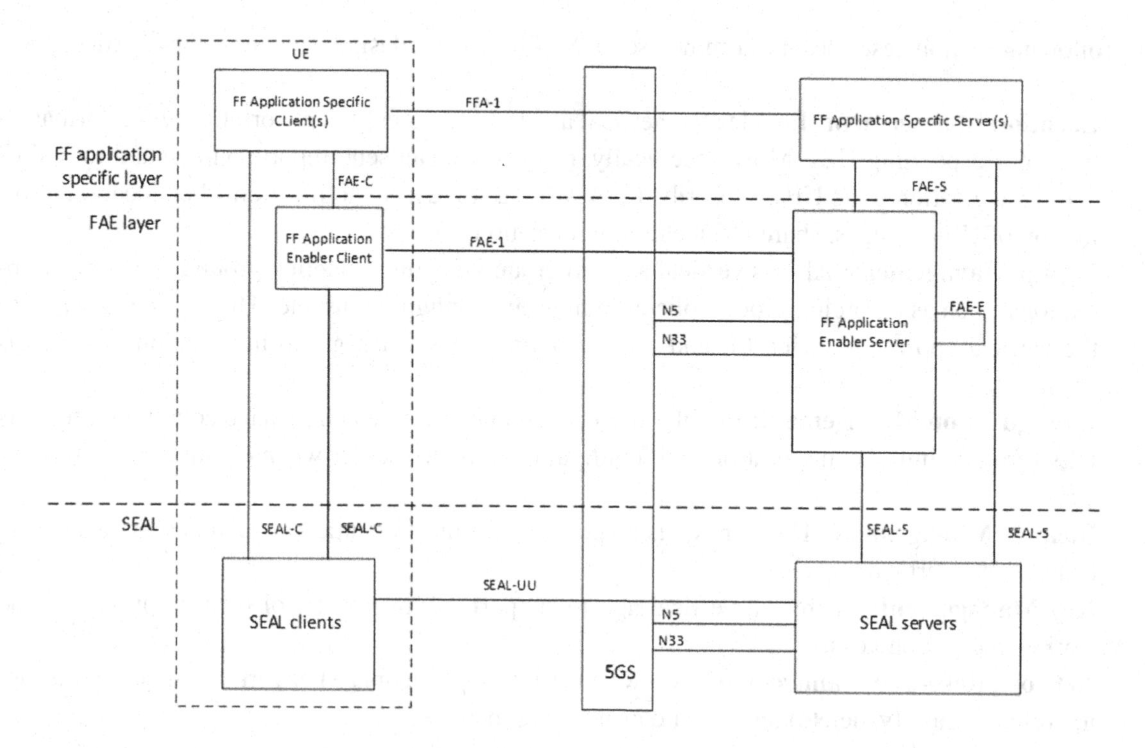

- **Vertical Application Specific Server:** Provides server-side functionalities corresponding to a specific vertical app (e.g., a platooning server in V2X use case). As mentioned, vertical apps can act as API invokers, if CAPIF is adapted. Specifically, the vertical app's server side represents the invoker (3GPP, 2021).
- **VAE Client:** Provides the client-side support functions for a specific vertical app. Some of the functions include delivering application messages to vertical app clients, receiving monitoring reports from VAE server, and providing location information to VAE server.
- **VAE Server:** Provides the server-side support functions for a specific vertical app (e.g., communicate with the underlying network, provide service discovery, support resource adaptation, etc.).

VAE servers can be deployed either in a centralized manner, in which one VAE server supports one or more vertical app specific servers, or in a distributed manner, in which one or more VAE servers (with or without interconnection between them) support one vertical application specific server. Furthermore, the VAE server can be deployed either in a PLMN operator domain or in a vertical service provider domain. As mentioned, the VAE layer utilizes the capabilities of the underlying SEAL, thus it provides additional vertical specific capabilities to enable the applications. 3GPP has already specified the VAE architecture for Vehicle to Everything (V2X) services and Unmanned Aerial Systems (UAS). Work and studies are ongoing also for the FoF services (3GPP, 2021d), which is the sector that the specific chapter is focusing on. The above-mentioned frameworks comprise 3GPP's SA6 initiatives occupying a vital role between the 5GS and the vertical service providers. With CAPIF enabling a unified and harmonized Northbound API framework, improvements to the existing platforms such as SEAL which serves as a common service layer, and VAE enabling the creation of the vertical specific application, can be applied. Thus, new frameworks may arise in the future to enhance and support the ever-increasing requirements of the vertical applications.

5. NPN INFRASTRUCTURE

At the onset of the Industry 4.0 era, factories and manufacturing companies around the globe are focusing on reducing the operational costs and increasing their production numbers and effectiveness, due to shorter business and product life cycles. To that end, the progress towards a new generation of more efficient and novel ICT technologies, is deemed necessary in order to introduce sustainable and connected industrial systems, allowing the FoF concept to emerge. Indeed, several leading manufacturers are already engaged in the digital transformation era that will leverage connectivity, intelligence and allow for flexible automation.

The envisioned digital transformation to be applied to manufacturing processes will provide several advantages such as advanced analytics (based on measuring device specific parameters like vibration, temperature or noise levels), situational awareness and prediction of maintenance needs. In addition, the interconnection of the several machineries extends to using the most prominent communication technology, which can enable new features, including, for instance, flexibility of the topology and load balancing that controls the quality of the communication traffic. Moreover, within the industrial environments, the cooperation of robots and humans is intensified due to massive utilization of wirelessly connected sensors within the factory. Compared to common use cases from other vertical sectors (e.g., energy, media, smart cities), the use cases that are related to Industry 4.0 will induce strict requirements in terms of latency, reliability and high-accuracy positioning.

All these challenges related to the envisioned digital transformation, rely on efficient connectivity, high throughput and low latency. Through the years different wireless technologies have been proposed to support the industrial domain. The wireless technologies that have been proposed through the last years are generally classified into short range (i.e., ZigBee, Wi-Fi), long range (i.e., LoRa, Sigfox) and cellular networks (Sikimić et al., 2020). The technologies when used for Industry 4.0 do not seem to be able to provide the necessary requirements. Both short-range and long-range networks draw upon the unlicensed spectrum, but they cannot offer the critical capabilities, latency, and Quality of Service that are needed, with the exception of Wi-Fi. The alternative option of wired connections is limited to stationary objects and is impractical when it comes to connecting the large number of devices used in Industry 4.0 factories.

In this context, the concept of 5G Non-Public Networks (NPN) has emerged in order to support communication applications with heterogeneous and demanding requirements within the Industrial sector (Ordonez-Lucena et al., 2019). According to 3GPP, next generation mobile networks can be classified into public and Non-Public (3GPP, 2020a). The former refers to Public Land Mobile Networks (PLMN) which are typically led by Mobile Network Operators (MNOs) that provide their services publicly and they primarily operate in national scope. On the other hand, a NPN enables the deployment of a 5G System (5GS) that restricts its operability to private organizations, typically an industry vertical and offers private network services to end users acting within organization's premises.

As Ordonez et al. (2019) describes, NPNs can be divided into two main categories namely Stand alone and Public Network Integrated (PNI). A stand-alone NPN is an isolated private network that has zero interaction with PLMN. Within the environment of Industry 4.0, the utilization of a NPN permits a vertical to make use of an in-premise 5G network, and as a consequence the traffic that is related to this network will be limited within the premises' boundaries, without spanning to the public domain. This aspect offers the following advantages:

- Provision of QoS utilizing 5G network functions and service applications as close as possible to the devices and making use of advanced technologies like Time Sensitive Networking.
- Isolation for the public domain, which enables advanced security towards the NPN
- Efficient authorization and authentication of the devices within NPN, through the isolated network operations

On the contrary, PNI-NPNs rely on PLMN, meaning that a part or the whole of the functionality is provided by the MNO. The autonomy of a NPN in relation to a PLMN can be described based on the following characteristics:

- Use of a unique identifier for the NPN.
- Allocation of private spectrum to the NPN.
- Full deployment of a 5G system, meaning both Radio Access Network and Core Network, within the perimeter of an industrial environment.

PNI-NPN category supports a variety of deployment modes and architectural options, which in turn facilitate the heterogeneous use cases with different requirements within the Industry 4.0 ecosystem. The 3GPP specifications describe a variety of deployment modes related to 5G NPN. One approach is to set up a standalone network independently of the public network. In such case all network functionalities, including both the radio access network (RAN) and the control plane, are placed within the facility. Moreover, dedicated spectrum (licensed or unlicensed) shall be obtained from a mobile network operator for the standalone deployment case.

In a second deployment scenario, the NPN and service provider share the same radio-access network (RAN), with control plane elements and other network operations located at the NPN site, as illustrated in Figure 6b. This type of configuration allows to the traffic of the network to be routed locally within the NPN's physical premises, while other type of data is routed to the service provider's network if need be. A third major deployment is when the NPN is directly connected to a public network, as depicted in Figure 6c. The traffic stemming from both the public and private network is located off-site. By using the slicing concept, the network services and the resources between the public network and the NPN are virtualised and can be kept fully distinct and isolated. In this sense, each slice created on top of the physical infrastructure stands for a complete logical network consisting of network capabilities as well as associated resources which can provide specific end-to-end enhanced service capabilities (Kourtis et al., 2020).

6. FoF USE CASE

In the era of Industry 4.0, factories and manufacturers are pressured to increase their production and effectiveness by including new technologies and equipment while at the same time focusing on in-

Figure 6. Deployments models of NPN

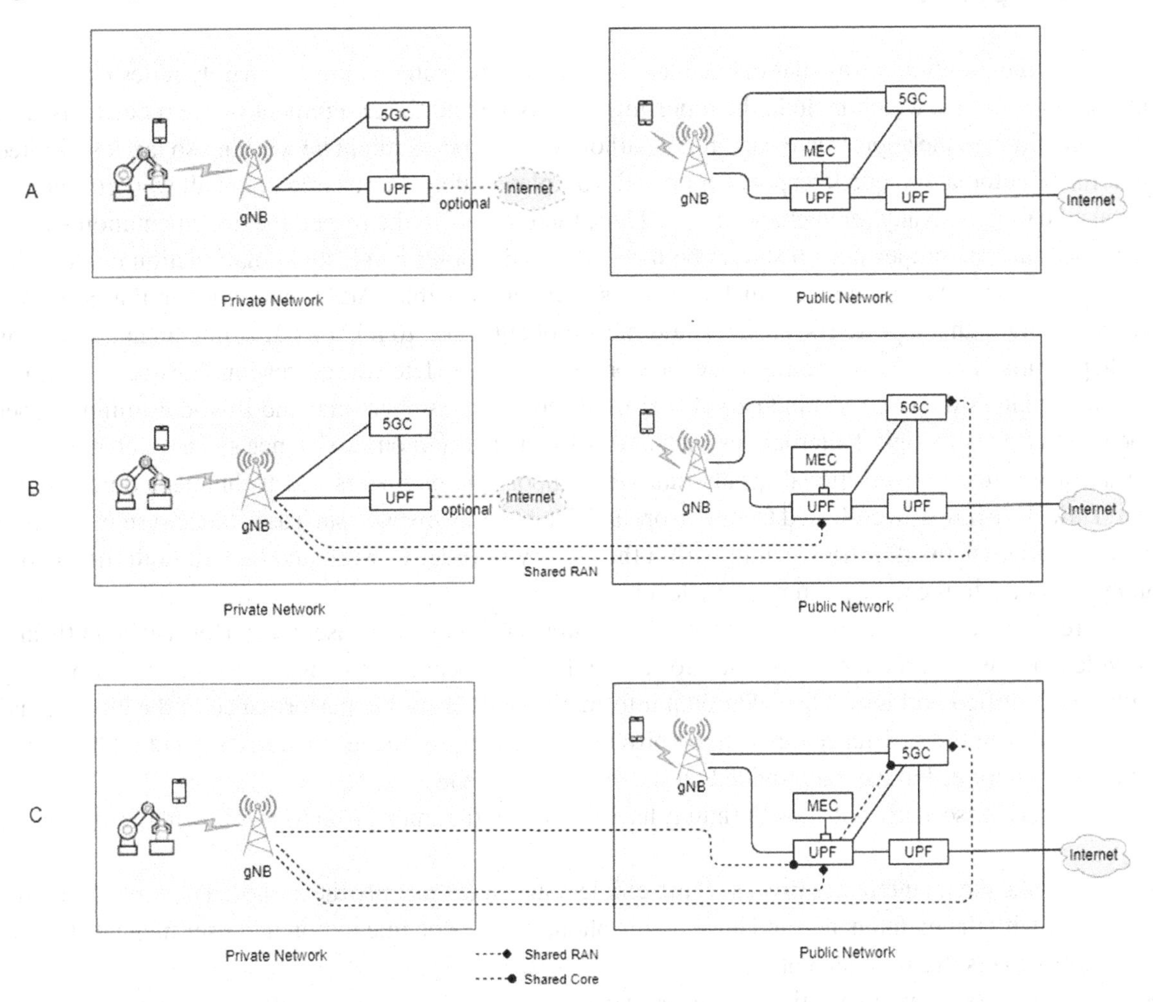

terconnectivity, automation, machine learning, and real-time data. To that end and as more and more manufacturing companies incline to that direction, thus forcing the Factory of Future (FoF) concept to continue its growth, there is an increasing need for new ideas that will further support that concept and bring fundamental changes to the core of manufacturing. While industry workers are directly affected by these changes and actively encouraged to interact with machines and collaborate using digital systems on an everyday level, human-computer interaction (HCI) is being placed at the heart of industry 4.0. Supporting the concept, chatbots are one of the most important applications of industry 4.0, combining artificial intelligence and HCI, being the perfect candidate for supporting employees' everyday work. At the same time, the emergence of the 5G system comes to support this evolution by promising a new generation of applications able to realize the innovative ideas towards Industry 4.0.

In the light of the above, the following sections describe a use case scenario focusing on how 5GC exposure of standard APIs and the concept of Vertical Application Enablers (VAE) can be utilised so as to build innovative applications that will be based on top of various vertical apps within the Industry 4.0 ecosystem.

6.1. The Opportunity

Currently, the traditional way that maintenance processes are being addressed mainly relies on custom internal procedures, which include the reporting of an issue and the assignment of the reported issues, in second time, to the appropriate personnel. Although this process might be working so far, it is limited in terms of automation, quick response time and workplace safety. For example, not all workers can undertake dealing with any maintenance issue. Thus, their access to the respective documentation should be denied and the proper person should be notified. On the other hand, some malfunction could be in high priority and thus require as quick a response time as possible. Additionally, given that a worker starts dealing with a reported issue, he would most probably have to rely on paper documentation. This could potentially be time consuming since the worker will have to fetch the corresponding documentation and then address the issue. As industry 4.0 aims at replacing cumbersome and time-consuming paper documentation with digital alternatives such as AI-driven autonomous assistance systems, chatbots can act as easy-to-use conversational agents that will support the engineers and technicians during daily workflows by using data collected from sensors and databases. However, such assistance systems require the appropriate network infrastructure within the factory in order to work and benefit both the factory and its workers to the maximum possible level.

More specifically, for the described scenario, chatbots need to locate users to perform efficiently and provide to the worker the most relevant information for the machinery that is located in close proximity with the identified problem. This additional information will boost the performance of the chatbot and the efficiency of the assistance that is to be provided. Such requirements are currently not addressed in traditional factories but can be provided by the 5G System (5GS).

The described scenario will most definitely have many benefits among which the following are expected:

- To make the maintenance process more efficient and safer, as only the authorized worker can proceed with the maintenance and have on his phone all the documentation and user manuals that he needs to resolve the incident.
- To optimize maintenance time and effectiveness, as only authorized employees will be allowed to address the reported issue.
- To introduce new ways of security monitoring and alerting of the factory environment.
- To enrich the value chain with a new means of communication and interactions which provide accuracy of information, ease of use, minimum delay of operations and versatile accessibility, leading to new business models and opportunities.

6.2. The Use Case Scenario

Leveraging the concept of the VAE and the APIs exposed by the 5GS, a use case scenario that can tackle the described scenario by proposing an innovative solution, using the advantages and capabilities of a chatbot, has been identified. The main idea is to establish a dedicated series of actions that will take place in a factory environment in order to realize the handling of maintenance scenarios via a chatbot platform. This use case targets the ways that maintenance scenarios are taking place within a factory. The goal is to introduce the chatbot application for reporting malfunctions and facilitating the course of actions needed to address and fix the reported issue. Additionally, a VAE will also need to be defined for the specification and realization of the use case.

Given a 5G non-public network that can be installed in a factory, a VAE will enable the use of a chatbot to help identify and solve possible malfunctions in a shorter time frame using a more user-friendly solution. Also, the 5G network can provide an Identification (ID) for all the connected workers of the factory and their relative location. It is also important to mention that the workers can be connected from any device of their choice (mobile phone or tablet device). At any point, a worker might encounter faulty equipment, and then he/she can use the chatbot app, that is installed and configured to their device, to report the issue. The overall procedure is as follows. Firstly, the VAE gets the location of the worker, for example the worker is located in Area A. After checking whether the specific area is safe, and thus not initiating an evacuation procedure, a second check takes place regarding the access status of the worker in "Area A". In case the person should not be in the specific area he/she is prompted to leave. If a worker is cleared a request for the manuals of the area can be made and further instructions to fix the damaged machinery. Finally, if the problem is not resolved the closest specialized technician will be notified to aid. In order to control accessibility to the areas and provide easy assistance with a friendly automated way are some of the benefits of this use case.

6.3. The Vertical Application Enabler

For the realization and specifications of the use case described in the previous section, there is a need for defining a VAE that will reside between the vApp and the 5G System in order to advance the functionalities that are offered to the worker, following the architecture described in Figure 7.

More specifically, the VAE, by using a REST API callback or making API calls to the 5G exposure interfaces, and receiving data from the internal database and files from the embedded file server, provides the following functionalities:

Figure 7. Service architecture

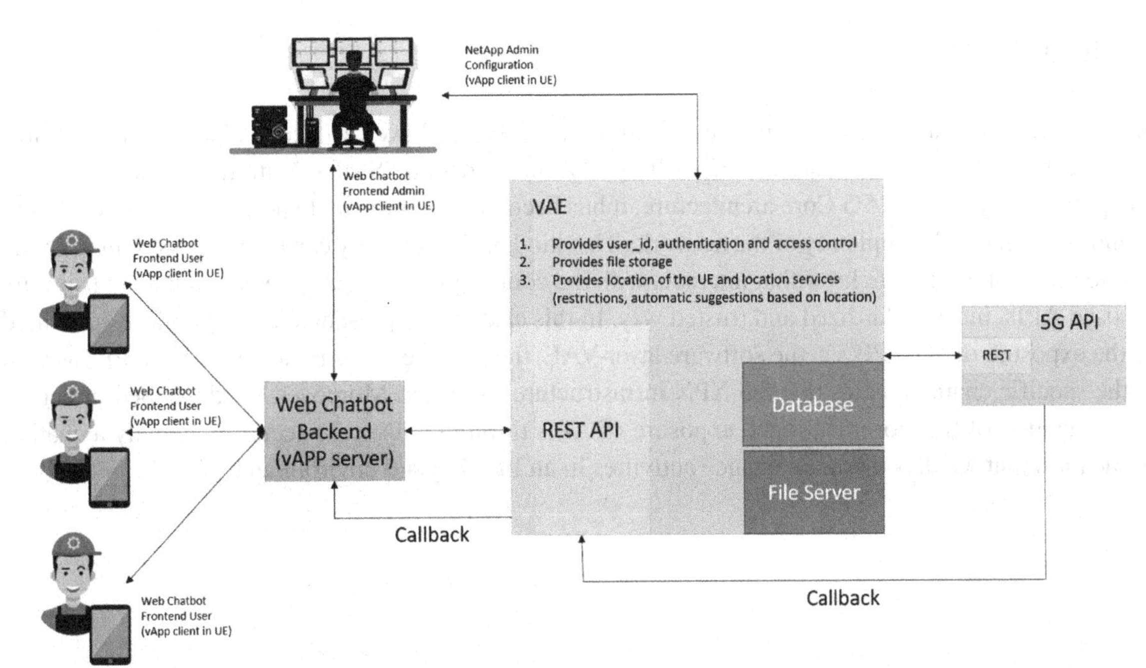

- Retrieves the User Equipment (UE) location via the 5GS. The VAE, based on the user ID, will request from the 5G system to retrieve the location info of the user.
- Correlates the UE-location provided by the 5GS to specific Factory areas. The VAE will have been properly configured in order to be aware of the factory areas and be able to map the retrieved 5G UE location to each of them. This functionality will allow the worker to easily choose only between the machines that are under this area while it also ensures the workers' safety since this area could be marked as dangerous.
- Prevents unauthorized personnel to perform actions that are not approved by the factory policy. Based on the factory area that has been spotted according to the UE location, the VAE performs an authorization process of the specific worker to proceed with the maintenance at the specific area. If the worker has not granted the appropriate level of authorization, then the worker in close proximity with the appropriate authorization level should be notified to fix the problem. Otherwise, clearance is given to the worker.
- Provides to authorized personnel, access to Maintenance documentation and Service manuals. The VAE, upon successful authorization and clearance of the worker to fix the issue, will provide access to its data storage in order to store and retrieve maintenance documentation and service manuals. Considering that, according to the use case scenario, storing input information and displaying documentation files, combined with the local deployment of the system, the VAE is also important to provide access to data storage.

The VAE as described above, is a standalone application that can be properly configured to connect with any compatible vertical application. Combining the VAE with a chatbot application is an effective solution that can follow the developments in Industry 4.0, taking advantage of the functionalities of the 5G System. The fact that chatbots are easy to learn and use only support that claim since it is easier to break into the traditional manufacturing processes and support the workers to their everyday tasks.

CONCLUSION

In the journey towards fully exploitable 5G infrastructures, we have reached the point where performance gains need to be made accessible to 3rd party innovators and SMEs. With the service composition provided by the new 5G Core architecture, it has become easier to create new services as well as to simplify the internal complexity of the networks This kind of accessibility can be performed through the development of a software layer that interacts with the control plane of a mobile network by consuming exposed APIs, in a standardized and trusted way. In this chapter we presented the capabilities provided by the exposure of 5G APIs via the software layer-VAE, the main services related to these APIs, as well as the specific characteristics that 5G NPN infrastructure can offer. Moreover, we described in details the realization of the aforementioned exposure of APIs through a use case scenario, namely a chatbot application that handles the maintenance activities in an Industry 4.0 environment.

ACKNOWLEDGMENT

Part of the work in this chapter has been funded by H2020 EU-funded projects EVOLVED-5G (GA No. 101016608).

REFERENCES

GPP TR 21.916. (2020a). *Services and System Aspects.* Release 16, v0.5.0.

GPP TR 23.745. (2021). *Study on application layer support for Factories of the Future in 5G network.* Release 17, v17.0.0.

GPP TS 23.222. (2021). *Common API Framework for 3GPP Northbound APIs.* Release 17, v17.4.0.

GPP TS 23.286. (2021). *Application layer support for Vehicle-to-Everything (V2X) services.* Release 17, v17.1.0.

GPP TS 23.434. (2019). *Service Enabler Architecture Layer for Verticals (SEAL); Functional architecture and information flows.* Release 16, v1.1.0.

GPP TS 23.434. (2021). *Service Enabler Architecture Layer for Verticals (SEAL).* Release 17, v17.1.0.

GPP TS 23.501. (2021). *System architecture for the 5G System (5GS).* Release 17, v17.0.0.

GPP TS 33.122. (2020). *Security aspects of Common API Framework (CAPIF) for 3GPP northbound APIs.* Release 16, v16.3.0.

GPPP. (2016). *5G Vision, The 5G Infrastructure Public Private Partnership: the next generation of communication networks and services* [Whitepaper]. https://espas.secure.europarl.europa.eu/orbis/document/5g-vision-5g-infrastructure-public-private-partnership-next-generation-communication

GPPP. (2021). *View on 5G Architecture* [White paper]. https://5gppp.eu/wp-content/uploads/2020/02/5G-PPP-5G-Architecture-White-Paper_final.pdf

Díaz Zayas, A., Caso, G., Alay, Ö., Merino, P., Brunstrom, A., Tsolkas, D., & Koumaras, H. (2020). A modular experimentation methodology for 5G deployments: The 5GENESIS approach. *Sensors (Basel), 20*(22), 6652. doi:10.339020226652 PMID:33233691

Ericsson. (2019). *Network Programmability, a new frontier in 5G* [White paper]. https://www.ericsson.com/en/blog/2019/1/network-programmability---in-5g-an-invisible-goldmine-for-service-providers-and-industry

Fragkos, D., Makropoulos, G., Sarantos, P., Koumaras, H., Charismiadis, A. S., & Tsolkas, D. (2021, September). 5G Vertical Application Enablers Implementation Challenges and Perspectives. In *2021 IEEE International Mediterranean Conference on Communications and Networking (MeditCom)* (pp. 117-122). IEEE. 10.1109/MeditCom49071.2021.9647460

Kostakis, P., Charismiadis, A. S., Tsolkas, D., & Koumaras, H. (2021, May). An Experimentation Platform for Automated Assessment of Multimedia Services over Mobile Networks. In *IEEE INFOCOM 2021-IEEE Conference on Computer Communications Workshops (INFOCOM WKSHPS)* (pp. 1-2). IEEE. 10.1109/INFOCOMWKSHPS51825.2021.9484528

Koumaras, H., Makropoulos, G., Batistatos, M., Kolometsos, S., Gogos, A., Xilouris, G., Sarlas, A., & Kourtis, M. A. (2021). 5G-enabled UAVs with command-and-control software component at the edge for supporting energy efficient opportunistic networks. *Energies*, *14*(5), 1480. doi:10.3390/en14051480

Kourtis, M. A., Anagnostopoulos, T., Kukliński, S., Wierzbicki, M., Oikonomakis, A., Xilouris, G., Chochliouros, I., Yi, N., Kostopoulos, A., Tomaszewski, L., Sarlas, T., & Koumaras, H. (2020, November). 5G network slicing enabling edge services. In *2020 IEEE Conference on Network Function Virtualization and Software Defined Networks (NFV-SDN)* (pp. 155-160). IEEE. 10.1109/NFV-SDN50289.2020.9289880

Ordonez-Lucena, J., Chavarria, J. F., Contreras, L. M., & Pastor, A. (2019, October). The use of 5G Non-Public Networks to support Industry 4.0 scenarios. In *2019 IEEE Conference on Standards for Communications and Networking (CSCN)* (pp. 1-7). IEEE. 10.1109/CSCN.2019.8931325

Sikimić, M., Amović, M., Vujović, V., Suknović, B., & Manjak, D. (2020, March). An overview of wireless technologies for IoT network. In *2020 19th International Symposium INFOTEH-JAHORINA (INFOTEH)* (pp. 1-6). IEEE. 10.1109/INFOTEH48170.2020.9066337

Tangudu, N. D., Gupta, N., Shah, S. P., Pattan, B. J., & Chitturi, S. (2020, September). Common framework for 5G northbound APIs. In *2020 IEEE 3rd 5G World Forum (5GWF)* (pp. 275-280). IEEE.

Tsolkas, D., & Koumaras, H. (2022). On the development and provisioning of vertical applications in the beyond 5G era. *IEEE Networking Letters*, *4*(1), 43–47. doi:10.1109/LNET.2022.3142088

World Economic Forum. (2019). *Fourth Industrial Revolution Beacons of Technology and Innovation in Manufacturing* [White paper]. https://www3.weforum.org/docs/WEF_4IR_Beacons_of_Technology_and_Innovation_in_Manufacturing_report_2019.pdf

KEY TERMS AND DEFINITION

Application Programming Interface (API): Is a set of rules, protocols, and tools for building software applications. APIs define how different software components should interact with each other, allowing for the exchange of data and functionality between different applications or systems. They provide a layer of abstraction between software components, making it easier for developers to create software that can work seamlessly with other applications or services.

CAPIF: Is a common API framework and its functionality is considered as a cornerstone in the realization of 5G openness, since it allows secure exposure of 5G core APIs to third party domains, and also enables third parties to define and expose their own APIs.

Factory of the Future (FoF): Is a concept that refers to the integration of advanced technologies such as artificial intelligence, the Internet of Things (IoT), robotics, and automation into manufacturing processes. The aim is to create a more efficient, agile, and responsive factory that can quickly adapt to changing customer demands and market conditions.

Industry 4.0: Refers to the fourth industrial revolution, which is characterized by the integration of advanced technologies and digitalization into manufacturing and other industrial processes. It builds on the previous revolutions, which introduced mechanization, mass production, and automation to industry. Industry 4.0 is driven by the increasing availability of digital data, connectivity, and computing power, which enable the use of advanced technologies, such as the Internet of Things (IoT), artificial intelligence (AI), machine learning (ML), and robotics, to optimize and automate industrial processes.

Network Exposure Function (NEF): Is a key component of the Service Based Architecture (SBA) adopted by the 5G Core (5GC) network. It provides a set of northbound APIs for exposing network data and receiving management commands. NEF facilitates the secure disclosure of network resources to third-party developers, such as network slicing, edge computing, and machine learning utilizing the 5G system.

Network Functions (NFs): Are software-based entities that provide various network-related services in a network infrastructure. These services include routing, switching, firewall, load balancing, and other network-related functions that are necessary for the proper operation of the network. NFs are typically deployed on standard computing platforms and virtualized environments to provide network services, which can be scaled up or down based on the network traffic demands.

Non-Public Network (NPN) Infrastructure: Refers to a private network infrastructure that is designed to provide secure and reliable communication services for industrial applications within Industry 4.0. NPN Infrastructure enables the deployment of vertical applications and the integration of various Industrial Internet of Things (IIoT) devices and sensors, which can be used to collect data and provide valuable insights for industrial processes. This infrastructure is isolated from public networks and provides dedicated connectivity, security, and quality of service, which is essential for the reliable and efficient operation of industrial applications.

Service-Based Architecture (SBA): Of 5G is the network architecture that underpins the 5G mobile networks. It is a new and innovative approach to designing and implementing network services, where network functions are developed as independent services that can be easily combined and configured to create customized services for specific use cases. SBA of 5G provides a flexible and modular framework for deploying and managing network services that is designed to meet the requirements of diverse applications and industries, including Industry 4.0.

Vertical Application Enabler (VAE): Is a software component or set of components that provides a set of functions to support the development of specific applications in a particular industry or vertical market.

Vertical Application Enablers (VAEs): Are a set of capabilities that enable the development and deployment of specialized applications for specific industries or use cases, such as smart cities, autonomous vehicles, and industrial automation. VAEs provide an abstraction layer between the underlying 5G network infrastructure and the application layer, allowing developers to access and utilize network resources and services in a simplified and standardized way.

Chapter 7
Internet of Things and Its Relevance to Digital Marketing

Shouvik Sanyal
Department of Marketing and Entrepreneurship, Dhofar University, Oman

Kalimuthu M.
Department of Commerce, Dr. N.G.P. Arts and Science College, Coimbatore, India

Thangaraja Arumugam
https://orcid.org/0000-0001-5496-7258
VIT Business School, Vellore Institute of Technology, Chennai, India

Aruna R.
VIT Business School, Vellore Institute of Technology, Chennai, India

Balaji J.
https://orcid.org/0000-0003-1898-9550
VIT Business School, Vellore Institute of Technology, Chennai, India

Ajitha Savarimuthu
https://orcid.org/0000-0002-0459-3208
Acharya Bangalore B School, Bengaluru, India

Chandan Chavadi
https://orcid.org/0000-0002-7214-5888
Presidency Business School, Presidency College, Bengaluru, India

Dhanabalan Thangam
https://orcid.org/0000-0003-1253-3587
Presidency Business School, Presidency College, Bengaluru, India

Sendhilkumar Manoharan
https://orcid.org/0000-0001-5116-8696
Dayananda Sagar University, Bengaluru, India

Shasikala Patil
Department of Management and Commerce, International School of Management Excellence, India

ABSTRACT

The internet of things characterizes a unified structure of internet-supported objects that can gather and send data through a wireless network with no intervention of humans. By this structure, the present business world is experiencing remarkable changes due to the irresistible potential derivable from the internet of things. This technology is already bringing significant changes across the industries, of which the digital marketing sector has the maximum benefits. This technology collects various data from the consumers through different forms of digital marketing platforms such as social media marketing, online marketing, electronic mail marketing, pay-per-click advertising. With these types of data, marketers

DOI: 10.4018/978-1-7998-9266-3.ch007

can generate some meaningful insights, develop interactions with customers, communicate with sellers and customers, and can also forecast the behavior and lifestyle of the customers. With this backdrop, the chapter has made an attempt to explain the relevance of the internet of things in digital marketing.

INTRODUCTION

One of the most important trends in digital marketing that all marketers and advertisers need to be aware of at present is data science. Because marketing plans that are based on data yield results that can be measured as data. According to Hal Varian, chief economist at Google and professor of information sciences mentioned about data science that, "the ability to take data- be able to understand it, process it, take out significance from it, to picture it, to disseminate it". Along with data science latest digital technologies join together and create tremendous changes in all industries, especially in the marketing sector by collecting and analyzing a huge quantity of data (Palmere, 2019). The quantity of data that can be evaluated is enormous, without difficulty accessible, and inestimable, with 2.5 quintillion bytes of data produced every day and increasing. Digital marketing make use of data science to find, collect, fragment, comprehend, and examine apparently arbitrary collections of 1s and 0s. The results are then transformed into useful outcomes. Although almost every business can profit from these insights, the digital marketing industry has a particularly bright future. Marketers have a greater responsibility than ever to provide targeted, intelligent advertising at a time when customers demand more customised content than ever. Thus digital marketers can succeed in their business process through data science. Understanding why data science is essential for marketing activities is especially crucial because the data science industry is both vast and profitable (Stellar Digital, 2022). Thus, data analytics' significance in digital marketing arises from its capacity to identify information that casual observers would otherwise overlook. The fundamental skill of data analytics is pattern recognition. Contemporary digital marketing strategies have been prepared with the help of big data, which is a vast collection of data from numerous sources. This data is processed by data analytics to give businesses useful information. Target marketing can be considered the best example of data analytics supporting business decision-making. A company can begin creating a community by creating rich, educational material that speaks to the fundamental beliefs of its target consumer group. When a user accesses material, the business collects information about that user. Then, it may create profiles for each person in its community, which will enable it to grow. Additionally, a company or a business can alter its pricing to be more competitive by gathering information about goods that are identical to or comparable to the product it sells. Having access to this information enables businesses to develop more effective sales tactics for large markets. Furthermore, buyers who were once loyal to another brand may be drawn in by decreased prices (Maryville University, 2019).Thus marketing industry has rejuvenated itself through digital technologies and shifted the marketing landscape from conventional to Digital. Digital marketing is playing a major role in the contemporary business world, and it cannot be undervalued or disregarded in the modern online business platform. This digital marketing operates with the help of various advanced digital technologies and it is also available in the hands of every individual. Thereby it spreads brand awareness and reaches the products to the consumers at a convenient time and place, thus this digital marketing is a tool for driving demand for the goods and services. irrespective of developed and developing countries, all the world countries are concentrating more on the digitalization process and gearing their attention to bring the country under the digitization process wherever possible, however, the process needs more time to take place and realize cent percent success. Even though the trade and commerce of the present world have been done through the digital marketing platform to a certain extent now (Skelia, 2020). In the technology-dominated business world, all the business houses concentrate more on doing business through the digital platform. If not, it is hard for the business houses to promote sales, create brand awareness and sustain in the competition. According to the market survey experts, more than 76 percent of the global populace now desire to do the shopping online, and 79 percent of the people look for various information such as consumers' preferences, investors' opinions, demand and supply of the markets through online, and around 51 percent of the buyers deciding their online purchases by going through the online reviews written by the existing buyers (Rajanarthagi, 2020). Digital marketing supports the business sector in managing its brand appropriately through diverse digital platforms such as company websites, blogs, social media, mobile applications, company pages, and so on. All these platforms are supporting equally for effective online business communications by way of social media marketing, e-mail marketing, content marketing, search engine marketing, online advertising, etc. Thus it could be understood that any type of marketing that utilizes electronic devices or technologies in their marketing operations is said to be digital marketing (Power Digital, 2017). Digital marketing is a new form of marketing pattern used in the promotion of goods or services through the use of digital technology, primarily the internet, mobile devices, and other digital media. While developing a digital business strategy, a digital marketer will use a set of actions that assist their business in achieving objectives through properly chosen online channels. A few examples of marketing channels are own websites, earned, and paid media (Venables, 2019).

TECHNOLOGIES THAT ARE SUPPORTING DIGITAL MARKETING

Digital marketing seems one of the profitable ways of doing business, as it is having the power to reach a vast consumer base within a short period. Since a lot of technological advancements support digital marketing, it helps business people to run their businesses profitably. The World Economic Forum stated that there are five major factors namely growth and development of smartphone communication, Artificial Intelligence (AI) enabled technologies, Internet-of-Things (IoT), big data analytics, blockchain technology, and cloud computing (Vandita, 2018). However, it is also expected that Virtual Reality (VR) will also join this landscape in the days to come. All these technologies are playing a major role to connect the business world through numerous touchpoints; it means, even though the internet supports these technologies, it doesn't have the power to control data related to business as well as consumers. More significantly, it is helping business people to generate holistically as well as determine operations in the future. Thus technologies are supporting a lot to the smooth flow of digital business. The following are some of the technologies that are supporting digital marketing (Iscoop, 2020).

Big Data Analytics: The introduction and the application of Cloud computing technology into marketing have supported a lot to all sorts of businesses irrespective of small or big to collect huge volumes of customer as well as business data, in part paving the way to big data. In recent times, companies are very keen to know more about their customers than ever before; to serve them by creating, accurate, as well as customized services (Fiftyfiveandfive.com, 2020).

Artificial Intelligence (AI): AI is a technology that operates along with cognitive as well as reasoning technologies built into the computers to simulate the customers' behavior, along with an algorithmic process. This technology helps business people to understand customers' mindsets accurately and thus enables them to take accurate and appropriate marketing as well as business decisions and predictions. Thus AI has the potential to unleash digital marketing operations fully (Douglas, 2019).

Chatbots: Chatbots technologies are a comparatively cheap as well as convenient approach for marketers to deliver the utmost customer service and retain customers. Because attaining new customers is more costly than retaining existing customers. Chatbots are more helpful to retain existing customers as they have been giving data-related responses and collecting customer requests. This technology can be incorporated into websites, applications, blogs, and social media platforms, to collect data to formulate appropriate marketing strategies (Douglas, 2019).

Voice Search: Voice search software has now become popular and it has been used more vibrantly nowadays. According to a report, it is found out of 4 billion google searches; one-third of the search is done through voice-enabled searches every day. This voice search has played a major role in the online platform, especially in the case of digital marketing and it is expected to change digital marketing strategies and practices in the upcoming days.

Virtual Reality and Augmented Reality: Both virtual reality (VR) and augmented reality (AR) have changed the business landscape drastically, recommend an instinctive try to the consumers before buying the products, thereby it enhances customers' experience, letting them explore a new product, connect with the brand, and purchase at the same time it enables them to go through a variety of senses and emotions.

Blockchain Technology: Blockchain technology is a comparatively novel one and it paves the way for digital transformation by collecting, storing, and distributing massive information technically. This technology supports a lot of digital marketing by providing transparent, secure, and accessible flow in terms of financial and customer as well as business data. Thus this technology improves the customers' digital marketing experience and supports businesses to invest money wisely and profitably in digital marketing campaigns. With the help of this technology, consumers can keep and sell their data to marketers and advertisers directly. Further, it also supports marketers to track and keep the public affianced in advertisements (Johannes, 2021).

Internet of Things and Its Enabled Technologies: Big data has become a magic word in the digital landscape. It is playing a major role in the business world along with IoT. The increased use of smart technologies and IoT is greatly transforming digital marketing. The application of IoT technology in the marketing field leads to enormous benefits; it also collects customers' data in many ways and used the same to predict the markets as well as consumers. Further, the increased usage of smart digital devices links the objects with the web, and thereby marketers can know about various consumer data, together with the likes and dislikes of consumers (Kiran, 2020).

INTERNET OF THINGS AND DIGITAL MARKETING

The term "Internet of Things" (IoT) effectively describes the network and communication between various devices that have been made possible by the internet and are employed in our daily lives. The data is then transmitted and collected using these devices via Bluetooth signals, beacons, and the cloud, respectively. IoT represents a new era in the creation of mobile applications. IoT gives customers a lot of importance. It interacts with users, studies their behavior, discerns their wants, and satisfies them all. Thus, it helps businesses to focus more on customers' needs and wants. It also has simplified numerous processes for users, including the purchasing process, which can now be completed with just a few clicks as opposed to several. With these advantages, the IoT is projected to leave its mark on the world and create new opportunities for digital marketing (Deshpande, 2021). It is an association or the arrangement of smart or digital devices connected with the help of the internet. The digital devices can be smartphones, tablets, laptops, smart wearables, computers, smart TVs, vehicles, smart home appliances, door locks, smartwatches, and heart and blood pressure monitors. All these digital devices are using sensors and actuators to gather and swap data from the users' surroundings with each other devices with the help of the internet (Louis, 2019). Sensors are playing a major role in the working of IoT, by collecting data from various digital devices and uploading those data on the internet, and sharing the same with interested parties. This type of process is also called device-to-device communication. IoT-enabled devices support swapping data but also help to collect, store and create a massive customer database because customers' data are playing a major role in the digital business landscape. Moreover, with the support of this database, marketers can get enhanced marketing insights and generate more exact and suitable digital marketing strategies. Thus the growth and development of IoT help digital marketers to use this platform for developing competent marketing campaigns (IoT for all, 2021). Same manner businesses can now collect more customer data and target them more effectively than ever thanks to the proliferation of IoT devices. Businesses can learn how to develop better products and services to satisfy their demands as they gather more data about their customers and their activities. Because IoT facilitates personalizing marketing messages based on the data gathered from each individual, business people may promote these goods and services more successfully (Brenner, 2020). The way that people shop online has already been altered by connected devices. Ask Alexa is the best example, to place an order for more when they run out of coffee or washing detergent. The prospects for direct marketing in smart speakers and digital assistants have been quite scarce thus far. It makes sense that wise product designers take a cautious stance. In a market with so many alternatives, annoying and obtrusive marketing messages would undoubtedly result in the premature demise of such products. However, this is a field that will undoubtedly present a lot more opportunities in the near future (Brenner, 2020).

The way that people shop online has already been altered by connected devices. Ask Alexa is the best example, to place an order for more when they run out of coffee or washing detergent. The prospects for direct marketing in smart speakers and digital assistants have been quite scarce thus far. It makes sense that wise product designers take a cautious stance. In a market with so many alternatives, annoying and obtrusive marketing messages would undoubtedly result in the premature demise of such products. However, this is a field that will undoubtedly present a lot more opportunities in the near future (Brenner, 2020). The way and means of communicating with customers become more personal than ever when using IoT in marketing. Targeting a specific audience and segmentation accuracy also reach new heights. It might even be regarded as a completely new branch of direct marketing. Businesses will be able to provide a good or service to every individual customer. The foundation for more complex tailored marketing can be built on this capability. Additionally, when a customer demonstrates a desire to learn more about goods and services, advertising becomes less grating. Another example of how the effectiveness of permission marketing is growing while the future of interruption marketing is becoming increasingly hazy. This trend may be facilitated by IoT technologies.

IoT technology provides a wealth of new communication opportunities for marketers. Companies already use social media extensively for product feedback. It affects how societal interests are formed and shapes the attitudes of every social media user. Last but not least, such engagement may affect a product's features. A virtual arena where a business and a user will have a closer connection will be developed when IoT technology and social media join. In addition to verbal communication, it will be feasible for a corporation to order a product from a customer and customize it to meet the customer's needs. IoT will transform how businesses interact with their current and potential customers and become an essential component of integrated marketing communications (Abashidze & Dąbrowski, 2016). Additionally, they can become more sophisticated, and only those businesses that will hold on to their market share will be able to make the best use of all communication channels IoT technologies are one of the most significant of them.

One of a marketer's main objectives is to gather marketing data. There are more options to target the intended audience, boost conversion rates, and promote brand awareness, and other factors, the more exact and thorough the consumer data. IoT technology makes it possible to gather high-quality, infinite amounts of consumer data. Age, gender, interests, social standing, income, preferences, buying and other forms of behaviors, location, and many other types of information can be included. Provide Opportunities for segmenting consumers more accurately. Since there can be a lot of contact between a customer, IoT devices, and the real environment, marketers can monitor even the most specific aspects of consumer behavior. The vast majority of shoppers now use their mobile devices to make purchases in stores, and a sizeable portion of these shoppers are willing to make touchless or one-click payments using their devices (Ng & Wakenshaw, 2017). The willingness of consumers to use wearable and mobile technology while shopping is growing. This fact is particularly intriguing for retailers because they can build statistical databases with information about every single purchase and consumer type that will later aid a marketing team in improving products, delivery methods, advertising strategies, targeting strategies, and many other things (Nguyen & Simkin, 2017).

The ability of IoT technology to track customer locations deserves special consideration. Knowing where and when consumers are at a given time is crucial since "place" is one of the important components in the 4Ps consists of a product, price, location, and promotion. A corporation can choose the other three elements, but a marketing team should know where to deliver a product. These elements are interdependent, so if even one of them is missing, the entire marketing chain may experience issues. Therefore, businesses make an effort to compile data from all sources, evaluate it, create customer profiles, and use it for profit. Although a lot of customers are naturally apprehensive about the possibility of someone tracking their whereabouts, a sizeable portion of customers are willing to provide this information to businesses in exchange for a perk or discount (Taylor et al., 2020). This situation needs to develop into the most alluring chance for marketers to encourage customers to share their data.

Undoubtedly, Nike+ is one of the most notable instances of IoT technologies being used for all facets of marketing. This is an excellent illustration of how to link a consumer's mobile phone, wearable technology, and the internet. All of these are now in use and already exist. Nike was able to provide users with a range of services by combining various components into a single mobile application. The application's primary purpose for users is to track jogging routes, gauge speed, and keep track of calories burned. The information is saved, and a user can keep track of development. Additionally, the app enables users to post findings on social media. Additionally, the app itself functions as a kind of social network where users may add one another to buddy lists, challenge one another, and view results. Regarding the application's commercial objectives, users can shop online for sportswear. Some pairs of running shoes have unique sensors inside that can track a wearer's run and send data to the gadget (Abashidze & Dąbrowski, 2016).

First and foremost, it's important to note that this software aids Nike's marketing staff in gathering priceless information about customers, like their age, gender, location, exercise preferences, routes, email addresses, etc. Most likely, a market research study would be unable to gather such accurate information.

What's more, this knowledge is free for Nike, unlike market research, which can be quite pricey. The program's integration with social media is another crucial feature since it enables the business to have a direct line of communication with its customers (Abashidze & Dąbrowski, 2016). Lastly, the application generates revenue because it allows for online product sales. The program is an outstanding example of IoT utilization in digital marketing because it combines all of its components.

WORKING STRUCTURE OF IoT

Just like Internet has changed the way we work and communicate with each other, by connecting us through the World Wide Web (internet), IoT also aims to take this connectivity to another level by connecting multiple devices at a time to the internet thereby facilitating man to machine and machine to machine interactions (Alec, 2020). Populace who came up with this idea, have also understand that this IoT environment is not restricted to a picky field, but it has various business relevance in areas of housing automation, automobile automation, industrial unit automation, health and medical, retail and wholesale, and more. Thus the IoT has occupied almost all sectors with the help of its four major components namely Sensors, Gateway, Cloud / Big data, and User Interface (Paul, 2018). The role played by these components is explained in Figure 1.

Sensors: The major role played by the IoT is collecting data from the users, and it is the primary work in the IoT process. Sensors are playing an important role in data collection by monitoring uses in various ways, and it is collecting data continuously from the objects, and surroundings under measurement. Smart devices such as smartphones, smart rings, smartwatches, and cars, have plenty of sensors and they observe and collect data. All these sensors are measuring various physical phenomena such as body heat, pressure, blood pressure, sound, heat, etc., converting them into data in digital form for

Figure 1. The role played by the components

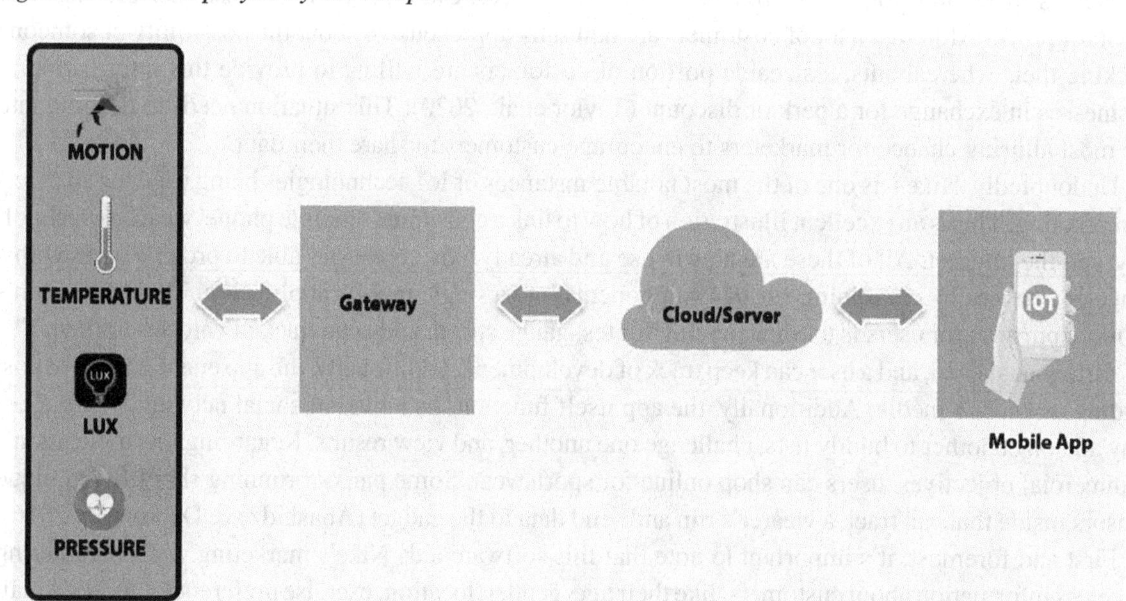

representation. There are types of sensors are in the practice nowadays such as GPS trackers, body and common temperature measurement sensors, humidity measuring sensors, blood pressure monitoring sensors, moisture monitoring sensors, smoke intensity measuring sensors, radiofrequency recognition tags, and so on. Apart from these, sensors are using a wi-fi network, Bluetooth, infrared technology, and zig bee to continue the data collection (Alex, 2021).

Gateway: Gateway is the second component of the IoT process, and it is acting like a bridge that joins IoT-enabled devices through a cloud. Wherein the IoT enabled devices to collect data from the targeted sources and convey them to the cloud, and it obtains the data from the cloud as well and sends it to the digital gadgets. Thus, the IoT gateway is a key component in the IoT system.

Cloud or Server: IoT-enabled devices generate and collect a huge volume of data. All these data need to be stored and analyzed, however, these activities are not possible in the physical storage, hence the cloud infrastructure used in this process with utmost data safety. Thus, this cloud technology seems to be the brain of IoT infrastructure. The analysis and interpretation of data are easy in the case of cloud technology and help to establish valuable as well as meaningful data. This data can be used as a predictive model for the end users (Altexsoft, 2020).

Mobile Applications or User Interfaces: The fourth component of an IoT network is user interfaces such as smart devices, and these user interfaces are distributing constructive data to the end-user, which is a noticeable part of the IoT network. The end-users may be business users or a new device, which means the D2D process. Thus the mobile applications or the user interface transforms the raw data into a processed one, and it exhibits diagrams, pictures, metrics, and markers. Thus mobile applications and user interfaces are acting as a monitor (Altexsoft, 2020).

ROLE OF THE INTERNET OF THINGS IN DIGITAL MARKETING

The diligent growth and development of IoT have been obtainable for decades; however, the role played by the IoT in the digital marketing domain is overwhelming and more obvious. For improved or shoddier, the IoT is disrupting digital marketing. IoT-enabled devices connect different types and numbers of devices into the daily life of the individual as well as business and it collects a huge volume of data and thereby help the marketers to utilize this data to generate necessary information about the consumers and to recognize patterns of customer communications, and forecast customers' buying behavior and way of life. This helps further to envisage customers' taste and preferences and connect it to buying intention (seo-alien.com, 2021). Thus IoT is helping a lot to the marketers, apart from this the following are the roles played by the IoT in digital marketing.

New Product Design Development: With the support of IoT-enabled technologies, marketers can collect huge volumes of data and thereby understand the taste and preferences of the customers, and what customers expect the products or services from their company or brands, this process will help marketers to design the new product as well as to perk up the quality of the product further. IoT not only facilitates augmenting the quality of the product but also aids to design and predict the demand, marketing plan, and strategies for new products. Thus IoT enabled devices support business houses to take better, more reliable and quick decisions (Letterbug, 2016).

Predicting Customers' Taste and Preferences and Buying Behavior: IoT-supported devices help marketers to comprehend the taste and preferences and buying intentions of customers from beginning to end and establish an apparent view about the customers and their buying passage. IoT technology also supports the assessment of the needs and wants of the customers, their buying patterns, what market trends influence the buying pattern of customers, and the area in that marketers need to modify the goods and services accordingly. Accordingly, IoT-enabled technologies present more precise data about consumers' tastes and preferences (Letterbug, 2016).

Big Data Analytics: The IoT-enabled smart devices such as touch screen types of machinery, smart TV, CCTV cameras, voice recognition technologies, and other smart devices support marketers to collect various data about customers and to design and implement an efficient marketing plan, strategies and campaigns. IoT-enabled devices collect data through different forms of sensors about consumers than conventional marketing research, and this data seems more accurate without human bias. Further, it is also expected that by 2020, about 45 zettabytes of data will be collected through IoT-enabled devices. As a result, big data offers more precise insights about consumers and helps marketers to diminish risk and augment business opportunities (smstudy.com, 2018).

Studying Customer Mindset: IoT-powered smart technologies help to read customers' mindsets about taste and preferences, buying behavior, and so on. These devices also provide a huge volume of data that will provide authentic, clear, precise insights about the customers. Data collected from IoT devices is utilized to comprehend the day-to-day lifestyle of the customers. In such a manner IoT enabled devices to help marketers to develop products and services based on composed data (smstudy.com, 2018).

Support to Track Customers' Data: IoT devices support tracking customers' data from IoT-enabled sensors and it will assist marketers to forecast customers' needs and wants towards goods and services. It also helps to predict the timings of sales and the real-time data about the customers and helps marketers to improve sales.

***H*elps to the Customization of Goods and Services:** Customization of goods and services is an efficient way to reach customers and make them buy the goods and services of a particular brand. IoT technology helps marketers to keep in touch with customers on an instantaneous basis. It also assists to reach customers more precisely, and it enhances the efficiency of marketing movements. IoT-enabled devices collect a range of data about the customers, and thereby support to design of the campaigns according to the capacity of the customers. Marketers can also design the best message at the right time to complete the buying process of the customers, based on the data collected (Diogo, 2021).

Improves the Satisfaction of the Customers: As modern marketing suggests, the satisfaction of the customers is an important duty of marketers. Customer satisfaction is a major part of any business, especially in the digital business world, customers need better customer service within a short period, and it is of paramount significance for marketers. This can be achievable with the help of IoT-enabled devices and their real-time data. Thus marketers can serve quickly to customers and respond to them immediately and quickly.

Better Customer Engagement: IoT enables smart devices can be used as a medium to engage with customers. Either one way or another way IoT-enabled smart device is helping a lot of marketers. As a result, marketers can understand the needs and wants of the customers, and serve them better through real-time engagement (Michael, 2020). Thus, IoT is a disruptive technology that has changed the way how businesses produce and distribute goods as well as sell their products and provide customer service. IoT devices have the potential to improve businesses and also bring with them a lot of commercial prospects.

INTERNET OF THINGS AND ITS DISRUPTION IN DIGITAL MARKETING

BI Intelligence research findings reveal that the number of IoT-enabled devices will be twofold by 2025. Gartner research institute also predicts that 25 billion IoT-enabled devices will be exited by 2030. The IoT-enabled technologies not only change the day-to-day individual lives, however, but a variety of IoT-enabled applications also support a lot to organizations to generate a new collection of business models. However, it's significant to comprehend how IoT technologies are disrupting the digital business landscape. Because this technology is having lots of advantages for various industries including, digital marketing. Dresner Advisory Services conducted a study and found that most of the business houses mentioned that IoT is not a significant one for the business process. But, in reality, IoT has the potential to transform the face of various businesses, especially digital marketing and marketing intelligence. Big data analytics and digital technologies lead to digital disruption and help to formulate new digital marketing strategies and introduction of new technologies. As the present world is dominated by digital technology, both business and individual use of IoT-enabled technologies will move forward and turn out to be a part of daily life. Similar to past technological developments, it's not easy to forecast accurately how entangled our routine business, as well as personal activities and thus technology, will change the future life (Analyticsinsight.net, 2018).

The introduction of IoT-enabled devices will generate extraordinary data with accuracy from all sorts of businesses; thereby it computes and quantifies the data to take meaningful insights. However, in the case of digital marketing, this technology presents vast chances to generate several marketing strategies and to adapt the same to various industries to find possible changes. In addition, this technology also helps system developers and companies to enhance data security and minimize the security risks associated with the huge volume of data stored in the storage. Whenever and wherever the business people, as well as individuals, can track, and analyze the stored data. Though, this technology influences various business landscapes, the following are how IoT-enabled devices are disrupting digital marketing (ytcstudios.com, 2019).

Enhancing Market Research: IoT-supported technologies are playing a major role in the case of marketing research. Even though marketing research is sophisticated in the earlier days, the accuracy of data collection, processing of data, and analysis of data are becoming so precise using this IoT, and it also supports digital marketing research people by making market research bigger, better, more accurate, and reliable. In the present marketing landscape, the role of IoT in business-to-business (B2b) enterprises becoming a norm already and consumers' admittance to IoT technologies will perform similarly for business-to-consumer (B2C) marketers. As a result, the occurrence of IoT-supported technologies in the residence will let marketers collect more data to collect and analyze. This rush forward in data collection and analysis will make the marketers rationalize and modernize the business processes and reach the target customers in all-new possible ways. Since the existence of a linkage between IoT-enabled devices and the cloud, a marketer can perform better than the previous time, and with the help of accurate predictive analysis of evolving needs and wants of the customers (Businessinsider. in, 2020).

IoT Helps to Read Consumers' Lifestyles: The IoT technologies don't access the data simply that a customer shared publicly on various platforms like web pages, review points, and social media, but it accesses the data of their day-by-day activities as a lifestyle. Thinking about the different types of IoT-supported devices; they can be used in various places by all types of customers like smartphones, coffee makers, smartwatches, home automation systems, and smart cars. All these smart devices are integrated with day-to-day lifestyle of the customers from morning to night sleep and collect different data set. All these data sets are much important for digital marketers, to access information about the lifestyle pattern of the customers, thereby it could be understood why the demand for the product increase or decrease, how customers utilize the products or services, and when it is being used, where it is being used and why is it being used. All these types of questions could be answered only through the IoT device's collected data (Bdigimarketer, 2019).

Impacting on the Search Engine Optimization (SEO) Industry: Individuals, who are specialized in SEO activities, will realize major changes in their sector, with the help of IoT, further these technology-enabled devices turn more advanced in this field. This technology helps to process the human language by its enabled devices and equipment, as users habitually pose questions more in nature when speaking in a contrast to when queries are typed into google search. According to Beth Kotz, the regular use of AI-enabled technologies like Siri from Apple, Alexa from amazon, and google home means that search engines attempt to realize the mindset of the customers as opposed to grinding in on a keyword. The web pages no longer are graded merely on the insertion of keywords but on the process of natural language that is controlled inside the content. In the case of marketers, this technology helps to move from keyword-supported SEO content to mindset-based scripts will be the main aspect of the IoT background (ytcstudios.com, 2019).

Generation of Big Data: IoT technologies help to access a variety of data from different touchpoints. From the digital marketing point of view, this is hugely supportive to generate a better understanding of the customers throughout their purchase. During the customers' purchase process, digital marketers have the chance to utilize this data to their benefit. Because the insinuation of IoT technology enables marketers to observe where customers are in the buying process, marketers can also create more touch-point for optimistic interaction with customers. With augmented contact with customers, marketers can be able to identify novel ways to communicate with the customers, to respond to queries related to the brand at and before purchase, rather than during or after purchase. On the other hand, it should also be important that the use of customers' data should be considering various security concerns (smstudy.com, 2018).

INTERNET OF THINGS AND DIGITAL MARKETING REVOLUTIONS

IoT technology is a game-changing one and it connects our personal as well as business lives with different IoT-enabled devices in a network system and helps to carry out manifold activities. Even the general public can be using it easily in everyday life. Apart from this various fields can also be benefited a lot from this technology the smart devices application. This technology also helps to use for real business state purposes and it can construct the business by the IoT networking. There are sectors such as health, retailing, logistics, supply chains, customer services; and so on those are using this technology to excel in their business operations. Because this technology is shifting the interacting mode with the surroundings since this technology is providing personalized as well as accurate services to the taste and preferences of the customers. IoT-enabled devices are influencing people and business processes in various ways such as customers' experience, tastes, and preferences, analysis of customers, prediction of markets, assess the trends of customers and markets, and so on. Thus, all marketers, online businesses, retailers, products, and services are transforming themselves by using IoT technologies to connect with clients. The IoT has transformed the conventional way of doing business into a modern and sophisticated one, and it has made easier shopping nowadays and provides a very good experience to customers (Chantal,

2016). Consequently, IoT-enabled technologies made some revolutionary changes in digital marketing. They are explaining as follows:

Development of the Markets for Digital Devices: If things unite something to the Internet, it can demonstrate important progressions in technology. In the same manner, combining business with Internet technology will change the lifestyle of customers to more peaceful, order and purchase goods and services through online stores from the place where customers are living or working. For enabling these marketing practices people should use digital technologies such as smartphones, smartwatches, tablets, laptops, and smart televisions, thereby it will lead to the augmented use of and availability of digital devices in the market. Thus these technologies enable customers to buy digital devices as well as goods and services from the market.

Data Revolution: IoT-enabled devices keep on observing and collecting a range of data from the users and saving it, analyzing it, and processing it, to get meaningful insights such as customers' behavior, buying patterns, tastes and preferences, and so on. Collecting and studying these data will help marketers to design new products and new marketing strategies to provide customized service as well as experience to the customers.

Revolutions in the Marketing Support: IoT-enabled devices support a lot to the marketing teams to collect a variety of data set to understand customers' preferences towards brands, price, shopping intention, and pattern, and thus data plays a major role in marketing and acts as a gold mine for the business people.

Augmenting the Role of the Agency: Conventional marketing agencies are concentrating only on products available in the market and computing profits. However, the role now has changed and developed and hence marketers need to become accustomed to the latest technologies. In the earlier days marketing activities were done through offline stores only, now the trend has changed to online and new targeted marketing and seems to be at all the online supply to the customers. It also entails executing a supple move towards marketing campaigns, initiatives, reduction of cost, and new product development. Thus IoT augments the role of agency in modern marketing.

Establishment of Customers' Security and Privacy: As all the brands and organizations are starting to watch their customers' daily purchase routines and collecting huge amounts of data, at this juncture customers' privacy should be under scrutiny. Hence tight security measures need to establish by various methods such as segmentation of the network, D2D authentication, and reinforced encryption methods to take care of IoT devices from being conciliation. Hence there is a need to put on the eye-centric to the rules and regulations of the technology and thereby more security and privacy have to establish to keep protect consumers as well as business data.

Leads to the Growth in E-Commerce: Internet technologies nowadays capture business as well as markets by the massive use of using Internet technologies from their computers, laptops, smartphones, and other smart devices. With the support of IoT-enabled devices, customers can expand their relations with e-commerce companies directly, and accordingly, e-commerce will be boomed.

Development of Omni Channels: With the support of IoT-enabled devices, several devices can be working with one user account in the form of omnichannel technology. After the arrival of IoT technologies into the business process, there is a development in the omnichannel business segment, and it has also augmented the usage of online business and stimulated customers to order goods and services from their residence or place of work.

Cost Saving: Utilization of IoT technologies into the business process helps to get better output and efficiency, efficient utilization of organizational assets, and enhancing the efficiency of the business process and thereby reducing the costs and increasing the profits of the organization. Moreover, IoT-enabled technologies also help marketers a lot to predict the market and demand for goods and services, thus maintenance costs can be trimmed down up to 40 percent (Deals Insight Team, 2020).

INTERNET OF THINGS AND THE FUTURE OF DIGITAL MARKETING

The growth and development of IoT, have changed the landscape of digital marketing and sustainability. Due to the incremental usage of IoT-enabled devices in all businesses and services, marketers are reaping lots of benefits with the data of the customers. With the help of the data captured by the IoT devices, marketers could understand the customers' purchase intention, purchasing patterns, and expectations on one side, and accordingly assess the market demand for the goods and services, opportunities for the business, and so on. Thus, IoT-enabled technologies are providing current, actionable, and up-to-date insights about customers and businesses; thereby marketers could predict the unprecedented opportunities available for businesses. It could also suggest that marketers develop strategies to promote the business, and gives insights about when to promote, where to promote, and how to promote the goods and services, and markets. Accordingly, IoT is creating a new pathway to the paradigm in digital marketing (Campaignsoftheworld.com, 2018). However, the following are some of how the future of digital marketing has to be changed by the IoT.

Determination of the Right Customers, Suitable Media, and Appropriate Message: In recent days marketers are benefiting a lot from industry 4.0 technologies such as AI, machine learning (ML), big data, and IoT. Though these many numbers of technologies are available, the role played by the IoT is limitless to digital marketing. The increasing number IoT enabled devices and their usage in the business have the potential to offer boundless services as well as benefits to the business community. As the IoT-enabled devices have multiple touch points and sophisticated usages, it's supporting marketers to collect a huge volume of data, and thus, process and fine-tune the collected data. Thereby it is offering meaningful, useful, and relevant insights about the customer and thus supporting the marketers to retain the existing customers by offering more value. In recent days all the stores have collected customers' contact details for sending offers as well as discount alerts to their mobile phones as well as emails. Apart from this the sellers can also able to send exact advertisements and sales promotion-related information to the customers, based on their previous shopping patterns and preferences. Thus IoT has changed the marketing pattern from traditional to digital already and it moves further with advancements to enhance the better customer experience, and offer distinctive opportunities for marketers to get close to customers through IoT-enabled interactions (Clarke, 2018).

Understanding What Customers Want: The irresistible as well predictive power of IoT-enabled devices supports digital marketers by supplying various insights about the purchase history and process of the customers, their decision points, and lifestyle. All these details can support digital marketers to develop new business strategies and thereby marketers can establish stronger as well as more profitable relationships with customers. This technology also supports marketers to understand the customers deeply and to predict the market accordingly. Based on that useful information motivates the customers to engage with the seller further. As the data have been collected from IoT-enabled devices, it also helps the marketers to predict the needs and wants of the customers, accordingly the marketers can do investments in prognostic insights to make the customers' purchases frequently. Such a kind of dedicated, requirement-based marketing helps to save businesses (Antonis, 2019).

Automatic Business Promotion: The IoT-supported devices collecting the data do prognostic analytics to empower the marketers to think and act more tactically to develop the brand, augment customer retention, and increase customer loyalty. However, this technology helps to develop digital algorithms through the collected data and assists to automate the business promotion processes. Based on the searches done by the customers, the system itself designs and prepares the advertisement banners and displays on the web pages of the online stores. Since IoT-linked devices automate their process, marketers are never far from time, technology, and the machines themselves making and applying the marketing decisions. Thus the IoT-enabled devices free up marketers from promotion-related worries and help them to move with innovative as well as competitive business strategies (Alpesh, 2021).

CONCLUSION

The growth and development of information and communication technologies have changed the world a lot. However, the introduction of industry 4.0 technologies has taken the world to the next level. These technologies have introduced advancements in various industries, of which digital marketing is one of the most beneficial sectors. This technology has transformed the way of doing business, and it renders support to marketers as well as customers. It has to switch people from offline to online and makes them love and uses the technologies in their personal life as well as purchasing. This technology helps a lot to acquire and retain customers, and make them continue with the existing brand. It can turn each machine to interact with people with the help of a smart device. An IoT-enabled device collects and shares data about customers' location, purchase patterns, buying behavior, purchase frequency, and so on. Thus, marketers can harvest benefits and turn them into useful insights to make stronger marketing campaigns. IoT has modernized the marketing world, it connects customers with marketers in an interrelated aspect. Due to this reason, marketers nowadays start to invest more in the IoT and its related. Marketers, who are utilizing IoT in their marketing as a model, have got an edge over competitors or rivalries. Thus the role of IoT in digital marketing is really about thinking outer the box and promoting technologies in the correct way to connect with genuine customers most authentically. Thus IoT has promised marketers to enable digital marketing campaigns most realistically with the proper customer engagement strategies. Since the IoT-enabled technologies offer flawless interfacing and data absorption over the internet on to community cloud, marketers will have capable to incorporate IoT into the digital marketing process and transforming how customers are engaged.

REFERENCES

Abashidze, I., & Dąbrowski, M. (2016). *Internet of Things in marketing: opportunities and security issues*. Management Systems in Production Engineering.

Alec, J. (2020). *The 4 Stages of IoT Architecture*. https://www.digi.com/blog/post/the-4-stages-of-iot-architecture

Alpesh, P. (2021). *The Internet of Things and Its Impact on Digital Marketing*. https://theseoplatform.co.uk/blog/the-internet-of-things-and-its-impact-on-digital-marketing/

Altexsoft. (2020). *IoT Architecture: the Pathway from Physical Signals to Business Decisions*. https://www.altexsoft.com/blog/iot-architecture-layers-components/#:~:text=IoT%20Architecture%3A%20the%20Pathway%20from%20Physical%20Signals%20to%20Business%20Decisions&text=IoT%20solutions%20have%20become%20a,enterprises%2C%20connected%20devices%20are%20everywhere.&text=This%20article%20describes%20IoT%20%E2%80%94%20the,its%20architecture%2C%20layer%20to%20layer

Analyticsinsight.net. (2018). *How the Internet of Things Is Transforming Digital Marketing*. https://www.analyticsinsight.net/how-the-internet-of-things-is-transforming-digital-marketing/

Bdigimarketer. (2019). *The growing Need for Ethics in Digital Marketing-Bdigimarketer*. https://medium.com/@bdigimarketer1/the-growing-need-for-ethics-in-digital-marketing-bdigimarketer-56da482c032a

Brenner, M. (2020). *Marketing and the Internet of Things*. https://marketinginsidergroup.com/strategy/marketing-internet-of-things/

campaignsoftheworld.com. (2018). *How the Internet of things is Transforming Digital Marketing*. https://campaignsoftheworld.com/technology/how-the-internet-of-things-is-transforming-digital-marketing/

Clarke. (2018). *How Internet of Things (IoT) Will Transform the Future of Marketing Businesses*. https://www.branex.ca/blog/how-iot-transform-future-marketing/

Dan, S. (2020). *Ethical Marketing: 5 Examples of Companies with a Conscience*. https://www.wordstream.com/blog/ws/2017/09/20/ethical-marketing

Deals Insight Team. (2020). *The Big Revolution in B2B Marketing-Internet of Things*. https://www.dealsinsight.com/the-big-revolution-in-b2b-marketing-internet-of-things-iot/

Deshpande, I. (2021). *What Is IoT And Why Should Marketers Care About It?* https://www.spiceworks.com/tech/iot/articles/what-is-iot-for-marketers/

Digital, S. (2022). *What Is The Role Of Data Science In Digital Marketing?* https://www.stellardigital.in/blog/what-is-the-role-of-data-science-in-digital-marketing/

Diogo, S. (2021). *5 Ways in which IoT will Impact Digital Marketing*. https://www.youlead.agency/blog/5-ways-in-which-iot-will-impact-digital-marketing

Douglas, K. (2019). *10 Modern Technologies That Are Enhancing Digital Marketing*.https://martech.zone/modern-technologies-impacting-digital-marketing

Fiftyfiveandfive.com. (2020). *6 Digital Marketing Technologies to Help You Raise Your Game*. https://www.fiftyfiveandfive.com/6-digital-marketing-technologies/

Grizhnevich. (2021). *IoT Architecture: Building Blocks and How They Work*. https://www.scnsoft.com/blog/iot-architecture-in-a-nutshell-and-how-it-works

Gudema. (2019). *7 Marketing Technologies Every Company Must Use*. https://hbr.org/2014/11/7-marketing-technologies-every-company-must-use

India. (2016). *The Internet of Things revolution and the transformation of Digital Marketing*. https://www.cyberclick.net/numericalblogen/the-internet-of-things-revolution-and-the-transformation-of-digital-marketing

IoT Dunia.com. (2020). *How does IoT work? - Explanation of IoT Architecture & layers.* https://iotdunia.com/iot-architecture/

IoT for all. (2021). *How IoT Data Can Improve Digital Marketing Outcomes.* https://www.iotforall.com/how-iot-consumer-data-affect-digital-marketing

Iscoop. (2020). *The Internet of Things in Marketing: The Integrated Marketing Opportunity.* https://www.i-scoop.eu/internet-of-things-iot/internet-things-marketing/

Johannes, B. (2021). *Ways IoT Is Changing Digital Marketing.* https://iotmktg.com/ways-iot-is-changing-digital-marketing/

Kiran, B. (2020). *Top 7 Ways IoT is Transforming Digital Marketing in 2020 and Beyond.* https://medium.com/@kiran.bhatt270897/top-7-ways-iot-is-transforming-digital-marketing-in-2020-and-beyond-a855101b6607

Letterbug. (2016). *The Role of IoT in Digital Marketing.* https://suyati.com/blog/the-role-of-iot-in-digital-marketing/

Michael, B. (2020). *Marketing and the Internet of Things.* https://marketinginsidergroup.com/strategy/marketing-internet-of-things/

Neocleous. (2019). *Digital Marketing and the Internet of Things.* https://www.baseelement.digital/en/digital-marketing-blog/digital-marketing-the-internet-of-things

Ng, I. C., & Wakenshaw, S. Y. (2017). The Internet-of-Things: Review and research directions. *International Journal of Research in Marketing, 34*(1), 3–21. doi:10.1016/j.ijresmar.2016.11.003

Nguyen, B., & Simkin, L. (2017). The Internet of Things (IoT) and marketing: The state of play, future trends and the implications for marketing. *Journal of Marketing Management, 33*(1-2), 1–6. doi:10.1080/0267257X.2016.1257542

Palmere, T. (2019). *How IoT Data Can Improve Digital Marketing Outcomes.* https://www.iotforall.com/how-iot-consumer-data-affect-digital-marketing

Paul, S. (2018). *4 Stages of IoT Architecture Explained In Simple Words.* https://medium.datadriveninvestor.com/4-stages-of-iot-architecture-explained-in-simple-words-b2ea8b4f777f

Power Digital. (2017). *The Internet of Things and The Future of Digital Marketing.* https://powerdigitalmarketing.com/blog/the-internet-of-things-and-the-future-of-digital-marketing/#gref

Rajanarthagi. (2020). *What is IoT? & Role of IoT in digital marketing.* https://gecdesigns.com/blog/role-of-iot-in-digital-marketing

Seo-alien.com. (2021). *The Role of IoT in Digital Marketing.* https://www.seo-alien.com/social-media-marketing/the-role-of-iot-in-digital-marketing/

Skelia. (2020). *10 Ways IoT is Changing Digital Marketing in 2020.* https://skelia.com/articles/10-ways-iot-is-changing-digital-marketing-in-2020/

Smstudy.com. (2018). *How IoT will help Digital Marketers in near future?* https://www.smstudy.com/article/how-iot-will-help-digital-marketers-in-near-future

Tahir, A. (2016). *7 Fundamental Ethics of Social Media Marketing*. https://www.business2community.com/social-media/7-fundamental-ethics-social-media-marketing-01571504

Taylor, M., Reilly, D., & Wren, C. (2020). Internet of things support for marketing activities. *Journal of Strategic Marketing*, *28*(2), 149–160. doi:10.1080/0965254X.2018.1493523

Vandita, G. (2018). *5 Ways IoT is Changing Digital Marketing*. https://www.martechadvisor.com/articles/iot/5-ways-iot-is-changing-digital-marketing/

Venables, M. (2019). *Optimizing Digital Marketing with Data Science*. https://towardsdatascience.com/how-data-science-is-shaping-digital-marketing-5a149443f90

ytcstudios.com. (2019). *Digital Marketing in the age of IoT.* https://www.businessinsider.in/advertising/ad-agencies/article/digital-marketing-in-the-age-of-iot/articleshow/72055063.cms

ytcstudios.com. (2020). *The Convergence of Marketing & the Internet of Things*. https://www.ytcstudios.com/blog/2020/8/6/the-convergence-of-marketing-amp-the-internet-of-things-iot

Chapter 8
The Role of Digital Twin in Accelerating the Digital Transformation of Smart Cities:
Case Studies in China

Poshan Yu
Soochow University, China & Australian Studies Centre, Shanghai University, China & EBU Luxembourg, Luxembourg

Hongyu Lang
Independent Researcher, China

Jericho I. Galang
Public-Private Partnership Center of the Philippines, Philippines & Ateneo de Manila University, Philippines

Yifei Xu
Tongji University, China

ABSTRACT

Cities are crucial carriers of economic prosperity and social development. With the approaching Society 5.0, the digital twin city has become the mainstream model for the construction of new smart cities. Digital twins, along with the internet of things, fifth-generation wireless systems, and artificial intelligence technologies, offer great potential in the transformation of the current urban governance paradigm toward smart cities. This chapter will explore the imperative role of digital twin in accelerating digital transformation of smart cities. From the perspective of the overall policy, the industry, and application of the said technology, this chapter gives insight into the development trend of digital twin cities. Additionally, through case studies of the two most outstanding digital pilot regions in China, the chapter investigates the role of digital twin in accelerating digital transformation of smart cities such as the technical approaches on city information modeling (CIM) and building information modeling (BIM).

DOI: 10.4018/978-1-7998-9266-3.ch008

INTRODUCTION

The world is at an unprecedented level of urbanization. The United Nations (2018) estimates that 68 percent of the world's population will live in urban areas by 2050. As a result of the rapid growth in the rate of urbanization, cities face huge challenges in meeting the housing, infrastructure, transport, and energy needs of their urban populations, and urgently require new ideas and methods to solve these issues (World Economic Forum, 2022).

Digital transformation is an inevitable choice for urban governance. Each major technological breakthrough has redefined the world landscape. Mergel et al. (2019) interviewed governments in several countries and the majority of the respondents noted that digital transformation is driven by the external environment (83%) rather than by internal pressures. Modern information and communication processes become an important force for social evolution (Mergel et al., 2019). The revolution has greatly increased the productivity of cities. The efficiency of these cities attracted labor, and the concentration of labor in turn increased the productivity of the cities.

Over the past 20 years, the concept of smart cities has become almost globally known and people have started to think about innovative ways of developing smart cities. Zheng et al. (2020) reviewed 7,840 studies on smart cities between 1990 and 2019. The authors noted that sustainability and sustainable development have become popular topics not only among scholars, especially in the fields of environmental economics, technology and science, urban planning, development, and management, but also among urban policy makers and professional practitioners. Caragliu and Del Bo (2019) consider smart cities as complex systems that involve symbiotic connections between people, institutions, technologies, organizations, the built environment, and physical infrastructure. Smart cities use data and technology to improve efficiency and sustainability, and to enhance the quality of life and the experience of working in a city (Albino et al., 2015). With information and communications technology (ICT), smart cities can be applied more effectively to support economies and societies. Bifulco et al. (2016) see ICT as a set of tools for the governance and management of cities and metropolitan areas.

The digital twin is the inevitable goal of digital transformation. The concept of the digital twin was introduced by Professor Grieves in 2003 during a product lifecycle management course at the University of Michigan. The digital twin consists of three key components: 1) the physical product, 2) the virtual product and 3) the connections that link them. Digital twins fully describe potential or actual physical manufactured products from the micro-atomic level to the macro-geometric level (Grieves & Vickers, 2017). Digital twin technologies deeply integrate hardware, software and Internet of Things (IoT) technologies to enrich and refine virtual entities.

Digital twin technology has been developed over nearly two decades and is used not only in manufacturing but also in many other industries and public life. Digital Twin Cities (DTCs) is a new mode of urban development and management based on DTCs technology, and its concept is gradually becoming clear.

Table 1 summarizes the development process of DTCs in recent years. From the early stages of conceptualization and experimentation to the emergence of large-scale pilot projects and the establishment of standards and frameworks, DTCs have become an increasingly important aspect of urban planning and management. The evolution of DTCs has been marked by a focus on data integration, citizen engagement, and sustainability.

China's DTC construction market has great vitality. According to statistics, the total investment scale of China's new smart cities in 2020 was about RMB 2.4 trillion (or 0.38 trillion USD). China's City Information Modelling (CIM) construction projects have shown a trend of rapid growth year on year, with

Table 1. The development of the concept of digital twin city

Time	Author	Contribution
2005	Grieves	A set of digital models is constructed in the virtual space to interact with physical entities and fully describe the trajectory of physical entities throughout their life cycle.
2012	Glaessgen & Stargel	Inspired by National Aeronautics and Space Administration's (NASA's) Apollo programme, NASA researchers E.H. Glaessgen and D.S. Stargel defined the digital twin as the integration of multidisciplinary and multiscale simulation processes by making full use of physical models, sensors, operational history, and other data.
2017	Mohammadi & Taylor	From the perspective of an urban platform, the Georgia Institute of Technology proposed that the digital twin of a smart city is an intelligent, IoT-enabled, data-rich urban virtual platform that can be used to replicate and simulate changes that occur in real cities to improve the resilience, sustainability, and livability of cities.
2018	China Academy of Information and Communications Technology (CAICT)	The concept of a "digital twin city" was first proposed by CAICT and used in the planning and construction of smart cities.

Source: WoS and Sciencedirect

the number of projects growing from two (2) in 2018 to 72 in 2021 (as of September 2021), and the total amount of investments is also climbing year on year accordingly. In August 2021, the World Economic Forum and the CAICT jointly solicited DTC cases. According to the data collected for this project, more than half of projects were invested at the RMB 10 million (1.6 million USD) level, 89 percent were invested at the RMB 1 million level or more, and the average investment of the projects reached RMB 28 million.

With over 26 million inhabitants, Shanghai is the most populous city in China and the third most populous city in the world. Beijing-based digital twin specialist 51World has successfully created a complete virtual clone of the city - a total of 3,750 square kilometers - in the Unreal Engine (Weir-McCall, 2020).

As recently reported on the building information modeling (BIM), 51World has individually modelled 20 landmarks, including the Oriental Pearl and the Shanghai Centre Tower, and used data from satellites, drones, and sensors to generate digital versions of countless other buildings, roads, waterways and green spaces using algorithms. Ultimately, the plan is to turn this model into a true digital twin, which will be constantly updated in almost real time (Weir-McCall, 2020).

Figure 1 depicts the DTC Model of Shanghai, created by 51World. The model showcases the various aspects of the city's infrastructure, including transportation, energy, and environmental systems, as well as social and economic indicators. Through its use of advanced technology and data analytics, the model serves as a valuable tool for urban planning and management.

Through Figure 2, we can see a range of global DTC cases, highlighting how cities worldwide are adopting and implementing digital twin technologies to improve their urban planning and management. These cases include a diverse range of cities, from established urban centers to emerging smart cities, each utilizing digital twin technologies to enhance their efficiency, sustainability, and livability.

LITERATURE REVIEW

1. How Are Digital Twins Being Used in Smart Cities?

Digital Twins are finding success in a wide range of domains relevant to cities, each one potentially brings in different data and has different challenges to implementation.

Figure 1. Digital twin city model of shanghai created by 51world
Source: The BIM.

Figure 2. Cases of global digital twin cities
Source: World Economic Forum.

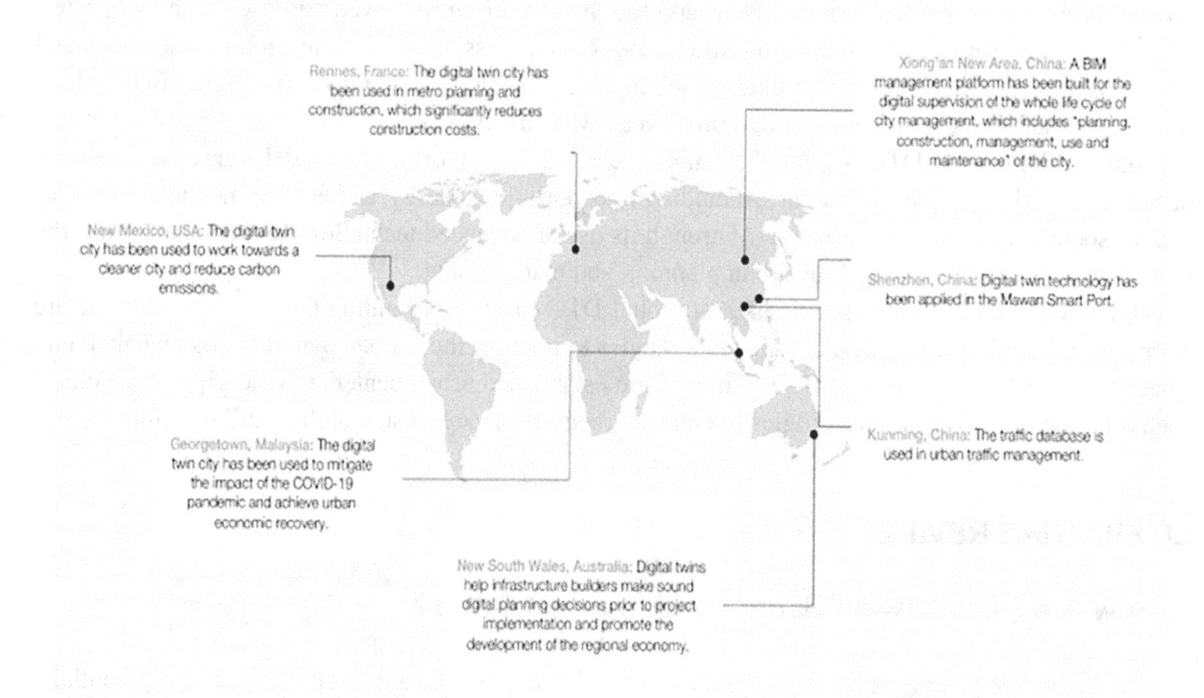

Table 2 presents a comprehensive overview of the applications of digital twin technology in different areas of smart cities, including transportation, energy, environment, public safety, and healthcare. The table demonstrates the wide range of use cases for digital twin technology, from improving traffic flow and reducing carbon emissions to enhancing emergency response and optimizing healthcare delivery.

DTCs aims to improve logistics, energy consumption, communications, urban planning, disasters, architecture, and transportation. In this section, we list several key technologies in DTCs: measurement and mapping technology, BIM technology, IoT, 5G, collaborative computing, blockchain, and simulation. The above technologies play different roles in DTC. Surveying and mapping technology is the basis of static data collection of urban buildings. BIM technology is the foundation of asset and infrastructure management. The IoT and 5G are fundamental to the effective collection of dynamic data and feedback. Blockchain technology is based on trust mechanisms for transactions, logistics, and human behavior. 5G's collaborative computing is efficient and basic real-time response. Simulation technology is the basis of policy support, planning and early warning mechanism.

A. Surveying and Mapping Technology

The DTC focuses on instantaneity and accuracy, which requires highly automated mapping and surveying technology. Surveying and mapping in a DTC are the fundamental technologies that make it possible to deliver data integrated holographic results (2D/3D, aboveground/underground and indoor/outdoor) in real time. Surveying and mapping technology in cities has two segments: 1) surveying the topography, environment, and spatial structure of the city; and 2) mapping this information into an integrated system

Table 2. Applications of DT in different areas of smart cities

Type	Description	Critical Data / Technologies	City Departments
Urban Planning / Zoning / Development	Both greenfield and older cities are better managing their developments by using the data to better visualize the impact of new buildings on the areas around them. By managing lands use, economic opportunities, and even the physical impact on wind and shadow they are able to answer questions from the public in a much more detailed way.	Land use, building footprints and details, shadow and wind patterns	GIS, Planning, Zoning board, Economic Development
Asset & Infrastructure Management	Asset management and especially preventative maintenance are a pressing need for cities that need to make their infrastructure dollars go further. Modeling when infrastructure needs service, or where new infrastructure can have desired outcomes becomes easier with a high quality platform.	Asset tagging and geolocation, IoT sensors on infrastructure	Parks, Public Works, Transportation
Environmental & Climate Monitoring	Sensors that measure noise, air pollution, temperature, water levels, and flood risk are being used to look at both quality of life issues(noise complaints) and scenario planning around climate change and sea -level rise. Cities where this is a factor are better able to plan development and see how critical infrastructure is impacted.	IoT sensors, topography, water, climate models	Environment, Facilities/Property Management, Parks, Public Works
Public Safety and Crowd Dynamics	Some cities are using surveillance as well as scenario planning around critical events to understand if something goes wrong how to deploy resources such as police and how to move people out of harms way efficiently and safely.	Optical sensors, video, GPS	Police, Fire, EMS, Emergency Management
Transportation and Pedestrian Planning	Understanding the transit needs and congestion of crowds are both important parts of vision zero and complete streets initiatives. These models can help for planning the streetscape and planning transit investments. In the case of parks and other public facilities, pedestrian information can be critical to understanding usage	Ped counting, optical sensors, ridership, traffic counts and light timing	Transportation, Parks, Public Safety
Facilities Management	Security, HVAC, occupancy, and lighting information is helping cities manage the extensive facilities they maintain such as schools, libraries, police stations, etc. These use cases will often bring a near -real time component in to complement the planning/maintenance usage	Building details, HVAC, security, ped counting, optical sensors, utilities	Facilities/Property Management
Energy Usage and Solar Deployment	From reducing carbon footprints to finding strategic places to add solar panels to buildings, cities are using the tech to better plan deployments and opportunities for cost savings	Utility, building details, solar radiation	Environment, Facilities/Property Management

Source: IDC Government.

based on geographic information system (GIS). In the surveying part, four technologies are considered: 1) tilt photography, 2) unmanned aerial vehicle (UAV), 3) 3D laser scanning, and 4) global positioning system (GPS). In the mapping part, two technologies are considered: 1) real-world 3D reconstruction technology and 2) multi-source geographic data processing technology.

In recent years, with the maturity of tilt photography and UAV technology, real-time and accurate acquisition of local ortho, tilt, or Lidar point cloud data in cities can greatly reduce the workload of field mapping. These unmanned facilities make full use of their respective advantages to make comprehensive measurements of urban entities from the land, sky, rivers, and underground space while implementing iterative data updates. Gao et al. (2017) studied the rapid acquisition and processing method of large-scale topographic map data which relies on the UAV low-altitude aerial photogrammetry system. El Meouche et al. (2016) used a UAV and other classical land surveyor instruments to investigate the possibility of exploiting and certificating it. The result indicated that the precision is satisfactory, with a maximum error of 1.0 cm on ground control points and 4 cm for the rest of the model. New-generation surveying technology, which focuses on the use of UAVs, is now widespread and operational for several applications. Yao et al. (2019) provided a summary of existing examples of UAV-based RS in agricultural, environmental, urban, and hazards assessment applications, etc. In comparison with traditional topography, aerial photogrammetry provides faster data acquisition and processing and generates several high-quality products, with an impressive level of detail in the outputs at a lower cost.

3D laser scanning technology is a low-cost technology using a high-speed laser scanning measurement method and laser ranging principle to accumulate large-area dense point cloud data. These data include three-dimensional coordinates, reflectance, and texture on the surface of the measured object. With the processing algorithm, this technology would create a 3D model of the measured object, as well as various map data, such as lines, areas, and volumes, promptly. With the rapid progress in urban planning, the cost of lightweight 3D laser scanning technology has decreased sharply, thus promoting the use of UAVs. Li et al. (2019) constructed a measurement system to acquire global coordinate point cloud data by combining 3D laser scanning and GPS, the positional accuracy in the global coordinate data was better than 10 mm.

The real-world 3D reconstruction technology is the key to mapping, based on the results of remote sensing and the fusion of multi-source data. In a city, this technology obtains 3D point clouds, 3D models, and real imagery through automated processes. The extraordinary location accuracy and geometric accuracy separate this technology from others. Zhou et al. (2019) proposed a method to improve the reconstruction effect of a 3D structure and model's visualization. Ma and Liu (2018) systematically summarized the up-to-date achievements and challenges for the applications of 3D reconstruction techniques in civil engineering and proposed key future research directions to be addressed in the field. Laser point cloud 3D construction technology will reflect the advantages of the automatic construction of 3D models in the DTC, and it has a very real, detailed, specific, and impactful visual experience.

B. Building Information Modeling Technology (BIM)

Many studies have been conducted to create BIM or city information modeling as the digital infrastructure to support various smart city programs (Chen et al., 2018). In a DTC, digital twin technology maps the physical city into a mirror city that can be disassembled, duplicated, transferred, modified, deleted, and operated repeatedly.

The identity technology is the basis of operations and maintenance in the DTC, which gives "identity information" to the physical objects in the digital world. Identity technology will certify the unique identification of each entity in the city and its "digital identification" is the only identification in the BIM system. After identity processes, indexing, positioning, and loading related information of the objects in the asset database quickly become possible. With identity technology, the BIM system can recognize each entity using several approaches, such as surveying and mapping technology. An information modeling system provides all the required tools and automation to achieve end-to-end communication, data exchange, and information sharing among identified collaborators. Xue et al. (2018) sought to understand the status quo and development trajectory of such radio-frequency identification (RFID)-enabled BIM systems and, finally, provided five-step guidelines for linking RFID to BIM. Accordingly, virtual 3D models, created through the BIM process, delivered as physical assets, monitored in real-time, and managed using building management systems, can adopt IoT designs and services.

C. 5G-Enabled Internet of Things

The IoT technology, including the acquisition control and the perception process, is the basis for collecting dynamic data and relaying feedback. On the one hand, the acquisition technologies make communication between IoT infrastructure and entities possible through the sensor network. On the other hand, the perception technologies read the original data and transfer it to a machine-readable edition. However, both technologies rely on the IoT infrastructure.

The concept of the IoT dates to the 1990s. It can be defined as a global infrastructure interconnecting physical and virtual things through interoperable information and communication technologies (Alavi et al., 2018). With the development of digitization, the conversion of analog information makes it easier to store, access, share, and process the information (Tilson et al., 2010), and the confluence of cost reduction and processing speed increase on sensors implies that the world has moved into the era of the IoT. The sensor is the key part of the IoT, which measures some property of the environment and translates the input into an electrical signal. To realize the system-level representation of a complex asset, where the physical entity has a dynamically coupled representation in the digital world, it is necessary to improve real-time sensors technologies. With the advancements in communications technologies, the vision of a vast array of interconnected sensors on machines, people, and products, coupled with intelligent controllers, becomes affordable (Olsen & Tomlin, 2019). Gubbi et al. (2013) present a cloud-centric vision for the worldwide implementation of the IoT and propose that the IoT has stepped out of its infancy and is the next revolutionary technology in transforming the Internet into a fully integrated future Internet. Advances in IoT technology monitoring techniques have led to the invention of various sensors that can be effective indicators. It has brought about improvements in the quality of real-time data, and there are many examples of the applications of real-time simulation to electrical systems including grids, power electronics, and control systems (Das & Saha, 2018).

In recent years, IoT has made it possible to monitor and manage the urban services of bicycle pool sharing and public parking, as well as water consumption, air and noise pollution, and traffic. Bibri (2018) suggests that sensors can be classified according to the type of energy they detect as signals: location sensors (e.g., GPS, active badges), optical/vision sensors (e.g., photo-diode, color sensors, IR, and UV sensors), light sensors (e.g., photocells, photodiodes), image sensors (e.g., stereo-type camera, infrared), sound sensors (e.g. microphones), temperature sensors (e.g., thermometers), heat sensors (e.g., bolometer), electrical sensors (e.g., galvanometer), pressure sensors (e.g., barometer, pressure gauges), motion sensors (e.g., radar gun, speedometer, mercury switches, tachometer), orientation sensors (e.g., gyroscope), physical movement sensors (e.g., accelerometers), biosensors (e.g., pulse, galvanic skin response measure), vital sign processing devices (heart rate, temperature), wearable sensors (e.g., accelerometers, gyroscopes, magnetometers), and identification and traceability sensors (e.g., RFID, Near-Field Communication or NFC). They generate business opportunities for companies or entrepreneurs that develop services for citizens or public authorities through a common platform (Grimaldi & Fernandez, 2019). The emergence of the digital twin offers a glimpse into the potential of IoT-enabled operations. The IoT technology in the digital twin is not simply astatic digital representation of the physical object, but a dynamic digital representation of the object in use. It is important to create a digital world in the virtual space that maps and interacts with the physical world. Therefore, establishing a global full-time IoT perception system with multi-dimensional and multi-level accuracy is a critical foundation for the DTC. By assigning unique digital identities to various urban entities, the IoT perception infrastructure would be able to match the one-to-one twinning world easily. The DTC needs a large-scale deployment of multifunctional information facilities and intelligent gateways that support various distance communication protocol standards. With IoT, current authentication and access control solutions may induce a very high load on the core network and cause network failures (Behrad et al., 2020). They claim that 5G is expected to support a set of many requirements and use cases, such as handling connectivity for a massive number of IoT devices. Subsequently, the 5G-based IoT facilities aggregate and process the information collected by the sensors and upload it to the ubiquitous sensing and intelligent facility management platform.

2. Inevitable Trend of Digital Transformation for Smart Cities

Digital transformation is an inevitable choice for urban governance. Each major technological breakthrough has redefined the world landscape. Cities have become increasing smarter in the last two decades, using pervasive ICT to monitor activities in the city (Neirotti et al., 2014). Data can then be generated from a wide variety of activities in the city, such as traffic and transportation (Menouar et al., 2017), power generation, utilities provisioning, water supply (Parra et al., 2015) and waste management (Medvedev et al., 2015).

The increased data available from smart cities, artificial intelligence, data analytics and machine learning allows for the creation of a digital twin that can update and change as the physical equivalents change (Kaur et al., 2020).

3—Smart cities

Digital twin

2—Buildings

infrastructure

An ideal digital twin should be identical to its physical object and have a complete, real-time dataset of all information about the object/system. As the complexity of the object/system increases, the digital twin may only be identical in the relevant domain and only have the real-time data needed to support any desired simulation. The accuracy and usefulness of a digital twin depends on its level of detail and the comprehensiveness of the data available. Undoubtedly, it is inevitable trend of digital transformation for smart cities through using the advanced technology of digital twin.

Figure 3 showcases the Manufacturing Process Digital Twin Model, which has experienced an impressive CAGR[1] of 230% from 2018 to 2021. The model provides a comprehensive view of the manufacturing process, allowing for real-time monitoring and optimization.

Figure 3. Manufacturing process digital twin model
Source: Deloitte University Press.

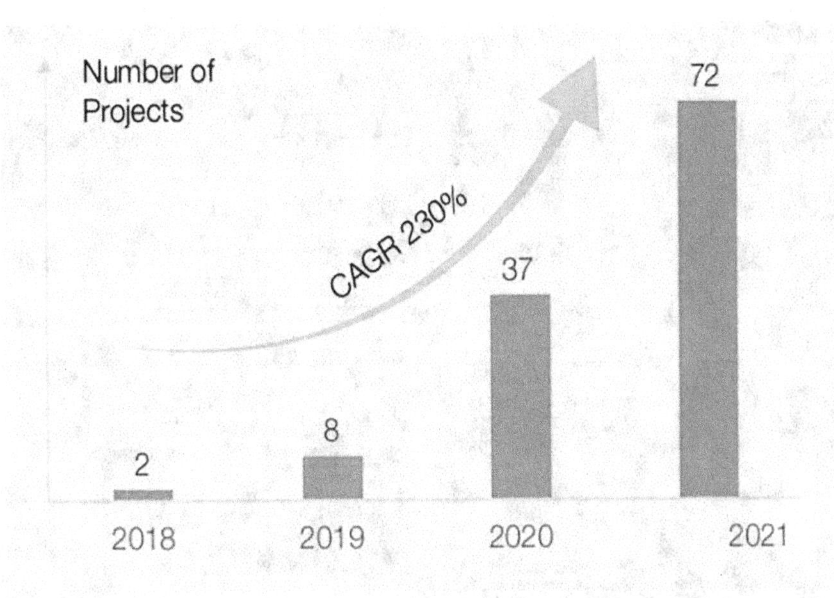

RESEARCH METHODS

1. Data Visualization With CiteSpace

CiteSpace software, which was jointly developed by Chen from the School of Information Science and Technology at Drexel University and the WISE Laboratory at Dalian University of Technology, is an information visualization software that measures and analyzes research data and then illustrates this analysis using knowledge maps. Before the analysis project was established in CiteSpace, the CNKI data were transformed; however, the data exported from the WoS core database did not need to be transformed. The CiteSpace information visualization software then developed a keyword network visualization map that indicated the time distribution zone, the high-frequency keywords, the emerging words in the special soil subgrade field, the most popular topics, and the research turning points, thereby revealing the evolution, the research frontier, and the development trends. Then, quantitative analyses of the countries/regions, organizations, source journals, and keyword co-occurrences were conducted, from which a comprehensive report on special soil subgrade was developed.

This chapter first analyzed the timeline of visualization with keyword of "digital twin cities" (Figure 4) and found that the research on digital twin is mainly focused between 2020 and 2021, which shows that digital twin is the hot spot of social concern and the direction of future urban development. We can also see application of big data and deep learning in the red timeline, together with digital twin in the green timeline and IoT in the purple timeline.

Then, we analyzed the keywords (Figure 5 and Figure 6) and could visually see the fact that the digital twin is the inevitable goal of digital transformation.

The analysis of Figures 5 and 6 shows that smart manufacturing is imperative in the process of digital transformation for smart cities. In the manufacturing industry, researchers have proposed various

Figure 4. Timeline visualization with keyword of "digital twin cities"
Source: Web of Science.

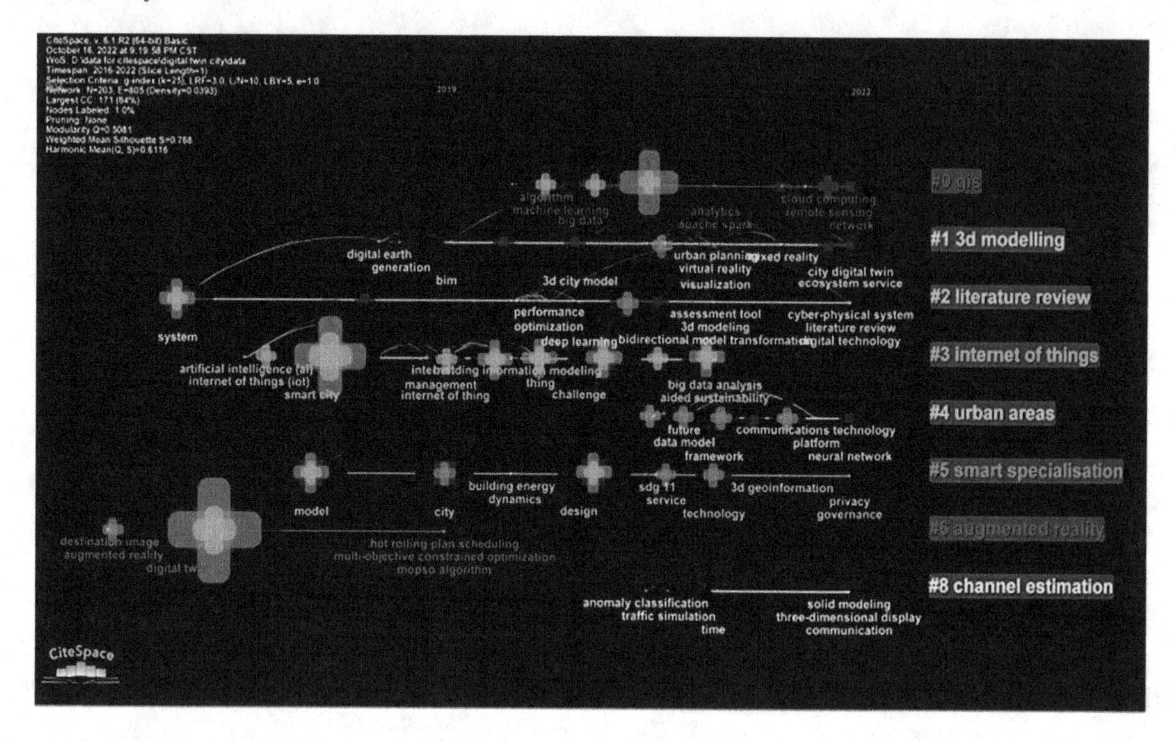

Figure 5. Cluster visualization with keywords of "digital twin city" and "smart city"
Source: Web of Science.

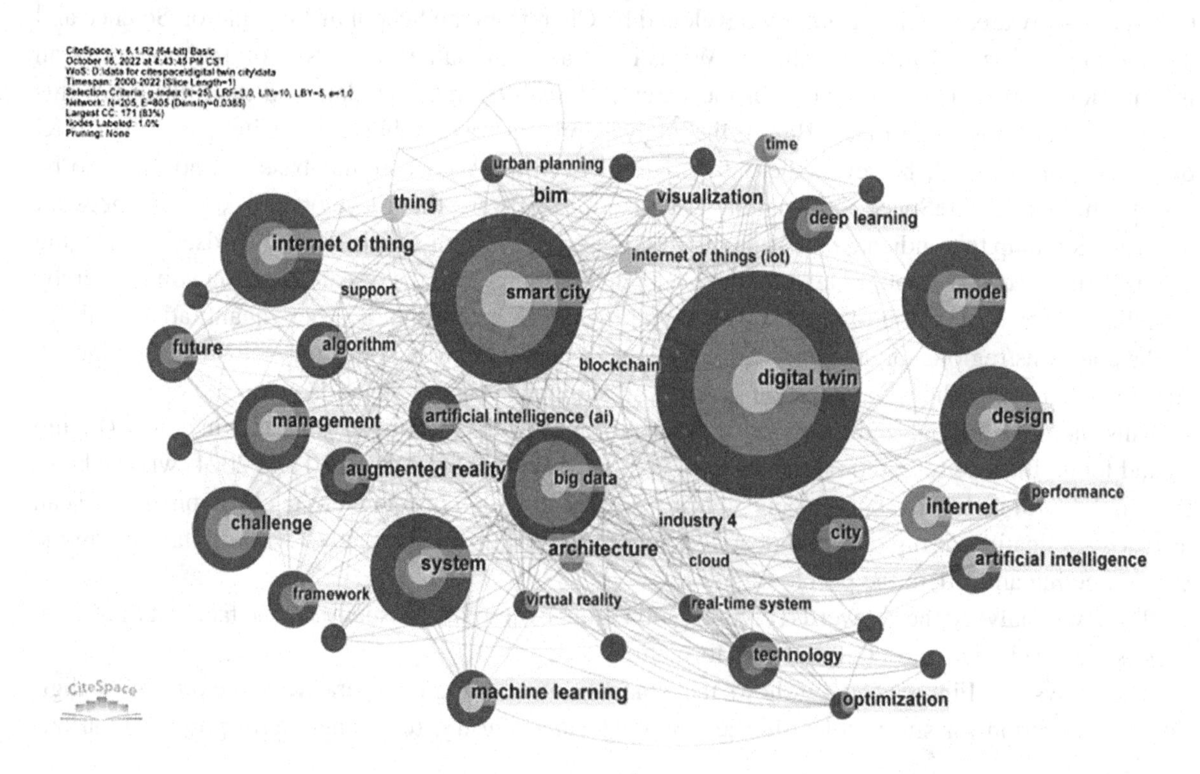

Figure 6. Cluster visualization with keyword of "digital twins city"
Source: Web of Science.

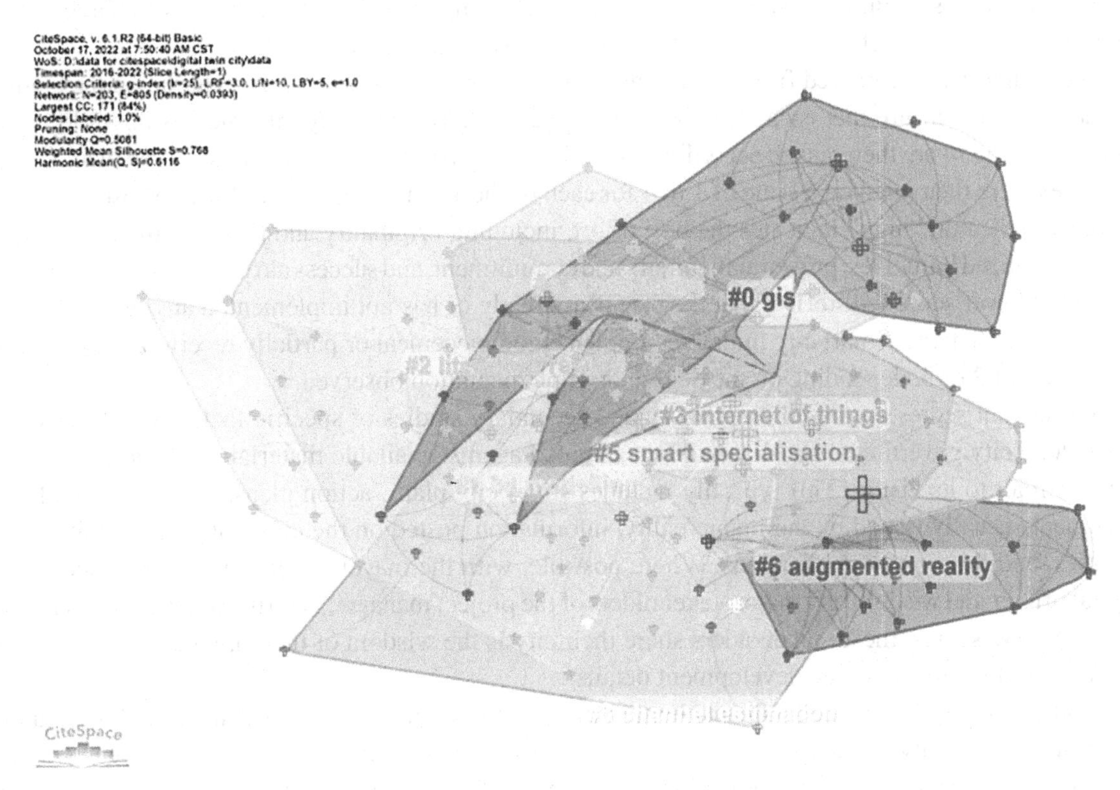

frameworks of digital twin-based shop-floor smart production management methods or approaches. Digital twins have been proven to be a practical method for integrating the physical world and the virtual world of manufacturing. Tao et al. (2019) present a new method for product design based on digital twin technology.

Zheng et al. (2019) studied a digital twin case of a welding production line and described the details of the implementation scheme, application process, and the effects of this case. The digital twin becomes the basis for simulation-driven assist systems as well as control and service decisions in combination (Bibri, 2018).

2. Case Studies in China

The Singapore-based Eden Strategy Institute released its second annual 2021 Top 50 Smart City Governments report on March 30, 2021. This year's report ranks 235 cities around the world. Nine (9) Chinese cities are on the list -- Shanghai, Beijing, Chengdu, Taipei, Shenzhen, Hangzhou, Guangzhou, Hong Kong, and Chongqing. Four (4) cities -- Chengdu, Guangzhou, Hangzhou, and Chongqing are on the top 50 smart city governments list for the first time. It marks the acceleration of urbanization, technological development, and smart city development in mainland China (Dong et al., 2022).

The top 50 Smart City Governments ranking requires a robust approach to ensure that as many potential smart cities around the world are considered and comprehensively evaluated as possible, with a clear focus on the role of city governments in driving smart city development.

The primary data gathered from the solicitation of proposals helped to complete the understanding of each city and ensure that each city government was as fairly and fully informed as possible in this study. The cities were then ranked according to the following 10 key metrics in Table 3 (Pan et al., 2022):

Cities were then rated on a scale of 1 to 4 for each of the 10 main indicators, from lowest to highest. For each indicator, "high" indicates the best effort, including originality and resourcefulness, multiple institutionalized initiatives, proven authenticity and commitment, and success attributable to that indicator. Instead, a lower score will reflect that the city is not ready or has not implemented any relevant plans, policies or initiatives. Smart city initiatives implemented piecemeal or partially receive an "intermediate" rating of 2 or 3, depending on the degree of implementation observed.

Subsequent scores are based on primary and secondary studies of specific factors in that city. For example, a city government's vision will be evaluated against available materials and sources that explain or refer to its vision. This typically includes smart city plans, action plans, strategy or road map documents usually issued by the municipality, information posted on the city's official website, press releases, and third-party publications. Where possible, with the mayor, chief innovation officer, chief digital officer and wisdom city major stakeholders of the project manager, in virtual interviews, to further validate these scores, these stakeholders share their city is the wisdom of the implementation of urban planning and its unique urban development details.

From Figure 7, the most significant changes in this year's Smart City Government rankings can be attributed to each city's management of COVID, as well as their resilience, adaptability, foresight, and initiative in anticipating and responding to major urban challenges. The Smart City Government ranking

Table 3. Ten indicators in evaluating smart cities

Number	指Indicator标	Interpretation of Indicators解释
1	VISION	A clear, well-defined, and updated strategy to develop a "smart city"
2	LEADERSHIP	Dedicated City leadership that steers smart city projects without major discontinuities
3	BUDGET	Sufficient and sustainable funding for smart city projects
4	FINANCIAL INCENTIVES	Financial incentives to effectively encourage private sector participation (e.g. grants, rebates, subsidies, competitions) specifically in smart city projects and initiatives
5	SUPPORT PROGRAMMES	In-kind programmers to encourage private actors to participate (e.g. incubators, events, networks, awards) specifically in smart city projects and initiatives
6	TALENT READINESS	Programmers to equip the city's talent with smart skills
7	PEOPLE CENTRICITY	A sincere, people-first design of the future city, with no systematic disregard for human life or basic human rights
8	INNOVATION ECOSYSTEMS	A comprehensive range of engaged stakeholders to sustain innovation and partnerships where the city government was instrumental in sustaining and catalyzing them
9	SMART POLICIES	A conducive policy and regulatory environment for smart city development (e.g. data governance, IP protection, urban design)
10	TRACK RECORD	The government' s experience in catalyzing successful smart city initiatives, with no systematic or widespread failure

Source:Singapore-based Eden Strategy Institute

Figure 7. Top 50 smart city governments
Source: Singapore-based Eden Strategy Institute (2021).

also reflects the extent to which city governments can manage and best support the needs and aspirations of their citizens, amid civil unrest, political unrest, and growing concerns about climate change mitigation throughout 2020.

Competition for the top 50 has intensified as city governments continue to develop their ability to promote smart city development. Eighteen of these new cities made it into the Smart City Government top 50 list, and have shown significant success in building urban resilience, promoting environmental sustainability, and leveraging multiple stakeholders to fund and support local initiatives. Many Chinese cities, including Hangzhou, Guangzhou, Chengdu, and Chongqing, have made significant efforts to boost innovation and support talent development respectively, the report said. Many European cities, such as Frankfurt, Zurich, Oslo, and Rotterdam, have also made significant progress in reducing carbon emissions and achieving greater environmental sustainability through interventions in transport, civic education, and energy use.

As Figure 8 shows, the most famous DTCs in China are Shanghai and Beijing. In Beijing, Xiongan New Area is the best DTC construction. In Shanghai, Lin-Gang Special Area is the best DTC construction. The following section mainly discusses these two areas.

Xiongan New Area: The World's First Digital Twin City

The concept of DTC was first proposed in the "Planning Outline of Xiong'an New Area in Hebei", which "insists on simultaneous planning and construction of the digital city and the real city", creating

Figure 8. Smart city government score of Shanghai (left) and Beijing (right)
Source: Singapore-based Eden Strategy Institute (2021).

Smart City Government Score		
VS	VISION	3.0
LS	LEADERSHIP	3.0
BG	BUDGET	2.1
FI	FINANCIAL INCENTIVES	3.1
SP	SUPPORT PROGRAMMES	4.0
TR	TALENT-READINESS	3.0
IE	INNOVATION ECOSYSTEM	4.0
SM	SMART POLICIES	3.0
PC	PEOPLE-CENTRICITY	2.1
TD	TRACK RECORD	4.0

Smart City Government Score		
VS	VISION	3.0
LS	LEADERSHIP	3.0
BG	BUDGET	3.0
FI	FINANCIAL INCENTIVES	3.1
SP	SUPPORT PROGRAMMES	1.1
TR	TALENT-READINESS	3.1
IE	INNOVATION ECOSYSTEM	4.0
SM	SMART POLICIES	2.9
PC	PEOPLE-CENTRICITY	2.1
TD	TRACK RECORD	4.0

the concept of "digital twin city", and the BIM management platform for planning and construction of Xiong'an New Area is the concrete implementation of this requirement (Qiaoling et al., 2021).

The construction of the BIM platform for the planning and construction of the Xiongan New Area is the concrete implementation of this requirement. The technical approach of City Information Modeling (CIM) is in the exploration stage, and one of the ideas is that CIM itself is formed by the aggregation of various BIM, and CIM and BIM will be integrated and connected continuously. From the concept of DTC to facilitate construction of CIM platform is also a major application exploration of the Ministry of Housing and Urban-Rural Development to promote the construction of CIM platform (Qiaoling et al., 2021).

The technical approach of CIM is currently in the exploration stage, and one of the views is that CIM itself consists of a variety of BIM aggregation, and CIM and BIM will be continuously integrated and connected. From the concept of DTC. The construction of CIM platform is also a major application exploration of the Ministry of Housing and Urban-Rural Development to promote the construction of CIM platform. It is also a major application exploration direction of the Ministry of Housing and Urban-Rural Development to promote the construction of CIM platform.

The Xiongan New Area's digital city is planned and built simultaneously with the physical city. It is a digital intelligent city where the virtual and the real interact and the twin grow together.

Shanghai Lin-Gang New Area: A Demonstration of Digital Transformation

Lin-Gang Special Area lay emphasis on applications of digital twin in keeping cities safe. With rapid development of modern industrialization, Lin-Gang New Area has formed a situation where there are many tower cranes and dense buildings. Zhang Danli, deputy general manager of Liuzi Technology (A famous technology company specializing in digital twinning), said, "For these construction projects, safety production is the lifeline, and the digital twin intelligent construction platform we developed targets possible risks and hidden dangers in the construction process from the very beginning and prescribes the right medicine one by one, allowing the digital twin to escort safety."

The digital twin is characterized by real-time. Through more than two hundred locators and cameras installed in the building, all kinds of data will be transmitted to the server in time through the 5G network, allowing elements, people, and supervision to be visualized through the digital twin. Figure 6 illustrates building data visualization through digital twin in Shanghai Lin-Gang Special Area (Wang & Deng, 2022).

These seemingly simple devices are actually given an important mission: the locator installed on the side of the guardrail will alert through the loudspeaker when workers are close and notify the project leader; AI intelligent camera, which can automatically identify unsafe behaviors and hidden dangers on site and send instructions to the site data center; workers' helmets are equipped with special positioning chips that can transmit workers' positions in real time; sensors are also designed and installed at various locations on the tower crane to transmit motion trajectories to the server in real time, realizing the whole life cycle management of equipment installation, acceptance, operation and maintenance, operational dynamics, and maintenance.

Figure 9 showcases the Digital Twin Smart Construction- Building Database, which provides a detailed digital replica of buildings, allowing for real-time monitoring and optimization of construction processes.

Lingang New Zone of Shanghai Free Trade Zone (FTZ) has issued the key points of urban Digital Development in Lingang New Zone in 2022 (Key points). The Key points emphasized that the construction of smart transportation infrastructure will be accelerated, digital infrastructure support such as smart security will be firmly established, and digital enabling platforms will be laid out for digital twin cities. It is mentioned in the Key points that a DTC basic platform should be built in Lingang New Area, a DTC data resource pool should be built, and the space-time base of the international innovation synergy zone should be built. Provide digital element expression, dynamic 3D presentation, intelligent decision support, simulation and simulation of digital twin enabling.

To further improve the work promotion mechanism, the Key points emphasized that Lingang new area "digital twin demonstration area" three-year action plan (2022-2024) will be released, focusing on the international innovation synergy area, building a future DTC of integration and growth, striving to become the "international digital capital" demonstration area. At the same time, the construction of digital twin ecosphere. The "International Digital Twin Ecological Conference" was held, and "Lingang Digital Twin City" Vision Plan and "Lingang Digital Twin City" Ecological Partnership Cooperation Plan were released to promote the environmental construction of "Lingang Digital Twin City" ecologi-

Figure 9. Digital twin smart construction-building database
Source: Chinese Academy of Digital Twin (2020).

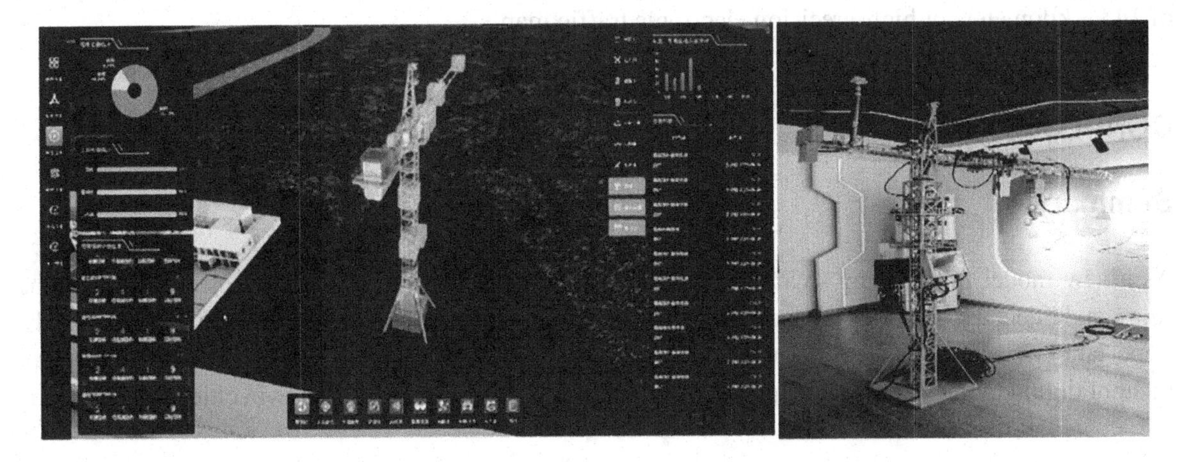

cal partnership. Lingang New Area will promote the construction of digital twin demonstration zones, application innovation, and platform opening.

First, Lin-Gang Special Area will speed up the construction of smart transportation infrastructure, strengthen digital infrastructure support such as smart security, and lay out digital enabling platforms for digital twin cities (Wang & Deng, 2022). The suppose are as follows (Lingang New Area Management Committee, 2020):

1. Speed up data center construction (Lingang New Area Management Committee, 2020).

Accelerate the construction of large-scale high-level cloud data centers such as China Telecom public cloud computing Power Center and Information Flying Fish Public Service Center and promote the acceptance and operation of the second phase project of China Mobile IDC Base in Lingang. Lingang New Area will promote the operation of new Internet exchange centers, improve the regulatory system, and network architecture, and increase the scale of access traffic. By the end of this year, Lingang New Area will have more than 50 enterprises connected to the Internet, and the peak exchange traffic will exceed 100 GIGA Bytes.

2. Improve 5G outdoor signal coverage (Lingang New Area Management Committee, 2020).

Over the course of the year, an additional 500 5G base stations were built, and fixed-mobile integrated communications facilities were built on a trial basis. Core urban areas will be further covered by 5G, while non-dense urban areas will basically be covered continuously.

3. Construct Road intelligent traffic perception facilities (Lingang New Area Management Committee, 2020).

Based on the standard requirements of Shanghai Road Traffic Management Information System, the traffic perception equipment of main roads and key areas in the central city of Lingang New Area will be built, the networking capacity of traffic signal machines will be optimized, the "base number" of real-time traffic trips will be grasped, and the hardware foundation will be laid for the construction of intelligent traffic integrated management platform. It plans to build 419 sets of video composite detectors at 97 points, upgrade and connect 83 single point signal machines, and build 40 roads with a total of 143.8 kilometers of high-precision electronic traffic map.

CONCLUSION

Summary

As a new model for building smart cities, the DTC will reshape the city governance structures and rules, and inject a continuous momentum for the development and transformation of cities. Many of the important cities around the world have launched plans to build digital twin cities. The rapid development of digital twin technologies has also made it possible to construct digital twin cities. By analyzing the existing digital twin technologies, based on the development of smart cities, this paper proposes a pattern of digital twins.

A self-perceiving, self-determining, self-organizing, self-executing, and adaptive platform for urban operation and maintenance is constructed through surveying and mapping technology, IoT perception, collaborative computing, simulation, and deep learning. The development of DTCs will create a new management pattern that can trace past events and explore frontier directions, which will be a trend for future research on DTCs.

As a complex system in which the physical world of a city and the virtual space of the network correspond, map, and interact with each other, DTC plays an important role in promoting the development, construction, and governance of smart cities and can bring important value to the development of smart cities in the future.

Digital twin cities can innovate the way cities are governed. As the abundance of urban functions and increasing urban governance problems, digital twin cities have brought the innovation management methods, such as digital planning using the simulation technology for urban planners to present the characteristics of different solution, and make all aspects of cities through the application of data technology and through digital twin present become more dynamic.

DTCs could improve residents' lives. With population expansion, traffic congestion, housing difficulties, resource shortage and other "urban diseases" increasingly serious, sustainable urban development is facing a major challenge. DTC can provide real-time response, virtual reality interaction, travel navigation services which will make citizens to have more convenience in terms traveling, telemedicine, and digital education which will also make public services more accessible to more people and make cities more livable.

DTCs can help urban sustainability. Sustainable development is the theme of urban development. Sustainable development can be realized only when the resources relied on by cities can be maintained continuously, without compromising future generations, and if cities veer away from environmental disasters and energy shortage. Through intelligent analysis and spatial deduction, digital twin can optimize resource allocation, adjust the distribution of population and industry, improve the carrying capacity of the environment, and promote the development of clean energy using a new distributed integrated energy system, enabling cities to achieve the goal of "double carbon" more quickly.

Theoretical Implications

With the global focus on the digital twin technology, industrial ecology is expected to explode. With the continuous introduction of strategies related to Society 5.0, digital twin technology has received widespread attention from all parties. Garter has listed digital twin as one of the top ten (10) strategic technology trends for the year for four (4) consecutive years. According to IDC statistics and forecasts, the global digital twin market reached 5.22 billion USD in 2020 and is expected to reach 21.28 billion USD by 2024, with a compounded annual growth rate of 42.09 period during the period. The future Asia-Pacific market size growth rate will exceed that of other regions such as North America and Europe, becoming the high ground of the global digital twin emerging market. So, it is very theoretically valuable to study these two regions in China, which are models of digital twin cities, to explore the role of digital twin in accelerating digital transformation of smart cities.

ACKNOWLEDGMENT

The authors extend sincere gratitude to:

- Our colleagues from Soochow University, The Australian Studies Centre of Shanghai University, The European Business University of Luxembourg and Krirk University as well as the independent research colleagues who provided insight and expertise that greatly assisted the research, although they may not agree with all of the interpretations/conclusions of this chapter.
- China Knowledge and Belt & Road Blockchain Association for supporting our research.
- The Editor and the International Editorial Advisory Board (IEAB) of this book who initially desk reviewed, arranged a rigorous double/triple blind review process and conducted a thorough, minute and critical final review before accepting the chapter for publication.
- All anonymous reviewers who provided very constructive feedbacks for thorough revision, improvement and fine tuning of the chapter.

REFERENCES

Alavi, A. H., Jiao, P., Buttlar, W. G., & Lajnef, N. (2018). Internet of Things-enabled smart cities: State-of-the-art and future trends. *Measurement, 129*, 589–606. doi:10.1016/j.measurement.2018.07.067

Albino, V., Berardi, U., & Dangelico, R. M. (2015). Smart Cities: Definitions, Dimensions, Performance, and Initiatives. *Journal of Urban Technology*, *22*(1), 3–21. doi:10.1080/10630732.2014.942092

Behrad, S., Bertin, E., Tuffin, S., & Crespi, N. (2020). A new scalable authentication and access control mechanism for 5G-based IoT. *Future Generation Computer Systems*, *108*, 46–61. doi:10.1016/j.future.2020.02.014

Bibri, S. E. (2018). The IoT for smart sustainable cities of the future: An analytical framework for sensor-based big data applications for environmental sustainability. *Sustainable Cities and Society*, *38*, 230–253. doi:10.1016/j.scs.2017.12.034

Bifulco, F., Tregua, M., Amitrano, C. C., & D'Auria, A. (2016). ICT and sustainability in smart cities management. *International Journal of Public Sector Management*, *29*(2), 132–147. doi:10.1108/IJPSM-07-2015-0132

Caragliu, A., & Del Bo, C. F. (2019). Smart innovative cities: The impact of Smart City policies on urban innovation. *Technological Forecasting and Social Change*, *142*, 373–383. doi:10.1016/j.techfore.2018.07.022

Chen, K., Lu, W., Xue, F., Tang, P., & Li, L. H. (2018). Automatic building information model reconstruction in high-density urban areas: Augmenting multi-source data with architectural knowledge. *Automation in Construction*, *93*, 22–34. doi:10.1016/j.autcon.2018.05.009

China Academy of Information and Communication Research. (2018). China Academy of Information and Communication Research.

Das, S., & Saha, P. (2018). A review of some advanced sensors used for health diagnosis of civil engineering structures. *Measurement*, *129*, 68–90. doi:10.1016/j.measurement.2018.07.008

Dong, F., Li, Y., Li, K., Zhu, J., & Zheng, L. (2022). Can smart city construction improve urban ecological total factor energy efficiency in China? Fresh evidence from generalized synthetic control method. *Energy, 241*, 122909. doi:10.1016/j.energy.2021.122909

El Meouche, R., Hijazi, I., Poncet, P. A., Abunemeh, M., & Rezoug, M. (2016). UAV photogrammetry implementation to enhance land surveying, comparisons and possibilities. *The International Archives of the Photogrammetry, Remote Sensing and Spatial Information Sciences, 42*(W2), 107–114. doi:10.5194/isprs-archives-XLII-2-W2-107-2016

Gao, Z., Song, Y., Li, C., Zeng, F., & Wang, F. (2017). Research on the application of rapid surveying and mapping for large scare topographic map by UAV aerial photography system. *The International Archives of the Photogrammetry, Remote Sensing and Spatial Information Sciences, 42*(W6), 121–125. doi:10.5194/isprs-archives-XLII-2-W6-121-2017

Glaessgen, E., & Stargel, D. (2012). *The digital twin paradigm for future NASA and U.S. air force vehicles*. Academic Press.

Grieves, M. (2005). *Product Lifecycle Management: Driving the Next Generation of Lean Thinking*. Academic Press.

Grieves, M., & Vickers, J. (2017). Digital Twin: Mitigating Unpredictable, Undesirable Emergent Behavior in Complex Systems. In F.-J. Kahlen, S. Flumerfelt, & A. Alves (Eds.), *Transdisciplinary Perspectives on Complex Systems: New Findings and Approaches* (pp. 85–113). Springer International Publishing. doi:10.1007/978-3-319-38756-7_4

Grimaldi, D., & Fernandez, V. (2019). Performance of an internet of things project in the public sector: The case of Nice smart city. *The Journal of High Technology Management Research, 30*(1), 27–39. doi:10.1016/j.hitech.2018.12.003

Gubbi, J., Buyya, R., Marusic, S., & Palaniswami, M. (2013). Internet of Things (IoT): A vision, architectural elements, and future directions. *Future Generation Computer Systems, 29*(7), 1645–1660. doi:10.1016/j.future.2013.01.010

Kaur, M. J., Mishra, V. P., & Maheshwari, P. (2020). The Convergence of Digital Twin, IoT, and Machine Learning: Transforming Data into Action. In M. Farsi, A. Daneshkhah, A. Hosseinian-Far, & H. Jahankhani (Eds.), *Digital Twin Technologies and Smart Cities* (pp. 3–17). Springer International Publishing. doi:10.1007/978-3-030-18732-3_1

Li, L., Cao, X., He, Q., Sun, J., Jia, B., & Dong, X. (2019). A new 3D laser-scanning and GPS combined measurement system. *Comptes Rendus Geoscience, 351*(7), 508–516. doi:10.1016/j.crte.2019.09.004

Ma, Z., & Liu, S. (2018). A review of 3D reconstruction techniques in civil engineering and their applications. *Advanced Engineering Informatics, 37*, 163–174. doi:10.1016/j.aei.2018.05.005

Medvedev, A., Fedchenkov, P., Zaslavsky, A., Anagnostopoulos, T., & Khoruzhnikov, S. (2015). *Waste Management as an IoT-Enabled Service in Smart Cities*. Paper presented at the Internet of Things, Smart Spaces, and Next Generation Networks and Systems, Cham, Switzerland.

Menouar, H., Guvenc, I., Akkaya, K., Uluagac, A. S., Kadri, A., & Tuncer, A. (2017). UAV-Enabled Intelligent Transportation Systems for the Smart City: Applications and Challenges. *IEEE Communications Magazine*, *55*(3), 22–28. doi:10.1109/MCOM.2017.1600238CM

Mergel, I., Edelmann, N., & Haug, N. (2019). Defining digital transformation: Results from expert interviews. *Government Information Quarterly*, *36*(4), 101385. doi:10.1016/j.giq.2019.06.002

Mohammadi, N., & Taylor, J. E. (2017). *Smart city digital twins.* Paper presented at the 2017 IEEE Symposium Series on Computational Intelligence (SSCI).

Neirotti, P., De Marco, A., Cagliano, A. C., Mangano, G., & Scorrano, F. (2014). Current trends in Smart City initiatives: Some stylised facts. *Cities (London, England)*, *38*, 25–36. doi:10.1016/j.cities.2013.12.010

Olsen, T. L., & Tomlin, B. (2019). Industry 4.0: Opportunities and Challenges for Operations Management. *Manufacturing & Service Operations Management*, *22*(1), 113–122. doi:10.1287/msom.2019.0796

Pan, T., Fan, Y., Shao, L., Chen, B., Chu, Y., He, G., Pan, Y., Wang, W., & Wu, Z. (2022). Multiple accounting and driving factors of water resources use: A case study of Shanghai. *Journal of Environmental Management*, *313*, 114929. doi:10.1016/j.jenvman.2022.114929 PMID:35421695

Parra, L., Sendra, S., Lloret, J., & Bosch, I. (2015). Development of a Conductivity Sensor for Monitoring Groundwater Resources to Optimize Water Management in Smart City Environments. *Sensors (Basel)*, *15*(9), 20990–21015. doi:10.3390150920990 PMID:26343653

Qiaoling, B., Tao, Y., Qiqing, H., & Ying, L. (2021). *Construction of a rule system for intelligent planning and construction management under the orientation of digital twin city: An example of BIM management platform for planning and construction of Xiongan New Area*. Academic Press.

Tao, F., Sui, F., Liu, A., Qi, Q., Zhang, M., Song, B., Guo, Z., Lu, S. C.-Y., & Nee, A. Y. C. (2019). Digital twin-driven product design framework. *International Journal of Production Research*, *57*(12), 3935–3953. doi:10.1080/00207543.2018.1443229

Tilson, D., Lyytinen, K., & Sørensen, C. (2010). Research Commentary—Digital Infrastructures: The Missing IS Research Agenda. *Information Systems Research*, *21*(4), 748–759. doi:10.1287/isre.1100.0318

Wang, J., & Deng, K. (2022). Impact and mechanism analysis of smart city policy on urban innovation: Evidence from China. *Economic Analysis and Policy*, *73*, 574–587. doi:10.1016/j.eap.2021.12.006

Weir-McCall, D. (2020). *51 World Created a Digital Twin of Entire Shanghai*. Retrieved from https://www.unrealengine.com/en-US/spotlights/51world-creates-digital-twin-of-the-entire-city-of-shanghai

World Economic Forum. (2022). *Digital Twin Cities: Framework and Global Practice*. Author.

Xue, F., Chen, K., Lu, W., Niu, Y., & Huang, G. Q. (2018). Linking radio-frequency identification to Building Information Modeling: Status quo, development trajectory and guidelines for practitioners. *Automation in Construction*, *93*, 241–251. doi:10.1016/j.autcon.2018.05.023

Yao, H., Qin, R., & Chen, X. (2019). Unmanned Aerial Vehicle for Remote Sensing Applications. *RE:view*, *11*(12), 1443.

Zheng, C., Yuan, J., Zhu, L., Zhang, Y., & Shao, Q. (2020). From digital to sustainable: A scientometric review of smart city literature between 1990 and 2019. *Journal of Cleaner Production, 258*, 120689. doi:10.1016/j.jclepro.2020.120689

Zheng, Y., Yang, S., & Cheng, H. (2019). An application framework of digital twin and its case study. *Journal of Ambient Intelligence and Humanized Computing, 10*(3), 1141–1153. doi:10.100712652-018-0911-3

Zhou, Y., Wang, L., Love, P. E. D., Ding, L., & Zhou, C. (2019). Three-dimensional (3D) reconstruction of structures and landscapes: A new point-and-line fusion method. *Advanced Engineering Informatics, 42*, 100961. doi:10.1016/j.aei.2019.100961

ADDITIONAL READING

Yu, P., Chang, X., & Mandizvidza, K. (2023). Development of New Energy Vehicles in Entrepreneurial Ecosystem Under the Carbon Neutrality Policy in China. In B. Marco-Lajara, J. Martínez-Falcó, & L. Millán-Tudela (Eds.), *Corporate Sustainability as a Tool for Improving Economic, Social, and Environmental Performance* (pp. 55–84). IGI Global. doi:10.4018/978-1-6684-7422-8.ch004

Yu, P., Chen, D., & Ahuja, A. (2022). Smart and Sustainable Economy: How COVID-19 Has Acted as a Catalyst for China's Digital Transformation. In S. Kautish & G. Kaur (Eds.), *AI-Enabled Agile Internet of Things for Sustainable FinTech Ecosystems* (pp. 106–146). IGI Global. doi:10.4018/978-1-6684-4176-3.ch006

Yu, P., Chen, J., Sampat, M., & Misuko, N. (2022). The Digital Transformation of Rural Agricultural Business Management: A Case Study of China. In S. Bilgaiyan, J. Singh, & H. Das (Eds.), *Empirical Research for Futuristic E-Commerce Systems: Foundations and Applications* (pp. 23–52). IGI Global. doi:10.4018/978-1-6684-4969-1.ch002

Yu, P., Ge, Y., Mandizvidza, K., & Mulli, J. (2023). How Can Small and Medium Enterprises in the Chinese Market Achieve Sustainable Development Goals Through Blockchain? In D. Taleb, M. Abdelli, A. Khalil, & A. Sghaier (Eds.), *Examining the Vital Financial Role of SMEs in Achieving the Sustainable Development Goals* (pp. 52–85). IGI Global. doi:10.4018/978-1-6684-4834-2.ch004

Yu, P., Gong, R., & Sampat, M. (2022). Blockchain Technology in China's Digital Economy: Balancing Regulation and Innovation. In P. Tehrani (Ed.), *Regulatory Aspects of Artificial Intelligence on Blockchain* (pp. 132–157). IGI Global. doi:10.4018/978-1-7998-7927-5.ch007

Yu, P., Gu, H., Zhao, Y., & Ahuja, A. (2022). Digital Transformation Driven by Internet Data Center: Case Studies on China. In D. Piaggesi, H. Landazuri, & B. Jia (Eds.), *Cases on Applying Knowledge Economy Principles for Economic Growth in Developing Nations* (pp. 203–230). IGI Global. doi:10.4018/978-1-7998-8417-0.ch011

Yu, P., Jiao, A., & Sampat, M. (2022). The Effect of Chinese Green Transformation on Competitiveness and the Environment. In P. Ordóñez de Pablos, X. Zhang, & M. Almunawar (Eds.), *Handbook of Research on Green, Circular, and Digital Economies as Tools for Recovery and Sustainability* (pp. 257–279). IGI Global. doi:10.4018/978-1-7998-9664-7.ch014

Yu, P., Liu, X., Mahendran, R., & Lu, S. (2022). Analysis and Comparison of Business Models of Leading Enterprises in the Chinese Hydrogen Energy Industry. In R. Felseghi, N. Cobîrzan, & M. Raboaca (Eds.), *Clean Technologies and Sustainable Development in Civil Engineering* (pp. 179–216). IGI Global. doi:10.4018/978-1-7998-9810-8.ch008

Yu, P., Liu, Z., & Hanes, E. (2022). Supply Chain Resiliency, Efficiency, and Visibility in the Post-Pandemic Era in China: Case Studies of MeiTuan Waimai, and Ele.me. In Y. Ramakrishna (Ed.), *Handbook of Research on Supply Chain Resiliency, Efficiency, and Visibility in the Post-Pandemic Era* (pp. 195–225). IGI Global. doi:10.4018/978-1-7998-9506-0.ch011

Yu, P., Liu, Z., Hanes, E., & Mumtaz, J. (2022). Integration of IoT and Blockchain for Smart and Secured Supply Chain Management: Case Studies of China. In S. Goyal, N. Pradeep, P. Shukla, M. Ghonge, & R. Ravi (Eds.), *Utilizing Blockchain Technologies in Manufacturing and Logistics Management* (pp. 179–207). IGI Global. doi:10.4018/978-1-7998-8697-6.ch010

Yu, P., Liu, Z., & Sampat, M. (2023). Enhancing the Resilience of Food Cold Chain Logistics Through Digital Transformation: A Case Study of China. In I. Masudin, M. Almunawar, D. Restuputri, & P. Sud-On (Eds.), *Handbook of Research on Promoting Logistics and Supply Chain Resilience Through Digital Transformation* (pp. 200–224). IGI Global. doi:10.4018/978-1-6684-5882-2.ch014

Yu, P., Lu, S., Hanes, E., & Chen, Y. (2022). The Role of Blockchain Technology in Harnessing the Sustainability of Chinese Digital Finance. In P. Swarnalatha & S. Prabu (Eds.), *Blockchain Technologies for Sustainable Development in Smart Cities* (pp. 155–186). IGI Global. doi:10.4018/978-1-7998-9274-8.ch009

Yu, P., Lu, S., Sampat, M., Li, R., & Ahuja, A. (2022). How AI-Enabled Agile Internet of Things Can Enhance the Business Efficiency of China's FinTech Ecosystem. In S. Kautish & G. Kaur (Eds.), *AI-Enabled Agile Internet of Things for Sustainable FinTech Ecosystems* (pp. 190–223). IGI Global. doi:10.4018/978-1-6684-4176-3.ch009

Yu, P., Shen, X., & Hanes, E. (2023). Promoting Responsible Research and Innovation in China's Hi-Tech Zones: Based on Case Studies of Zizhu Hi-Tech Zone, East Lake Hi-Tech Zone, and Guangzhou Hi-Tech Zone. In B. Marco-Lajara, J. Martínez-Falcó, & L. Millán-Tudela (Eds.), *Corporate Sustainability as a Tool for Improving Economic, Social, and Environmental Performance* (pp. 222–245). IGI Global. doi:10.4018/978-1-6684-7422-8.ch012

Yu, P., Weng, Y., & Ahuja, A. (2022). Carbon Financing and the Sustainable Development Mechanism: The Case of China. In A. Rafay (Ed.), *Handbook of Research on Energy and Environmental Finance 4.0* (pp. 301–332). IGI Global. doi:10.4018/978-1-7998-8210-7.ch012

Yu, P., Xu, S., Cheng, Z., & Sampat, M. (2023). Does the Development of New Energy Vehicles Promote Carbon Neutralization?: Case Studies in China. In A. Pego (Ed.), *Climate Change, World Consequences, and the Sustainable Development Goals for 2030* (pp. 109–131). IGI Global. doi:10.4018/978-1-6684-4829-8.ch006

Yu, P., Xue, W., & Mahendran, R. (2022). The Development and Impact of China's Digital Transformation in the Medical Industry. In M. Rodrigues & J. Proença (Eds.), *Impact of Digital Transformation on the Development of New Business Models and Consumer Experience* (pp. 97–128). IGI Global. doi:10.4018/978-1-7998-9179-6.ch006

Yu, P., Yu, M., & Sampat, M. (2022). Smart Management for Digital Transformation in China. In B. Barbosa, S. Filipe, & C. Santos (Eds.), Handbook of Research on Smart Management for Digital Transformation (pp. 411-438). IGI Global. doi:10.4018/978-1-7998-9008-9.ch019

Yu, P., Zhang, Y., Sampat, M., & Chen, Y. (2023). Research on Cross-Industry Digital Transformation Under the New Normal: A Case Study of China. In B. Marco-Lajara, J. Martínez-Falcó, & L. Millán-Tudela (Eds.), *Corporate Sustainability as a Tool for Improving Economic, Social, and Environmental Performance* (pp. 246–277). IGI Global. doi:10.4018/978-1-6684-7422-8.ch013

Yu, P., Zhao, Z., & Sampat, M. (2023). How Digital Twin Technology Promotes the Development of Smart Cities: Case Studies in China. In I. Vasiliu-Feltes (Ed.), *Impact of Digital Twins in Smart Cities Development* (pp. 198–227). IGI Global.

ENDNOTE

1 The compound annual growth rate (CAGR) is the annualized average revenue growth rate between two years, assuming that the growth is conducted at an exponential compound rate.

Chapter 9
Optical Networking Technologies for 5G Services

Baskaran S.
SASTRA University (Deemed), India

Srinivasan A.
https://orcid.org/0000-0003-1171-5573
SASTRA University (Deemed), India

Mardeni Bin Roslee
Multimedia University, Malaysia

ABSTRACT

The chapter provides information as a first step for individuals who are thriving to get a bird's eye view of the aspects underlying the optical networking in the context of 5G technology. Apart from capacity requirement challenges targeted by 5G coverage, it requires a lot of fibers to be successfully provisioned to achieve formidable performance goals of 5G such as diversified capacity requirements, availability, and coverage issues. The goals could be achieved by the underlying optical network with a greater number of interconnected fiber paths. In 5G, the requirements of reliable and ultra-low latency services required at the access side of a network shape up the research and evolution of underlying optical segments spanning from core to access part of the network. The reconfigurability and security issues of the present mode of optical communication need to be addressed, and the proposals given by the researchers are summed up. The chapter includes a general framework and theoretical concepts behind machine learning and software defined networking paradigms.

INTRODUCTION

The key challenges for the communication networks been thriving since few years have been to satisfy the burgeoning demands and enhance the capability for the network operations to integrate the state of the art technologies pertaining to different domains of wireless communications and highly promising optical communication technologies. Driven by mobility of devices, proliferation of IoT technology, the

DOI: 10.4018/978-1-7998-9266-3.ch009

rampant demand for communication resources has given rise to research areas that attempt to provide seamless interaction of radio access and optical communications. There has always been a requirement to cater to the needs of 5G services such as low latency to support real time communications coupled with energy efficiency and delivery of services that inherently involve traffic variability. With increasing dynamic traffic changes where frequent and complex reconfigurations of resources in networks spanning over multiple segments, the allocation of resources of varying capacities and highly dynamic utilization patterns are involved. To ensure satisfactory level of quality of service, the transport networks evolve with incorporation of automation for provision of connectivity and resource allocation.

Covering the aspects, the chapter presents a summary of optical networking technologies enabling 5G communications with focus on spatial multiplexing, propositions of optical distribution network in the front haul required by 5G services based on usage metrics and area of deployment, investigations regarding survivability and resilience issues pertaining to outage of fiber cables/ switching equipment in the back haul network. Excerpts on techniques deployed to ensure secure communication by sharing quantum keys to assist encryption, the enabling technologies to accomplish spectral multiplexing with high entropy by way of holographic techniques mimicking the Multiple Input Multiple Output(MIMO) are presented. The application of machine learning for management and control of optical communication network to ensure service guarantees invoking protection mechanisms have been covered. Software Defined Network aspects for control and management of dynamically varying connectivity, spectrum resources driven by the requirement of 5G services are also included.

SPACE DIVISION MULTIPLEXING IN OPTICAL NETWORKS

A highly flexible, adaptive high-capacity front haul with desirable power conservation becomes necessary attributes to be supported to cater to the requirements of 5G specifications. The optical spatial multiplexing is a promising member proposed to be included in the Next Generation Passive Optical Networks NGPON. Lagkas et al. (2020) have proposed a minimized energy consuming 5G front haul with an emphasis on assignment of optical resources. The work approaches spatial multiplexing incorporating the optical beamforming (OBFNs), Multicore fibers (MCFs) and Spectrally Spatially Flexible optical network propositions modelled as optimization problems and evaluated through simulations for their effectiveness.

In 5G architecture, the functionality of Baseband Unit (BBU) in a BBU pool is taken care of centrally by the Cloud Radio Access Network (C-RAN) by integrating the front haul, backhaul telecommunication infrastructures and radio access. Wireless connectivity for the mobile devices are provided by the Remote Radio Heads (RRHs). As a convergence of C-RAN, the traffic collected is delivered through the PON of the Optical Distribution Network (ODN) front haul. This enables the system to have high scaling capabilities with better adaptation to dynamic traffic situations. The Central Office (CO) where BBU pools are resident supports multiple operators and multiple services, while supporting Optical Line Terminals also. Through Remote Nodes (RNs), with dedicated fiber links the RRHs are reached. Other options of front haul design are also proposed to lower energy consumption, end to end delay and complexity by migrating the tasks carried out at the RRH to the Central Office. By adopting Digital Radio over Fiber (DRoF) evolved as eCPRI and Next Generation Front Haul Interface (NGFI), processing of wireless data at baseband and transmitted over optical fibers with addition of Medium Access Control layers processes to RRH sites.

A variant called Analog Radio Over Fiber (ARoF) eliminates signal processing and initial conversion of radio signals to baseband. The approach allows analog modulation of carrier with radio signals with inherent lesser complexity and lesser performance when larger distances are involved. In C-RAN architecture, the front haul needs to be supported with heavy load and as such Common Public Radio Interface could not be utilized since its inherent low scalability. Spatial Division Multiplexing based solutions for the Optical Distribution Network (ODN) promisingly support higher data rates. Further the authors have envisaged SDN solutions for the provisioning of enormous capacity required by the 5G front haul.

The spatial multiplexing specified by Richardson et al. (2013) in the form of integrated fibers enlisted as below into three categories as follows.

- Bundled Single Mode Fibers(B-SMFs) and multicore fibers with weak coupling
- MultiCore Fibers(MCFs) with strong coupling of spatial modes and Few Mode Fibers(FMFs)
- The coupled cores formed as uncoupled groups in integrated fibers and Few Mode MultiCore Fibers.

Even there are interference effects in the integrated fibers, the effects are of lesser hindrance to the system as the distances involved in the front haul are lesser. A multicore fiber (MCF) has 2 to 30 cores which may be coupled or spatially uncoupled. SMFs even though they have synchronization problems are promising options to support in the form of Bundled – Single Mode Fibers (B-SMFs)

The front haul ARoF aids in ensuring greater bandwidth by way of optical beam forming to targeted areas. Analog radio frequencies modulate optical carriers with transmission banks connected parallel to fiber cores. This arrangement encompasses Optical Beam Forming Networks (OBFN) between the cell sites and central office. Though there are three options available to place OBFN before the array of receivers, after the array of receivers of the cell site and at the cell site, the initial two options seem to be practicable for optical beam forming.

Regarding capacity allocation, the wavelength division multiplexing and time division multiplexing paradigms of NGPON exhibit enormous potential to support 5G front haul. Multicore fiber links with SMF or Few Mode Fibers allow greater granularity levels and permit capacity increase. If bandwidth increments are to be supported, on a per link basis the SDM spatial modes allow the requirement to be satisfied. By adopting time slots in transmission over a wavelength, finer granularity of required scale can be achieved. The blueSPACE project as discussed by Lagkas et al. (2020) envisages agile resource allocation based on the usage metrics and the area of deployment. The spatial and spectral dimensioning depends on the type of fibers with number of supported cores, the switching capabilities in terms of OBFN ports and distances involved.

The agility in provisioning resources depends on whether switching elements support spectral and spatial switching together or spectral, spatial switching separately and switching with respect to the ports or fiber cores. For each light path demand, the allocator decides on the available options to switch between ports, nodes, cores and wavelengths. These options of spatial multiplexing in the front haul could be used up to a distance of 5km. An optimized approach for allocation of resources, i.e., minimized usage of optical elements by evaluating the shortest routes. Power consumption minimization and effective utilization of physical resources is done by having the light paths aligned with common switching resources. The SDN controller resident in the Central Office takes care of dynamically activating and deactivating the ports/resources based on the traffic needs and available networking resources. This leads to power utilization to optimized levels. The routing demands are initiated after the allocation of radio resources. The ILP formulation for the said scenario goes as optical slots mapped to binary values of decision variables where the optical slots corresponds to wavelength channel of a fiber core connected between a pair of network nodes. Optimization decides on the assignment of resources such as port, core, wavelength to the traffic demand originated with pertinent constraints of the problem. The associated constraints are that an optical demand is permitted to utilize a specific optical resource. The second constraint accounted for is that for a demand, the sum of ingress spectrum allocations should equal the sum of egress spectrum allocations in the contexts of only spectral switching, only spatial switching, only port switching scenario. The work has made an emphasis on minimization of latency and power consumption for a targeted area coverage, where MCF options thrive for reduction in phase variations with spatial and spectral dimensioning done for optimized allocation of resources.

Multi – RAT (Multi Radio Access Technology) situation is dimensioned as converged SDM by Rommel et al. (2019), which demands support of highly latency sensitive live video interaction and one way streaming with comparatively lesser latency sensitivity and applications which are not sensitive to latency and these traffic scenario are mapped to different capacity requirements for fixed and mobile equipment. At the central office, Bandwidth/Bitrate Variable Transceivers (BVTs) are flooded with traffic by a set of BBUs. The BVTs remotely controlled by the CO through a control plane can be configured to maximum capacity. Wavelength Selective Switches (WSS) aid in to optically switch the inputs and outputs of BVTs based on wavelengths. M different signals can be wavelength multiplexed/demultiplexed based on the indications from the control plane. The WSS connected to Multicore Multiplexers (MCMs) which fan in /out devices on to different fiber cores.

The connectivity from the Central Office to the cell sites through Optical Distribution Network (ODN) go into different options as Spatially multiplexed sites, wherein for scaling up the capacity same wavelengths are used by the BVTs, sharing of light resources such as lasers among the BVTs so that the cost involved is kept low. Spectral multiplexing is performed by multiplexing wavelengths on to a single core. This entails usage of Arrayed Waveguide Grating (AWG) and another variant of spectral multiplexing uses Wavelength Selective Switches(WSS). In both of the options, dual wavelength duplex can be supported. Further cell site option with spectral and/or spatial multiplexing is used to utilize the network resources effectively. The optical switches residing at the central office becomes redundant if the BVTs are directly connected to WSSs. This leverages the integration of front haul to the data centres and enhances statistical multiplexing gains inherent with packet switching. For control of the BVTs, WSSs, the SDN is approached in a layered manner as child controllers of SDN governing the front haul and on top of it parent controller of SDN acting as an orchestrator. The packet flow across backhaul networks and network segments of NGFI are governed by the Openflow protocol. The front haul packet flow is managed by deploying SDN agents at the central office and cell sites. Configuration and monitoring of the 'blueSPACE nodes' is performed by SDN agents.

At the metro edge (ME), the WDM transponders deployed in two sets each handling 128 wavelengths form metro ring. Initial 128 wavelengths are launched into clockwise rings and further 128 wavelengths into counter clockwise rings. Of these rings formed in clockwise direction is maintained active. Along with this additional monitoring wavelength is launched into the ring. The entire set of wavelengths are grouped into three wavebands. The add/drop wavelength channels of access rings 1-13 are grouped as first waveband and as such 14-26 access rings' wavelengths are grouped as second waveband. The wavelengths of each waveband are demultiplexed into corresponding access rings. Since the metro network is formed as ring, the through channels are directed towards the metro edge. In the normal operating conditions, when there is no failure of components, the spans' communication are intact and the monitoring wavelength is detected at the metro edge. Under failure of fibers or nodes, monitoring wavelength is not detected at the metro edge and as a way of restoration of traffic, the second set of wavelengths are invoked across the counter clockwise protection rings. The reflective metro and access ring protection provides protection against failures on the metro span and as well in the access rings. This is accomplished by allocating separate monitoring channels for each access rings and metro rings. This variant also has groupings of wavelengths into wavebands as in R-MP but in different spans of the network. When failure in more than one access rings happen, the corresponding monitoring wavelengths are not received back at the metro edge. Traffic recovery is achieved by sending the transmission channels of the rings in both the directions. Analysis of the survivability schemes is done by the metrics of connection availability, i.e., probability of a connection being served at any point of time, failure impact factor as a measure of how traffic is impacted on failure occurrence, incremental cost as a measure of additional cost required to implement monitoring network components and in terms of yearly network consumption.

SURVIVABLE OPTICAL NETWORKS

In an infrastructure network or a backhaul network irrespective of the span covered by an optical network, there are inherent possibilities of outage of optical ports, optical fibers, and network elements due to natural disasters or fiber impairments. As the optical transport network handles enormous amount of traffic, a failure in the network elements reflects on a dramatic scale of number of users leading to loss of revenue. It leads to apparent magnification of the complexity as virtualization of network functions, BBUs, transponders in an optical transport network. A failure of an optical fiber propagates down the entire light path and require a survivability mechanisms of different QoS level needs that may be heterogeneous. The solutions proposed by Wong et al. (2017) in 16-18 consider overlay of mobile backhaul and front haul access networks over Passive Optical Networks (PON) in the form of hybrid passive optical access networks (HPCANs). The optical network units (ONUs) transition between active and sleep states, the sleep states invoke power down of transceivers to save power during idle conditions.

The main central office (MCO) along with the Optical Network Units (ONUs) by using monitoring modules detect failure of components or links. Once the loss of signal (LoS) of upstream transmissions are detected, protection switching is invoked to carry the affected traffic over the protection path. As per ITU-T(1998) G.775 specifications given by International Telecommunication Union[ITU], when there are no transitions for a duration of 175±75 contiguous intervals an LOS alarm is generated. This method of detection is not fairly applicable since the sleep mode also will create a situation of LOS. The works in Wong et al. (2016a), Wong et al. (2016b), and Machuca et al. (2016) mitigate the situation by providing survivable schemes.

Monitoring is done in these schemes by cw signals. A 5G network over a Manhattan, New York network for 5G is used to analyze three scenarios of brownfield, duct reuse and green field. The two schemes namely reflective metro ring(R-MP). Shahriar et al. (2019) proposed avenues for reducing network resource usage by incorporating a Virtual Network (VN) embedded solution. The network resources are optimized by tuning the guaranteed bandwidth in the event of failures and providing multipath survivable mechanisms. The tuning of guaranteed bandwidth is based on the fact that, while a path is operational, the entire bandwidth is not utilized by the services. A Bandwidth Squeezing Rate (BSR) is proposed as a measure of percentage of bandwidth available after a failure. The second technique survivable multipath mechanism is based on the idea that not all of the failures belong to the same virtual link VLink. The proposed techniques could reduce protection path resources. The multipath provisioning beats the performance of band squeezing technique only when disjoint paths are available and a reduction in gain results when there is dearth of disjoint paths. The adoption of Elastic Optical Networking (EON) by the researcher leverages in terms of fiber spectrum allocated, different modulation formats and FEC overheads. Mathematical modelling of the network is done as an ILP. The network is modelled as a substrate EON represented by undirected graph. The links are assumed to be bidirectional connecting adjacent nodes. The available spectrum on each link is subdivided into frequency slots of equal width. A set of k-Shortest paths are evaluated between nodes where the source- destination pairs demand connection for services. The transmission parameters such as symbol rate, modulation format and FEC overheads are configured for the set of k-Shortest paths.

With the virtual network (VN) formed, each virtual link requiring a bandwidth with a limitation placed by the parameter BSR based on reliability requirement. One approach mentioned is that the total bandwidth requirement is divided into multiple subrate paths. The traffic is split into subrate paths which are disjoint. For the k-Shortest path evaluated contiguous slices of spectrum are allocated along the precomputed paths with an objective to minimize the number of slices, subject to the constraints of resource availability and spectrum contiguity. The performance analysis is done taking the parameters, total number of spectrum slices required, protection overhead in terms of ratio of total bandwidth allocated on the split paths and accrual bandwidth demanded by the virtual link, maximum number of disjoint paths and maximum number of splits.

Survivability issues were investigated by Ayoub et al. (2020) as Survivable Virtual Network mapping in optical networks paradigm called Filterless Optical Networks (FON). The conventional usage of Reconfigurable Optical Add/Drop Multiplexers and Wavelength Selective Switches in switching architectures are replaced by passive wavelength splitters and wavelength combiners to provide way for cost effective architectures adopting the broadcast and select switching. The inherent spectrum wastage in FONs compared to the Wavelength Switched Optical Networks (WSON) poses challenges to establish loop free coverage. This is accomplished by forming fiber trees as to prevent laser-loop effects. The establishment of trees leads to complexities in ensuring survivability and demands excess transceivers to ensure several disjoint routes. The logical links formed in the virtual topology make sure that the physical link failures are thwarted and the logical topology is survivable in case of failures in an SVNM for a FON. As an instance, the virtual network is mapped i.e. an optical path corresponds to a virtual link. Two fiber tress of edge disjoint nature are considered and the interconnection of nodes that do not form part of same fiber are considered for trees. Additional transceivers are equipped as bridges to the trees to guarantee survivability. With an ILP formulation for fiber tree establishment and SVNM with an objective to minimize inter tree transceivers needed to be provided additionally. Analysis with a pre-establishment of fiber trees has been done by Ayoub et al. (2020). The cost involved with the survivability issues are benchmarked against (1) SVNM in the context of WSON and (2) Virtual Network mapping in the context of FON in the absence of survivability requirements. Joint optimization of survivability and fiber tree establishment proved promising to effective usage of spectrum and minimization of additional transceivers. The impetus on network management in case of component / fiber failures to provide resilience to failures beforehand of their occurrence by forecasting the scenario or invoking protection mechanisms after the occurrence of impairments.

QUANTUM KEY DISTRIBUTION

The 5G architectures do promisingly support lower latency connectivity, increased bandwidth requirement of customers along with the capability of meeting Quality of service requirements. In the converged architecture of 5G networks, the lower latency capability is realized by Ksentini and Frangoudis (2020) incorporating the resources closer to the edge nodes of the network leading to edge computing. The advantages of low latency networks are tapped in the form of usage of network for managing automated vehicles, IoT control and monitoring and realization of content delivery networks (CDN).

For any of the services, the encryption of data for secure communications is provided by the application layer as encryption sessions rather than encryption being an intrinsic aspect in the data. The security aspects in conventional communication systems relies on the end users or application developers as mentioned by Yousaf et al. (2017). As larger concentrations of traffic flow across metro nodes and aggregation switches, eavesdroppers may target to thwart the very purpose of security. The network operators could probe the avenues of secure communications across Layer 1 by providing intrinsic encryption in the form of Quantum Key Distribution (QKD). QKD is a technology meant for secure distribution of symmetric encryption keys. Cryptographic key exchange between two remote nodes / destinations are governed by laws of quantum physics. These keys can be securely used alongside keys generated by conventional algorithms of cryptography. Computational threats to secure keys by eavesdroppers could be avoided by utilizing QKD, wherein quantum states agreed upon by the users that are encoded on photons with Information Theoretic Security (ITS).

The security framework in the optical front haul of 5G network has been proposed by Zavitsanos et al. (2020) with a focus on Quantum Key Distribution (QKD) enabled assistance for secure communication. The idea behind the proposal is that the integration of quantum key delivery to the encryption engines of Advanced Encryption Standards (AES) – 256 through BB84 QKD link. The analysis of secure key rate were done for the shared fiber and the dedicated fiber configurations with a thrust on low latency and quantum keys availability. The researchers have found a technique that the dark fibers deliver quantum keys at much longer lengths than the front haul distance of optical fibers for point to point (P2P) and point to multipoint (P2MP) topologies. A quantum safe secure framework has been proposed for the optical front haul required for the 5G infrastructure.

The Advanced Encryption Standards(AES) known to be quantum resistant prompted by Bonnetain et al. (2019) is used to encrypt a message using distilled secret keys. Eriksson et al. (2019) have made extensive studies about the propagation of quantum keys along with conventional data channels carrying traffic of the order of Tbps across backbone optical networks. In the Passive Optical Networks (PON) segments, the edge nodes share quantum keys aiding secure transfer of traffic as given in Aleksic et al. (2013). To prevent contamination between quantum signal and classical front haul signals, multicore fiber(MCF) based solutions were proposed by the research community Kong et al. (2020) and Hugues-Salas et al. (2020).

A quantum secure front haul has been designed by Cho et al. (2020) in order to meet the security requirements specified by MACsec or IPsec and packet based switching of enhanced Common Public Radio Interface (CPRI) designed for delay sensitive Ethernet and IP. The dark fiber technology mentioned takes care of quantum keys distribution over the same length optical fiber links. The exchange of quantum keys over the front haul fibers interconnecting Radio Equipment Control(REC) and Radio Equipment(RE) at the terminal node of 5G network.

The quantum key distribution is suggested to take place in two approaches used alternatively as (1) dedicated one to one links interconnect nodes in P2MP stations, (2) a time division multiplexing technique can be adopted for sharing the dark fiber resources for key sharing amongst the users. The multiuser scenario to share the quantum keys involves deployment of an Alice station located at Base Band Unit(BBU) to share quantum keys to the quantum receiver at each of the terminal nodes in the 5G network. The number of terminal nodes 'N' defines the size of the 5G network with the security of each of the segments ensured by the respective quantum keys distributed / received at the edge nodes. Each receiving user is allocated with a randomly selected bits that the Alice station has transmitted through the feeder fiber. Security of each of the fiber segment is ensured by AES engines.

In Zhao et al. (2018) the resource allocation problem in optical networks which has additional ability in terms of security using QKD has been proposed. Traditional Communication data channel (TDChs) with additional Quantum Signal Channels (QSChs) along with Public Interaction Channels(PIChs) are also incorporated to have key synchronization. The ability of the QKD to identify the presence of the eavesdropper trying to gain information on the security key makes it eligible for the researchers Zhao et al. (2018) to introduce security aspects at different levels with several key updating periods and analysis made in two configurations of key updating periods. An algorithm for Routing, Wavelength and Time slot Assignment (RWTA) has been proposed by the researchers. The QKD protocol BB84 proposed by Gleim et al. (2016) uses polarization coding scheme along with decoy method for long distance networks. A series of 256 bit long keys as in AES are shared among Alice and Bob nodes via Quantum Signal Channels (QSChs) and Public Interaction Channels(PIChs)

For SDN based management of the network, a QKD plane is also introduced in the optical network architecture along with control plane and data plane. Sharing of spectrum resources in WDM technology among QSCh, PICh and TDCh is entrusted with QKD plane and data plane. In the RWTA analysis by the authors with inclusion of QKD feature, the secure key for TDCh needs to be updated periodically based on level of security requirements. When update interval is reduced security level increases. With more diverse intervals followed, between updation of security keys, the eavesdroppers face difficulty in inferring the update periods and secure keys.

Different levels of security with different update periods for keys are followed as the fixed updating periods, flexible updating periods. A time sliding window mechanism is adopted for data transfer and analysis of the RWTA has been performed with discrete event simulation with dynamic generation of connection requests. Blocking probability is evaluated with several updating intervals, different time sliding windows (TSWs) and varying number of wavelengths.

1. Attributes of QKD networks:

The QKD networks have the following requirements as given by Mehic et al. (2020).

a. **Key Rate:** The average key rate associated with a QKD link becomes important as sufficient key material is required for encryption and decryption procedures. The rate at which quantum keys are stored and the rate at which it is utilized for encryption and decryption influences network performance.
b. **Link Length:** The length of the fiber over which key material has to be generated becomes a constraint to the QKD link due to inherent optical fiber impairments such as absorption and scattering. The limits the quantum channels to be established over a short distance and current distances of about 100km can be covered with deployment of QKD links.
c. **Key Material Protection:** Privacy of key material in terms of the strong probability to be associated to make the keys unique as communication happens as hop-by-hop and evaluate for its privacy when stored and managed.
d. **Key Usage:** Additional amount of key generated means spending the resources of communication to a greater extent as hop-by-hop communication takes place across several nodes enroute the destination.

The QKD networks can be grouped as switched QKD networks and trusted repeater QKD networks and the later category of QKD networks are deployed currently to overcome distance constraint associated with QKD links and routing issues. A QKD protocol BB84 specifies generation of optical power and its detection at photonic levels, said in Lin et al. (2019). As QKD implementation involves sharing the existing network resources entails contamination of the QKD channel by the conventional data channel as several millions of photons are required for delivery of information. The researchers have proposed a newer dimension of incorporating QKD mechanisms into wavelength division multiplexing network and SDM networks. By WDM approach, the usage of different wavelengths for data transmission. The implementation scheme involved Coarse Wavelength Division Multiplexing with a data channel at 1.25Gbps operating at 1550nm wavelength and quantum channel at a wavelength of 1310nm and secret key rate of 3kbps.

The Single Mode Multi Core Fiber (SM-MCF) provides better isolation between adjacent spatial channels with better spectral efficiency as reported by Gan et al. (2018). The trench assisted MCF (TA-MCF) and non-trench assisted (NT-MCF) with different refractive index profiles are adopted with a boundary on inter-core cross talk. The realization of Space Division Multiplexing (SDM) of data channels and QKD channels has been reported in Dynes et al. (2016).

2. QoS in QKD networks:

QKD networks have characteristics as covering limited distance and the quantum key rate diminishing exponentially with distance. If sufficient key material is not available, link becomes unavailable. Another aspect is that as the communication happens on a hop-by-hop basis, there is lack of number of quantum repeaters.

In providing ITS communication as a goal of QKD, the network solution may not be suitable for all the scenario. For instance the network flooding specified in Elliott (2002) and Marhoefer et al. (2007) assumes all of the nodes to be trusted in hop-by-hop connectivity or a repetitive key distribution case. This makes the eavesdropper incapable of attacking at nodes and not able to gain information on the key shared via the link, but may attempt to disable the link. Thomas and Weil (2007) discuss that as multiple paths are utilized to share the secure key information, the routing information of the multiple paths need to be minimized. The routing packets need to be encrypted or authenticated and to be minimized to reduce the routing overheads.

In an SDN based QKD network as in Dahlberg et al. (2019), additional ITS layer needs to be incorporated for information critical infrastructures. Multi-tenancy requirements of an organization for instance may be provided secure Key as a service (Kaas), wherein the control layer of SDN architecture has key protocols (KP) mapped to virtual key pools at the application layer.

HOLOGRAPHIC TECHNIQUES

The idea behind holographic spectral multiplexing revolves around subdividing the spectrum into several subchannels as quoted in Khodaei and Deogun (2021). Accomplishing this task is by contiguous frequency slots, splitting wavelengths, spatial multiplexing of continuous wave laser spectrum specified by Essiambre and Tkach (2012) and spatial multiplexing of visible light spectrum specified by Winzer et al. (2018). The holographic spectral multiplexing proposed by Khodaei and Deogun is inspired by the

Multiple Input Multiple Output (MIMO) technology in RF communication domain to enhance spectral efficiency. The MIMO scheme has the RF wave propagate through sharing of physical medium. Through beam forming techniques decorrelation of virtual channels created by interference phenomenon is performed. An optimization of spectral beams is achieved by beam forming and enables MIMO to achieve comparably higher data rates.

The possibility of laser generators to generate femtosecond pulses provide scope to have holographic spectral multiplexing is proposed by Kieu et al. (2010). Line codes are generated as 2D hologram patterns since the availability of ultra-short pulses with high resolution. Hologram generation is done by dispersing spatially the frequency components by spectrum processing device (SPD) and forming 2D patterns with intensity variations. This method eliminates the lower spectral efficiency in the conventional PCM line coding applied in classical multiplexing schemes as TDM and WDM.

High entropy symbols are generated by the holographic technique and enables a block of data to be sent as single symbol. An analogous situation to array of sources and array of sensors in a MIMO is mimicked in a HSM as these points are replaced by pixels of hologram. Beam forming between the multiple receiver ends and sender is done in such a way that channel scattering is managed. In HSM implementation, chromatic dispersion effects are measured to study the channel status information (CSI) associated with the structure of ultra short pulses. A spectral diversity is created in the pulsed spectrum as there are different spectral components. MIMO channels are built as a matrix of elements reflecting the degree of scattering of the ultra short pulse. For the elements which are uncorrelated, form the HSM matrix to enable parallel transmissions of MIMO. Kahn et al. (1998) have proposed an array of imaging receivers which are pixelated rather than using conventional sensor detectors. In the typical set up mentioned, the visible light beams make an image to be incident on the receiver and the constituent beams are correlated. This limits the non-imaging MIMO systems from practical applications and has limited diversity gain as investigated by Shen et al. (2019).

The holographic spatial multiplexing set up consists in having a laser generator for ultra short pulses which propagate over a diffraction grating to spread the wavelength in spatial dimensions. Spatial light modulation (SLM) along with graded index lens(GRIN) is used to collimate the light beam. As the SLM can be electronically controlled to give required spectral width, the individual pixels can be invoked on or off or different stages of intensity modulation. SLM can be utilized to transmit subspectra with varying intensities.

MACHINE LEARNING IN OPTICAL NETWORKS

Machine learning is a domain of Artificial Intelligence (AI) that weaves linkages with statistical information, optimization and game theory. ML goes into categories as classification/clustering problem and regression problem. Machine learning techniques are applied in optical domain largely into the areas of fault detection, Software Defined Networking, physical layer design, quality of transmission estimation, nonlinearity compensation and optical performance monitoring. To provide reliable network communications to ensure service guarantees, in terms of Service Level Agreements (SLAs), proactive warning and invoking protection mechanisms beforehand deterioration of network components are necessitated. The conventional protection mechanisms fail to forecast the risks and recovery is initiated only after the occurrence of failure. This leads to have network designs with larger operating margins and failure to ascertain the cause of faults apart from inability to detect slow performance degradation of the system reported in Zhou and Xie (2016).

Proactive fault management with the help of ML has been followed with the retrieval of parameters such as power levels, gains of EDFAs, shelf temperature at the nodes and provides for data driven management of the network and to ensure reliable optical network operations. Discovery of faults in a proactive manner is accomplished by learning algorithms relying on historical data of fault occurrences and fault patterns over the network by data analytics. Yu et al. (2019) have proposed machine learning algorithm to locate fault in front haul optical networks which caters to 5G namely Deep Belief Network based Fault Location (DBN-FL). DBN-FL has been implemented as two phases of pre-training phase and fine-tuning phase which relies on classifier algorithms of machine learning.

False alarms are generated by the performance monitoring equipment, where in for the 5G network more numbers of nodes will be equipped as monitoring nodes. The pre-training phase in DBN involves unsupervised learning in a bottom-to-top approach to gather initial parameters pertaining to adjacent layers. A model generalization is obtained in the pre-training phase. The fine tuning phase consists in a supervised learning back propagation algorithm, wherein optimization of parameters from top-to-bottom approach is performed. The resulting multilayer model performs the task of alarm classification. The inherent incapability of DBN in its convergence rate during fine tuning phase due to non-availability of volumes of training vectors, 'latent features' of alarms are utilized by DBN-FL, to generate deep features and allow classification into real alarm or false alarm by analysis of differences among latent features. By DBN-FL the authors could locate single fault with an improved accuracy. In Hinton et al. (2006), the parameters learned by supervised learning in the pre-training phase of Restricted Boltzmann Machines(RBM) are transferred as initial parameters to unsupervised learning model. The back propagation algorithm is replaced by the LM algorithm in the fine tuning phase. By adopting these methods, training vectors are required less in numbers.

Autonomous decision making proposed in Morais (2021) is accomplished by AI system based on the observations of environment and actions invoked accordingly using a cognitive loop as in Mestres et al. (2017). Morais et al. (2021) have pointed that the machine learning algorithms learn from the historical data by analyzing the network behavior and makes decisions. The system gradually evolves to automatically make decisions and the decisions are translated into the status of network elements (NE) by Mestres et al. (2017). Linux's unified platform called Open Network Automation Platform (ONAP) which includes analytics, control and management aspects where artificial intelligence in multilayer networks could be introduced. The self-driving networks proposed by Atwal et al. (2018) based on dynamics in requirements, failures/outages of components reorganize themselves to adapt to the changes where ML can be applied with highlights as:

- The impending network failures due to anomalous network parameters that are detected and found out for their root cause and invokes preventive decisions.
- Prediction of traffic growth in terms of volume and allocation of resources based on dynamics accordingly.
- Performance of components over the entire span of light path by gathering the characteristics of optical components and switching systems.

SDN IN OPTICAL NETWORKS

The advancements and flexibilities in the 5G network in terms of Quality of Experience (QoE) quoted in Cvijetic (2014) in a heterogeneous networking environment with metrics for latency, capacity of the network, probability of outage and spectral efficiency necessitates management of 'device centric' entities evolving from a conventional 'cell centric' management. The Device to Device (D2D) communications, increased use of smart machines and sensors give rise to 5G in a need to provide any-to-any device connectivity with least latency. This scenario has given rise to developments of change in the control plane as well as in the data plane. The cloud radio access networks (CRAN) architectures emerged as promising avenues with the support of underlying optical networks to support low latency, high speed connectivity requirements of 5G. The dynamic changes in the signal formats, optical components used and management of optical spectrum and network control are discussed. A translation from Common Public Radio Interface (CPRI) and ability to support all WDM types provides a way for configuring a dynamic topology with the help of Software Defined Networking(SDN) as discussed in Cvijetic (2014). In a way similar to CPRI to Ethernet mapping, wherein variable rates are multiplexed via Ethernet, an analogous mapping of CPRI to OTN in the optical domain. Even though these mapping are reliable techniques, they lack bandwidth expansion required of in the CPRI. The further option is to introduce Fast Fourier Transforms applied at RRH so that the data modulated in frequency domain are propagated through optical links.

The PON architecture and the evolution in WDM to accommodate a plethora of Coarse WDM, colourless, dense and coherent WDMs to provide wavelengths by tuning and routing/switching to different wavelengths. Reconfigurations in the WDM in a decentralized manner is discussed. For D2D applications and the mobile front haul architecture with reconfigurable optical modules have been experimentally demonstrated in Cvijetic et al. (2014b). To dynamically invoke CoMP and D2D communication features, a open flow control is utilized to coordinate flow tables in control plane to Software Define optical switching elements. This introduces reconfigurable optical modules in front haul networks with dynamically changing topology along with the conventional fiber optic equipment. 10 Gbps data rate with latency less than 7 microseconds and a total link power budget of 29.6dB has been experimentally demonstrated. Cvijetic et al. (2014b) have discussed that Software Defined Networking based control offers a vendor agnostic way of controlling switching elements in optical and electrical domains with dynamic changes. In Cvijetic et al. (2014a) traffic flows are combined to device with an identifier name rather than using the techniques in IP addressing in SDN based control plane to ensure security and meet requirements of QoS policies.

A SDN based hybrid optical wireless network that incorporates virtualization, flexible and elastic spectrum slices and secure management of interfaces is discussed by Sarigiannidis et al. (2017). A flexible and agile network management has been discussed with a mention on SDN architecture and Openflow standard by Nunes et al. (2014) whereas Sarigiannidis et al. (2017) differs by integrating optical and wireless technologies by using SDN. A hypervisor is proposed that performs abstraction of physical SDN network divided into virtual networks logically isolated to provide slicing phenomenon in wireless and optical domains. Based on the arising tenant requirements dynamic control of the management plane is initiated. The underlying control plane has a repository of software libraries, abstractions and models for efficient usage of infrastructure. The interconnecting sublayer takes care of authentication and security aspects. By usage of Openflow protocol, data and control planes interact. Route evaluation, load balancing and resilience policies are governed by the decision-making modules of SDN paradigm.

Han et al. (2015) have done the proactive and reactive management of SDN based on the decisions made from the information base shared between control and management planes. Device modeling offered by the APIs which provide hardware abstractions with technology agnostic aspects of the underlying network resources.

The hybrid wireless optical architecture proposed will have control of the Wavelength Selective Switches and eNodes, LTE equipment integrate and controlled by SDN controllers. The WSS are connected to Central Office (CO) using optical fibers and the required optical amplifiers are also integrated with the SDN controllers. By configuring WSS- ADM through Openflow, the optical switching at core network has been optimized by Sun et al. (2013) to have flexible configurations.

In the optical SDN proposed by Yu et al. (2020) all information about the optical network components gathered online as well as offline are stored in postgreSQL database. The offline database consists of information of fiber links, types of devices, node locations and their IP addresses. The RyuSDN controller abstracts online data to assist the control plane in terms of topology reconfiguration, path backup and recovery information in case of failures. PostgreSQL saves in the Traffic Engineering Database (TED), the details of network information gathered offline. Path computation, control and management are coordinated by the TED and runs the Routing and Wavelength assignment(RWA) and controls the wavelength selective switches. With the port connection information, assigned ID, Optical wavelengths required and bandwidth the TED database is updated. With space switching becoming effective, TED and PostgreSQL are updated with the information regarding lightpath and physical topology.

CONCLUSION

An overview of the contributions towards Spatial Division Multiplexing, a way of enhancing the physical limit of fibers thwarting the capacity crunch have been discussed. The work accomplished by researchers to increase transmission capacity, offset the limits, and increase the node capacity involving Multicore fiber (MCF) have been reviewed. Holographic techniques in optical domain that mimic the highly spectral efficient MIMO in RF communication systems by exploiting the enormous spectral bandwidth contained in ultra-short optical pulses, by way of multiplexing also been reviewed. A secure way of distribution of keys for end-to-end lightpaths is affected by Quantum Keys for allocation of spectral resources as a way of effective utilization as the fiber resources have been included. The domain largely referred as Software Defined Networking which takes care of provision of network service starting computation of routes, allocation of resources have been discussed. The aspects of survivability of a long-haul optical network in cases of node or physical connectivity failures, resilience of the network, the research into the pertinent protection mechanisms are also furnished.

REFERENCES

Aleksic, S., Winkler, D., Franzl, G., Poppe, A., Schrenk, B., & Hipp, F. (2013). Quantum key distribution over optical access networks. *Proceedings of 2013 18th European Conference on Network and Optical Communications & 2013 8th Conference on Optical Cabling and Infrastructure (NOC-OC&I)*, 11–18.

Atwal, K. S., & Bassiouni, M. (2018). Deep SDN: connecting the dots towards self-driving networks. *Proceedings of*2018 *IEEE 37th International Performance Computing and Communications Conference (IPCCC).* 10.1109/PCCC.2018.8711025

Ayoub, O., Bovio, A., Musumeci, F., & Tornatore, M. (2020). Survivable Virtual Network Mapping in Filterless Optical Networks. *Proceedings of 2020 International Conference on Optical Network Design and Modeling (ONDM),* 1-6. 10.23919/ONDM48393.2020.9133008

Bonnetain, X., Naya-Plasencia, M., & Schrottenloher, A. (2019). Quantum Security Analysis of AES. *IACR Transactions on Symmetric Cryptology, 2,* 55–93. doi:10.46586/tosc.v2019.i2.55-93

Cho, J. Y., Sergeev, A., & Zou, J. (2020). Securing Ethernet-based Optical Front haul for 5G Network. *Journal of Cyber Security and Mobility, 9*(1), 91–110. doi:10.13052/jcsm2245-1439.913

Common Public Radio Interface. (n.d.). *ECPRI Interface Specification v1.2.* http://www.cpri.info/spec.html

Cvijetic, N. (2014). Optical network evolution for 5G mobile applications and SDN-based control. *Proceedings of 2014 16th International Telecommunications Network Strategy and Planning Symposium (Networks),* 1-5. 10.1109/NETWKS.2014.6958537

Cvijetic, N., Tanaka, A., Ji, P., Sethuraman, K., Murakami, S., & Wang, T. (2014). SDN and OpenFlow for dynamic flex-grid optical access and aggregation networks. *Journal of Lightwave Technology, 32*(4), 864–870. doi:10.1109/JLT.2013.2274991

Cvijetic, N., Tanaka, A., Kanonakis, K., & Wang, T. (2014). SDN-controlled topology-reconfigurable optical mobile fronthaul architecture for bidirectional CoMP and low latency inter-cell D2D in the 5G mobile era. *Optics Express, 22*(17), 20809. Advance online publication. doi:10.1364/OE.22.020809 PMID:25321284

Dahlberg, A., Oliveira Filho, J., Hanson, R., Wehner, S., Skrzypczyk, M., Coopmans, T., Wubben, L., Rozpedek, F., Pompili, M., Stolk, A., Pawełczak, P., & Knegjens, R. (2019). A link layer protocol for quantum networks. In *Proceedings of the ACM Special Interest Group on Data Communication.* ACM. 10.1145/3341302.3342070

Dynes, J. F., Kindness, S. J., Tam, S. W. B., Plews, A., Sharpe, A. W., Lucamarini, M., Fröhlich, B., Yuan, Z. L., Penty, R. V., & Shields, A. J. (2016). Quantum key distribution over multicore fiber. *Optics Express, 24*(8), 8081–8087. doi:10.1364/OE.24.008081 PMID:27137247

Elliott, C. (2002). Building the quantum network. *New Journal of Physics, 4,* 46. doi:10.1088/1367-2630/4/1/346

Eriksson, T. A., Hirano, T., Puttnam, B. J., Rademacher, G., Luís, R. S., Fujiwara, M., Namiki, R., Awaji, Y., Takeoka, M., Wada, N., & Sasaki, M. (2019). Wavelength division multiplexing of continuous variable quantum key distribution and 18.3 Tbit/s data channels. *Communications on Physics, 2*(1), 1–8. doi:10.103842005-018-0105-5

Essiambre, R. J., & Tkach, R. W. (2012). Capacity Trends and Limits of Optical Communication Networks. *Proceedings of the IEEE, 100*(5), 1035–1055. doi:10.1109/JPROC.2012.2182970

Gan, L., Shen, L., Tang, M., Xing, C., Li, Y., Ke, C., Tong, W., Li, B., Fu, B., & Liu, D. (2018). Investigation of channel model for weakly coupled multicore fiber. *Optics Express*, *26*(5), 5182–5199. doi:10.1364/OE.26.005182 PMID:29529725

Gleim, A. V., Egorov, V. I., Nazarov, Yu. V., Smirnov, S. V., Chistyakov, V. V., Bannik, O. I., Anisimov, A. A., Kynev, S. M., Ivanova, A. E., Collins, R. J., Kozlov, S. A., & Buller, G. S. (2016). Secure Polarization-Independent Subcarrier Quantum Key Distribution in Optical Fiber Channel Using BB84 Protocol with a Strong Reference. *Optics Express*, *24*(3), 2619–2633. doi:10.1364/OE.24.002619 PMID:26906834

Han, B., Gopalakrishnan, V., Ji, L., & Lee, S. (2015). Network function virtualization: Challenges and opportunities for innovations. *IEEE Communications Magazine*, *53*(2), 90–97. doi:10.1109/MCOM.2015.7045396

Hinton, G. E., Osindero, S., & Teh, Y. W. (2006). A fast learning algorithm for deep belief nets. *Neural Computation*, *18*(7), 1527–1554. doi:10.1162/neco.2006.18.7.1527 PMID:16764513

Hugues-Salas, E., Alia, O., Wang, R., Rajkumar, K., Kanellos, G. T., Nejabati, R., & Simeonidou, D. (2020). 11.2 Tb/s Classical Channel Coexistence with DV-QKD over a 7-Core Multicore Fiber. *Journal of Lightwave Technology*, *38*(18), 5064–5070. doi:10.1109/JLT.2020.2998053

(2020). Indoor Visible Light Communication Networks for Camera-Based Mobile Sensing. InShen, X. S., Lin, X., & Zhang, K. (Eds.), *Encyclopedia of Wireless Networks* (pp. 1–7). Springer International Publishing. doi:10.1007/978-3-319-32903-1_270-1

International Telecommunication Union. (1998). *Loss of signal (LOS), alarm indication signal (AIS) and remote defect indication (RDI) defect detection and clearance criteria for PDH signals*. ITU-T G.775.

Kahn, J. M., You, R., Djahani, P., Weisbin, A. G., & Tang, A. (1998). Imaging diversity receivers for high-speed infrared wireless communication. *IEEE Communications Magazine*, *36*(12), 88–94. doi:10.1109/35.735884

Khodaei, A., & Deogun, J. (2021). *Optical MIMO Communication Using Holographic Spectral Multiplexing of Pulsed Ultrashort Laser*. doi:10.48550/arXiv.2106.13896

Kieu, K., Jones, R. J., & Peyghambarian, N. (2010). Generation of Few-Cycle Pulses From an Amplified Carbon Nanotube Mode-Locked Fiber Laser System. *IEEE Photonics Technology Letters*, *22*(20), 1521–1523. doi:10.1109/LPT.2010.2063423

Kong, W., Sun, Y., Cai, C., & Ji, Y. (2020). Impact of Modulation Formats and Bandwidth on Quantum Secured 5G Optical Fronthaul over Multicore Fiber. *Proceedings of the CLEO 2020*. 10.1364/CLEO_AT.2020.JTu2A.25

Ksentini, A., & Frangoudis, P. A. (2020). Towards Slicing-Enabled Multi-Access Edge Computing in 5G. *IEEE Network*, *34*(2), 99–105. doi:10.1109/MNET.001.1900261

Lagkas, T., Klonidis, D., Sarigiannidis, P., & Tomkos, I. (2020). 5G/NGPON Evolution and Convergence: Developing on Spatial Multiplexing of Optical Fiber Links for 5G Infrastructures. *Fiber and Integrated Optics*, *39*(1), 4–23. doi:10.1080/01468030.2020.1725184

Lin, R., Udalcovs, A., Ozolins, O., Tang, M., Fu, S., Popov, S., Ferreira da Silva, T., Xavier, G. B., & Chen, J. (2019). Embedding Quantum Key Distribution into Optical Telecom Communication Systems. *Proceedings of 2019 21st International Conference on Transparent Optical Networks (ICTON),* 1-4. 10.1109/ICTON.2019.8840512

Linux Foundation. (n.d.). *Open Network Automation Platform (ONAP).* https://www.onap.org/

Machuca, C. M., Wong, E., Furdek, M., & Wosinka, L. (2016). Energy consumption and reliability performance of survivable hybrid passive optical converged networks. *Advanced Photonics 2016.* doi:10.1364/NETWORKS.2016.NeTu3C.2

Marhoefer, M., Wimberger, I., & Poppe, A. (2007). Applicability of quantum cryptography for securing mobile communication networks. In Long-Term and Dynamical Aspects of Information Security: Emerging Trends in Information and Communication Security. Nova Science Publishers.

Mehic, M., Niemiec, M., Rass, S., Ma, J., Peev, M., Aguado, A., Martin, V., Schauer, S., Poppe, A., Pacher, C., & Voznak, M. (2020). Quantum Key Distribution: A Networking Perspective. *ACM Computing Surveys, 53*(5), 41. doi:10.1145/3402192

Mestres, A., Rodriguez-Natal, A., Carner, J., Barlet-Ros, P., Alarcón, E., Solé, M., Muntés-Mulero, V., Meyer, D., Barkai, S., Hibbett, M. J., Estrada, G., Máruf, K., Coras, F., Ermagan, V., Latapie, H., Cassar, C., Evans, J., Maino, F., Walrand, J., & Cabellos, A. (2017). Knowledge-defined networking. *Computer Communication Review, 47*(3), 2–10. doi:10.1145/3138808.3138810

Morais, R. M. (2021). On the suitability, requisites, and challenges of machine learning. *Journal of Optical Communications and Networking, 13*(1), A1–A12. doi:10.1364/JOCN.401568

Nunes, B.A.A., Mendonca, M., Nguyen, X.N., Obraczka, & Turletti, T. (2014). A survey of software defined networking: past, present, and future of programmable networks. *IEEE Communications Survey & Tutorials, 16*(3), 1617–1634. doi:10.1109/SURV.2014.012214.00180

(2016). Optical performance monitoring for fiber-optic communication networks. InZhou, X., & Xie, C. (Eds.), *Enabling Technologies for High Spectral-Efficiency Coherent Optical Communication Networks.* John Wiley & Sons.

Richardson, D. J., Fini, J. M., & Nelson, L. E. (2013). Space-division multiplexing in optical fibres. *Nature Photonics, 7*(5), 354–362. doi:10.1038/nphoton.2013.94

Rommel, S., Perez-Galacho, D., Fabrega, J. M., Muñoz, R., Sales, S., & Tafur Monroy, I. (2019). High-Capacity 5G Front haul Networks Based on Optical Space Division Multiplexing. *IEEE Transactions on Broadcasting, 65*(2), 434–443. doi:10.1109/TBC.2019.2901412

Sarigiannidis, P., Lagkas, T., Bibi, S., Ampatzoglou, A., & Bellavista, P. (2017). Hybrid 5G optical-wireless SDN-based networks, challenges and open issues. *IET Networks, 6*(6), 141–148. doi:10.1049/iet-net.2017.0069

Shahriar, N., Taeb, S., Chowdhury, S. R., Zulfiqar, M., Tornatore, M., Boutaba, R., Mitra, J., & Hemmati, M. (2019). Reliable Slicing of 5G Transport Networks with Dedicated Protection. *Proceedings of 15th International Conference on Network and Service Management (CNSM),* 1-9. 10.23919/CNSM46954.2019.9012711

Sun, G., Liu, G., Zhang, H., & Tan, W. (2013). Architecture on mobility management in OpenFlow-based radio access networks. *Proceedings of IEEE Global High Tech Congress on Electronics,* 88–92. 10.1109/GHTCE.2013.6767247

Thomas, W., & Weil, P. (Eds.). (2007). Lecture Notes in Computer Science: Vol. 4393. *Quantum network coding*. Springer Berlin. doi:10.1007/978-3-540-70918-3

Winzer, P. J., Neilson, D. T., & Chraplyvy, A. R. (2018). Fiber-optic transmission and networking: The previous 20 and the next 20 years. *Optics Express*, *26*(18), 24190–24239. doi:10.1364/OE.26.024190 PMID:30184909

Wong, E., Grigoreva, E., Wosinska, L., & Machuca, C. M. (2017). Enhancing the survivability and power savings of 5G transport networks based on DWDM rings. *Journal of Optical Communications and Networking*, *9*(9), D74–D85. doi:10.1364/JOCN.9.000D74

Wong, E., Machuca, C. M., & Wosinska, L. (2016). Survivable architectures for power-savings capable converged access networks. *Proceedings of 2016 IEEE International Conference on Communications,* 1-7. 10.1109/ICC.2016.7510719

Wong, E., Machuca, C. M., & Wosinska, L. (2016). Survivable hybrid passive optical converged network architectures based on reflective monitoring. *Journal of Lightwave Technology*, *34*(18), 4317–4328. doi:10.1109/JLT.2016.2593481

Yousaf, F. Z., Bredel, M., Schaller, S., & Schneider, F. (2017). NFV and SDN - Key Technology Enablers for 5G Networks. *IEEE Journal on Selected Areas in Communications*, *35*(11), 2468–2478. doi:10.1109/JSAC.2017.2760418

Yu, A., Yang, H., Yao, Q., Li, Y., Guo, H., Peng, T., Li, H., & Zhang, J. (2019). Accurate Fault Location Using Deep Belief Network for Optical Front haul Networks in 5G and Beyond. *IEEE Access : Practical Innovations, Open Solutions*, *7*, 77932–77943. doi:10.1109/ACCESS.2019.2921329

Yu, J., Gutterman, C., Minakhmetov, A., Sherman, M., Chen, T., Zhu, S., Zussman, G., Seskar, I., & Kilper, D. (2020). Dual Use SDN Controller for Management and Experimentation in a Field Deployed Test bed. *Proceedings of 2020 Optical Fiber Communications Conference and Exhibition (OFC),* 1-3. 10.1364/OFC.2020.T3J.3

Zavitsanos, D., Ntanos, A., Giannoulis, G., & Avramopoulos, H. (2020). On the QKD integration in converged fiber/wireless topologies for secured, low-latency 5G/B5G front haul. *Applied Sciences (Basel, Switzerland)*, *10*(15), 5193. doi:10.3390/app10155193

Zhao, Y., Cao, Y., Wang, W., Wang, H., Yu, X., Zhang, J., Tornatore, M., Wu, Y., & Mukherjee, B. (2018). Resource Allocation in Optical Networks Secured by Quantum Key Distribution. *IEEE Communications Magazine*, *56*(8), 130–137. doi:10.1109/MCOM.2018.1700656

KEY TERMS AND DEFINITIONS

Baseband Unit (BBU): A telecommunication network device used to interpret and process baseband signals in telecom systems. Baseband signals are the original frequencies of transmissions prior to being modulated. A BBU typically connects to transceivers – Remote Radio Units (RRU) through wired or wireless connections.

Common Public Radio Interface (CPRI): Wireless interface specification for transport, connectivity and control communication between BBUs and RRUs of cellular wireless networks. An I/Q interface protocol for communication between towers and base stations for various standards such as GSM, WCDMA, LTE, etc.

Multiple Input Multiple Output (MIMO): Usage of multiple transmitters and receivers to transfer more data at the same time. MIMO technology uses spatial dimensions and smart antennas thus utilizing diversity and multipath to achieve higher speeds of data transfer.

Optical Beam Forming Networks (OBFN): Communication networks which have the ability to focus, shape and steer the light signals. Capable of manipulating or indirectly change a radio frequency on an optical carrier.

Passive Optical Network (PON): Fiber optic technology to cater to multiple end pints from a single source. The word "Passive" signifies the usage of an unpowered splitter(passive) which transmits data from a service provider to multiple customers, each customer being connected with the splitter through optical fiber cables. A GPON is capable of high-speed Gigabit transmissions.

Radio Access Network (RAN): A telecommunication network infrastructure that connects individual user devices to other parts of a network through a radio link. RAN includes base station and antennas that cover a specific region according to its capacity, design, and propagation.

Routing and Wavelength Assignment (RWA): In optical Wavelength Division Multiplexing (WDM) networks, with an aim for maximizing the number of optical connections, wavelengths are assigned for connection requests, i.e., light paths are assigned to carry traffic over optical carriers according to specified routing schemes.

Software-Defined Networking (SDN): A networking approach that envisages communication with underlying hardware infrastructure using software based controllers or Application Programming Interfaces to direct traffic on the network.

Space Division Multiplexing (SDM): A multiplexing technique used to transmit several independent channels separated in spatial dimensions, adopted in MIMO wireless communications and fiber optic communications.

Wavelength-Selective Switches (WSS): In optical networks, reconfiguration of traffic and bandwidth sharing over the optical layer is taken care of by the WSS. A WSS uses a diffraction grating to separate wavelengths and a switch section with an array of optical ports.

Wavelength-Switched Optical Networks (WSON): WSON are Wavelength Division Multiplexing (WDM) networks which consist of switching elements that perform switching depending on the wavelength of signals transported through an optical fiber.

Chapter 10
A Summary on 5G and Future 6G Internet of Things

Kok Yeow You
https://orcid.org/0000-0001-5214-7571
Universiti Teknologi Malaysia, Malaysia

ABSTRACT

This chapter comprehensively surveys the six aspects of the 5G and future 6G internet of things (IoT). First, most of the 5G- and 6G-IoT usage scenarios and key performance indicators are summarized in the forms of tables, pictures, and diagrams to facilitate readers to understand and compare current and future IoT technologies more easily and quickly. Second, 5G- and 6G-IoT access networks, protocols, and standards were briefly analyzed and compared, such as coverage, transfer data speed, energy consumption, operating frequency, and the number of device connectivity. Third and fourth, the impact of 6G-IoT on society's daily life and industry operation, as well as its underlying research were described. Fifth, five types of 6G-IoT challenges were analyzed and discussed in detail in this chapter, namely transmission path loss at THz, wireless network coverage, transfer data rate, latency, security, privacy protection, and energy-efficient and reliable devices/services. Finally, the latest nine IoT business models are described and summarized in tabular form.

INTRODUCTION

Around 2008 and 2009, the 'Internet of Things' term began to be mentioned in academia and industry. The 'Internet of Things' was normally represented by short form of 'IoT' and originally defined as (International Telecommunication Union, 2012):

A global infrastructure for the information society, enabling advanced services by interconnecting (physical and virtual) things based on existing and evolving interoperable information and communication technologies.

Over the past decade, a great deal of studies and research related to the Internet of Things (IoT) has been conducted, as well as commercial IoT products already exist in the market. Therefore, academic

DOI: 10.4018/978-1-7998-9266-3.ch010

circles, industry circles, and manufacturers now have a clearer understanding of the infrastructure form and business model of the IoT compared to the initial proposal. As a result, recently, more specific and explicit terms have been used in the definition of the IoT (Riazul Islam *et al.*, 2015, IoT for all, 2021, Wikipedia, 2021):

A physical object that is embedded with sensors, processing ability, software, and other technologies, and that connects and exchanges data with other devices and systems over the Internet or other communications networks.

or

A system of interrelated computing devices, mechanical, and digital machines provided with unique identifiers (UIDs) and the ability to transfer data over a network without requiring human-to-human or human-to-computer interaction.

or

A device embedded with electronics, software, sensors, actuators, and network connectivity that are capable of covering a variety of protocols, domains, and applications, which include the automotive industry, public safety, emergency services, and medical field.

In short, IoT is an infrastructure where the use involves people, devices, and services connected by wireless communication technology. For instance, the implementation of future sixth generation wireless technology (6G) in IoT assisted by artificial intelligence (AI) features is called 6G-IoT.

At present, the IoT applications have been widely implemented in the industrial field and our daily lives, covering eight main infrastructures, namely smart wearables, smart home, smart city, smart agriculture, smart vehicle, smart health care, industry automation, smart energy as shown in Figure 1. In 2018, the utilize of the IoT is expanded through the evolution of fifth generation wireless technology (5G). When 5G wireless technology is used, predictions and preliminary studies related to future 6G wireless technology have already begun, with operating frequencies above 95 GHz to 3 THz. For example, since 2018 until Jan 2023, up to thousands of studies and recommendations related to 6G communications and IoT have been documented (Huang et al., 2019, Letaief et al., 2019, Saad et al., 2019, Zhang et al., 2019a, Zhang et al., 2019b, Zong et al., 2019, Akhtar et al., 2020, Alsharif et al., 2020, Ian et al., 2020, Lee et al., 2020, Michailidis et al., 2020, Sekaran et al., 2020, Barakat et al., 2021, Chen and Okada, 2021, Dao et al., 2021, De Alwis et al., 2021, Dinh et al., 2021, Guo et al., 2021, Imoize et al., 2021, Ji et al., 2021, Jiang et al., 2021, Padhi & Charrua, 2021, Schroeder, 2021, Spyridis et al., 2021, You et al., 2021).

A comprehensive review of 6G-IoT can be found in several literatures (Kim, 2021; Chen & Okada, 2021; Barakat et al., 2021; Guo et al., 2021; Padhi & Charrua-Santos, 2021; Pattnaik et al., 2022; Dinh et al., 2022; Hosseinzadeh et al., 2022; Pajooh et al., 2022; Qadir et al., 2022). Figure 2 shows the number of publications indexed in the Web of Science (WoS) platform using the search keywords of '6G' and '6G IoT', respectively. Most of the documents are published by IEEE, Elsevier, Springer Nature, Wiley, MDPI, and other publishers. Based on WoS results using the search keyword '6G IoT', a total of 794 indexed publications are recorded from 2018 to January 2023, of which IEEE publishers contributed 526 (66%). On the other hand, Elsevier, Springer Nature, MDPI, and Wiley had 61 (7.7%), 56 (7.1%),

55 (6.9%), and 17 (2.1%) publications, respectively. Obviously, studies or research outputs related to the topic of 6G-IoT is increasing every year, especially from 2020 onwards.

Previously, the author has been conducted survey on three sub-topics of recent and future IoT, namely 5G-IoT protocols and standards, 5G- and 6G-IoT application categories, and their access networks (You, 2022a, 2022b). In this chapter, the author attempts to extend the survey of IoT comprehensively, such as the topic of the 6G-IoT challenges and IoT business models. Due to the 5G-IoT infrastructure has been launched since 2018, most of the information related to 5G-IoT is not new and well known. Hence, apart from 5G-IoT protocols, standards, and application categories, most of the survey topics focus on the future of 6G-IoT. Furthermore, this chapter summarizes all views, recommendations, and expectations of other researchers on future 6G-IoT. From the comparative analysis, readers can better understand the development trend of 6G-IoT in various aspects.

Figure 1. IoT applications
Source: You (2022b)

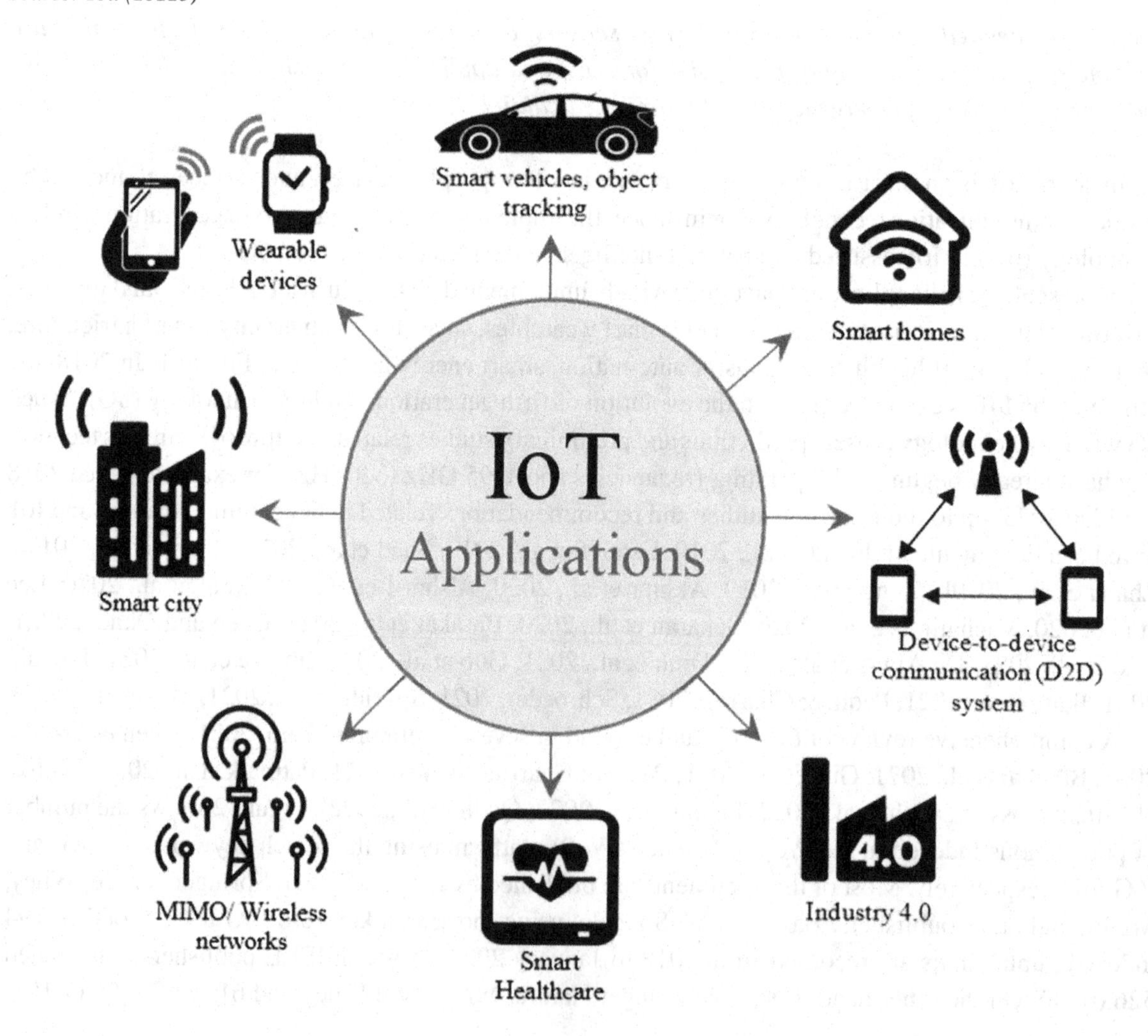

Figure 2. Indexed documents in WoS platform based on '6G' and '6G IoT' keywords

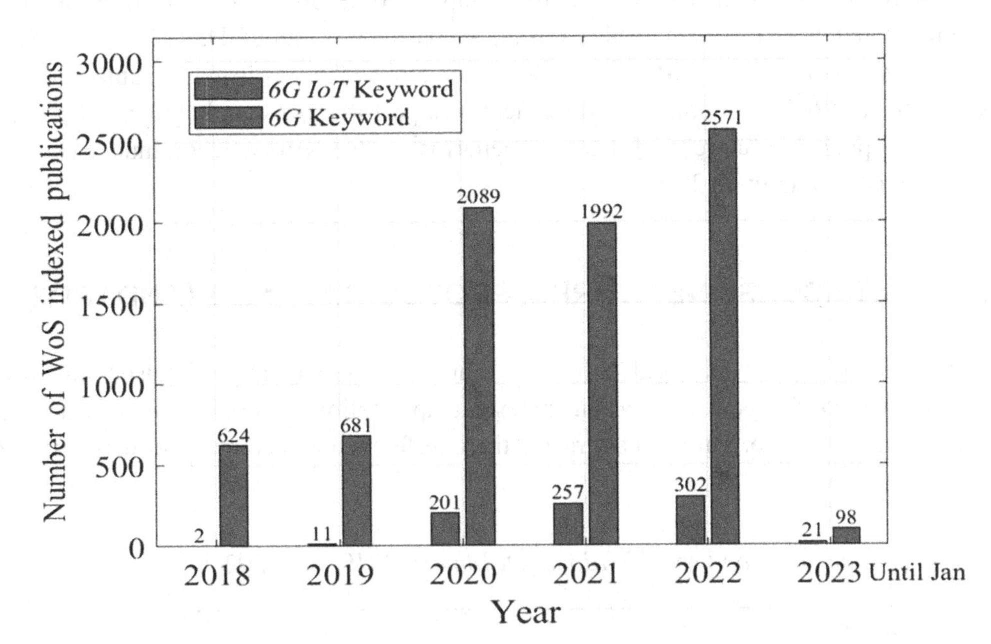

5G- AND 6G-IoT USAGE SCENARIOS AND KEY PERFORMANCE INDICATORS (KPI)

The International Telecommunication Union (ITU) has defined three main uses for fifth-generation (5G) wireless technology, namely Enhanced Mobile Broadband (eMBB), Ultra-reliable and Low-latency Communications (uRLLC), and Massive Machine Type Communications (mMTC), respectively. The eMBB aims to achieve the people's demand for an increasingly digital lifestyle, and focuses on services that have high requirements for bandwidth, such as high definition (HD) videos, virtual reality (VR), augmented reality (AR), and broadband IoT. On the other hand, the uRLLC and mMTC are designed for critical IoT and massive IoT, respectively. The uRLLC refer to using the network that requires uninterrupted and robust data exchange, such as assisted and automated driving, as well as remote management. While, the mMTC is used to connect to a large number of low power, low cost devices, which have high scalability and increased battery lifetime, in a wide area, such as smart city and smart agriculture as shown in Table 1 (DeTomasi, 2019). The applications of eMBB, uRLLC, and mMTC scenarios are shown in Figures 3, 4, and 5, respectively. The available industrial IoT (IIoT) subcategories, applications, technologies, and used protocols are listed in Table 2 (Oza, 2020).

As mentioned earlier, the use of devices connected to the internet will increase (massive connectivity) and reach the state of 'Internet of Everything (IoE)' in the near future. Therefore, the three 5G scenarios require to be strengthened and separated into more sub-scenarios for more diversified applications, namely 6G application scenarios. Starting in 2019, several 6G application scenarios have been proposed (Saad et al., 2019; Zhang et al., 2019b; Zong et al., 2019; Letaied et al., 2019; Gui et al., 2020; Dang et al., 2020). For instance, Saad et al. (2019) extended 6G applications to four scenarios, namely MBRLLC, muRLLC, HCS, and MPS. On the other hand, Zhang et al. (2019b) proposed five 6G usage scenarios, such as FeMBB, eURLLC, LDHMC, umMTC, and ELPC. In addition, Zong et al. (2019) enhanced the available three 5G usage scenarios suitable for 6G applications, re-named as uMUB, uHSLLC, and uHDD. While Letaief et al. (2019) named 6G usage scenarios as CAeC, EDuRLLC, and COC. Dang et al. (2020) is given as BigCom, SuRLLC, 3D-InteCom, and UCDC, as listed in Table 3. Besides, the comparison of 5G and 6G requirements of specifications, performances, and applications is tabulated in Table 4. In fact, predicted future 6G-IoT usage scenarios and 6G KPIs will continue to evolve and be revised in the years leading up to 2030.

5G- AND 6G-IoT ACCESS NETWORKS, PROTOCOLS, AND STANDARDS

Due to wireless technology not only used for human-to-human communication but also human-to-device or devices-to-device (D2D), as well as various types and applications of IoT has been used in the same environment at the same time. This will aggravate the interference and coupling effects between wire-

Table 1. 5G usage categories and specifications (DeTomasi, 2019; You, 2020b)

5G Services	Target:
Enhanced Mobile Broadband **(eMBB)**	High data rate, large data applications, massive devices, user capacity.
	Features:
	1. Transfer all data all the time. 2. Cover 2 billion people on social media. 3. Support 500 km/h mobility. 4. Peak data rate: 20 Gbps for downlink & 10 Gbps for uplink.
	Main Applications:
	Visual seen in Figure 3. (a) Fixed wireless, (b) Ultra high definition (UHD) video, (c) Video call, (d) Mobile cloud computing, and (e) Virtual reality (VR) /Augmented reality (AR)
Ultra-Reliable, Low Latency Communications **(uRLLC)**	**Target:**
	Fast and highly reliable, perfect coverage and uptime, strong security.
	Features:
	1. Ultra-high reliability (99.9999% reliability). 2. Ultra-responsive. 3. Data rate from 50 kbps to 10 Mbps. 4. Less than 1 ms air interface latency and 5 ms end-to-end latency.
	Main Applications:
	Visual seen in Figure 4. (a) Vehicle-to-vehicle, (b) Industrial automation, (c) Public safety, and (d) Remote surgery
Massive Machine Type Communications **(mMTC)**	**Target:**
	Massive connection density, energy efficiency, reduced cost per device
	Features:
	1. Cover 30 billion 'things' connected. 2. Low cost and low energy consumption. 3. Connectivity density of 10^5 to 10^6 devices/km^2 4. 1 to 100 kbps/device. 5. 10 year battery life.
	Main Applications:
	Visual seen in Figure 5. (a) Wearables, (b) Health care monitoring, (c) Smart home/city, and (d) Smart sensors

Figure 3.eMBB applications

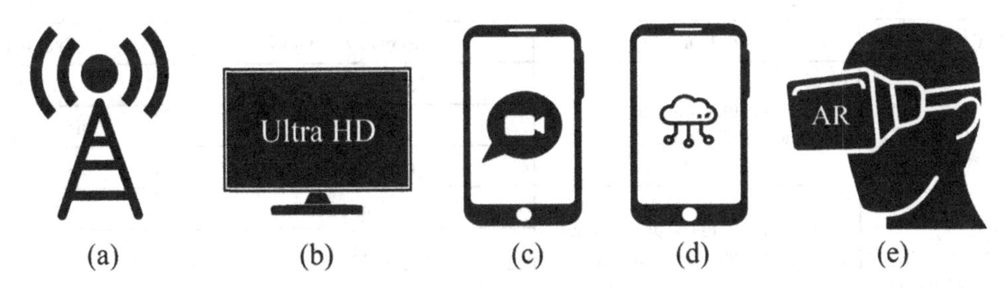

Figure 4.uRLLC applications

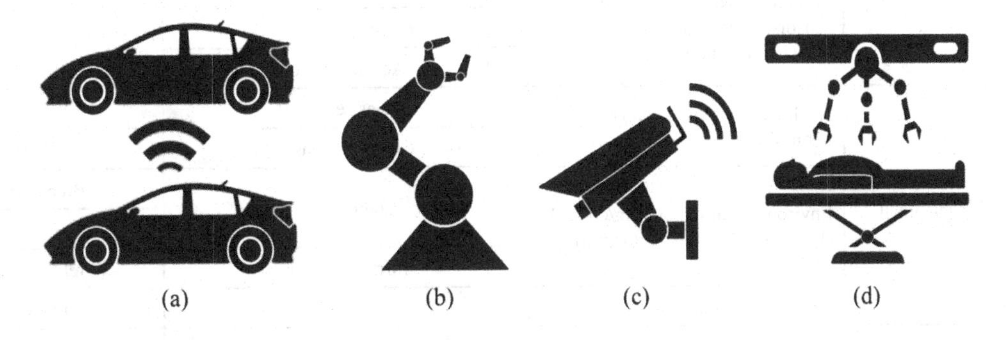

Figure 5.mMTC applications

less signals and cause data to be lost, voice quality may degrade, and the working range and battery life of the device may be reduced. It is impossible to allocate a separate frequency spectrum to each user, device, and application. Hence, various wireless protocols and standards employ a variety of communication techniques to help the communication operation peacefully coexist despite sharing the same frequency bands.

The IoT technologies can be categorized into three groups of coverage distance range applications as tabulated in Table 5. In addition to the coverage distance, IoT users can also select an appropriate wireless access protocol based on its baud rate, energy consumption, and cost required for particular IoT applications. Recently, these are two common unlicensed technologies used in IoT, namely Wireless Fidelity (Wi-Fi) and LoRaWAN (Long-range wide-area network) (Qorvo & RFMW, 2020). In fact, Wi-Fi is the unprecedented leader in broadband connectivity and its generations/protocols are shown in Table 6. Recently, the new generation of WiFi protocol (Wi-Fi 6) can benefit IoT hardware by improving battery performance, better outdoor operation, and expanding range (Qorvo and RFMW, 2020). On the

Table 2. Potential industrial internet of things (IIoT) (Oza, 2020; Chettri & Bera, 2020; You, 2022b)

<table>
<tr><th>Industrial IoT Categories</th><th>Applications</th><th>Sensory Devices</th><th>Protocols</th></tr>
<tr><td rowspan="10">Process Monitoring/ Predictive Maintenance</td><td rowspan="3">Machine health monitoring (precision CNC, conveyor belt)</td><td>Temperature sensor</td><td>LPWAN</td></tr>
<tr><td>Camera</td><td>WirelessHART</td></tr>
<tr><td>Humidity sensor</td><td>ISA100.11a</td></tr>
<tr><td rowspan="4">Asset monitoring (Hydraulic hose, pipeline, wellhead, steam trap, corrosion/structural integrity, seismic monitoring, tank level)</td><td>Pressure sensor</td><td>Cellular</td></tr>
<tr><td>Level sensor</td><td rowspan="6">Zigbee</td></tr>
<tr><td>Gas sensor</td></tr>
<tr><td>Proximity sensor</td></tr>
<tr><td rowspan="3">Remote visualization (Force sensors, laser measurement devices, cameras)</td><td>Acoustic sensor</td></tr>
<tr><td>Chemical sensor</td></tr>
<tr><td>Accelerometer</td></tr>
<tr><td rowspan="6">Facility Management</td><td rowspan="2">Health and safety monitoring (Emissions/Toxin)</td><td>Gas sensor</td><td>Wi-Fi</td></tr>
<tr><td>Chemical sensor</td><td>Cellular</td></tr>
<tr><td rowspan="3">Environmental monitoring/control (Lighting, HVAC, smart metering)</td><td rowspan="2">Light sensor</td><td>Bluetooth</td></tr>
<tr><td>ISA100.11a</td></tr>
<tr><td>Infrared sensor</td><td>WirelessHART</td></tr>
<tr><td>Perimeter security</td><td>Camera</td><td>LPWAN</td></tr>
<tr><td rowspan="4">Inventory Management</td><td rowspan="4">Asset Tracking (RTLS)</td><td>Bluetooth beacons</td><td>Bluetooth</td></tr>
<tr><td>RFID</td><td>Wi-Fi</td></tr>
<tr><td>Camera</td><td rowspan="2">UWB</td></tr>
<tr><td>Infrared sensor</td></tr>
<tr><td rowspan="4">Fleet Management</td><td rowspan="4">Delivery truck tracking, passenger car tracking, route development</td><td rowspan="4">GPS module</td><td>Cellular</td></tr>
<tr><td>LPWAN</td></tr>
<tr><td>NB-IoT</td></tr>
<tr><td>LTE-M1</td></tr>
</table>

Terminology: **RFID** — Radio Frequency Identification, **LPWAN** — Low-Power Wide-Area Network, **GPS** — Global Positioning System, **CNC** — Computer Numerical Control, **HVAC** — Heating, Ventilation, and Air Conditioning, **RTLS** —Real Time Location Systems, **UWB** — Ultra-wideband

other hand, LoRaWAN is used for long-range and low-energy connectivity. However, still a lot of IoT connectivity technologies use cellular networks and licensed spectrum under 3rd Generation Partnership Project (3GPP) access standards, such as Extended Coverage GSM IoT (EC-GSM-IoT), Long Term Evolution-Machine Type Communication (LTE-M), Narrowband Internet of Things (NB-IoT), 4G LTE, and 5G New Radio (NR). Compared with non-3GPP standards (LoraWAN and SixFog), 3GPP standards are mainly used for long-range high-quality mobile voice and data services.

As more and more IoT devices are connected to 3GPP standards, for instance, the china's government announced a policy that aims to reach 600 million NB-IoT nodes by 2020 compared to only 20 million NB-IoT based connections in 2017 (International Telecommunication Union, 2018a). Thus, here, author would like to briefly introduce what 3GPPs are. In fact, the 3GPP is an European Telecommunications Standards Institute (ETSI) formed in December 1998, partnered with other standard development organi-

Table 3. A comparative analysis of 5G and 6G usage scenarios (Saad et al., 2019; Imoize et al., 2021; Zhang et al., 2019b; Zong et al., 2019; Letaied et al., 2019; Gui et al., 2020; Dang et al., 2020)

Services/Usage Scenarios						
Recent 5G	Future 6G					
	Saad et al. 2019 Imoize et al., 2021	Zhang et al., 2019b	Zong et al., 2019	Letaief et al., 2019	Gui et al., 2020	Dang et al., 2020
Enhanced Mobile Broadband (eMBB)	**Mobile Broadband Reliable Low Latency Communication (MBRLLC = eMBB + uRLLC)**	**Further-Enhanced Mobile Broadband (FeMBB)**	**Ubiquitous Mobile Ultra-Broadband (uMUB)**	**Contextually Agile eMBB Communications (CAeC)**	**Mobile Broad Bandwidth & Low Latency (MBBLL)**	**Enhanced Mobile Broadband Plus (eMBB-Plus)**
i) Mobile cloud computing	i) XR/AR/VR	i) Holographic verticals & Society			i) Holographic teleconferencing	i) Conventional mobile communications
ii) Ultra high definition (UHD) video	ii) Autonomous vehicular systems	ii) Full-sensory digital sensing & reality			ii) AR, & VR	**Big Communications (BigCom)**
iii) Fixed wireless	iii) Autonomous drones	iii) Tactile/ Haptic internet				i) Large network coverage of urban and remote areas
iv) Video call	iv) Legacy eMBB & uRLLC	iv) UHD/SHD/ EHD videos				**Three-Dimensional Integrated Communications (3D-InteCom)**
v) XR/AR/VR						i) Satellite UAV
						ii) Underwater communications
Ultra-Reliable, Low Latency Communications (uRLLC)	**Massive Ultra-Reliable, Low Latency Communication (muRLLC = uRLLC + mMTC)**	**Extremely Ultra-Reliable, Low Latency Communications (eURLLC)**	**Ultra-High Speed with Low Latency Communications (uHSLLC)**	**Event Defined uRLLC (EDuRLLC)**	**Massive Low Latency Machine Type (mLLMT)**	**Secure Ultra-Reliable Low-Latency Communications (SURLLC)**
i) Industrial automation		i) Fully automated driving			i) Large scale industrial IoT (IIoT)	i) Industrial communications: Variety of robots, machine tools, and conveyor systems
ii) Vehicle-to-vehicle	i) Classical IoT	ii) Industrial internet			ii) Smart manufacturing	ii) Military communications
iii) Public safety	ii) User tracking	**Long-Distance and High-Mobility Communications (LDHMC)**			iii) Automatic transportation	iii) Vehicular communications
	iii) Blockchain & DLT					
iv) Remote surgery	iv) Massive sensing					
	v) Autonomous robotics					
		i) Deep-sea sightseeing				
		ii) Hyper-HSR				
		iii) Space travel				

continued on following page

Table 3. Continued

Services/Usage Scenarios						
Recent 5G	Future 6G					
	Saad et al. 2019 Imoize et al., 2021	Zhang et al., 2019b	Zong et al., 2019	Letaief et al., 2019	Gui et al., 2020	Dang et al., 2020
Massive Machine Type Communications (mMTC)	**Human-Centric Services (HCS)** i) BCI	**Ultra-Massive Machine Type Communications (umMTC)**	**Ultra-High Data Density (uHDD)**	**Computation Oriented Communications (COC)**	**Massive Broad Bandwidth Machine Type (mBBMT)**	**Unconventional Data Communications (UCDC)**
i) Wearables	ii) Haptics iii) Empathic communication	i) Internet of Everything (IoE)			i) Tactile Internet of Things (IoT)	i) Holographic ii) Tactile
ii) Health care monitoring	iv) Affective communication	**Extremely Low-Power Communications (ELPC)**				iii) Human bond communications
iii) Smart home/city	**Multi-Purpose 3CLS and Energy Services (MPS)**	i) Internet of Bio-Nano-Things				
iv) Smart sensors	i) CRAS ii) Telemedicine iii) Environmental mapping & imaging iv) Some special cases of XR services					

Terminology: **UHD/SHD/EHD** — Ultra-High-Definition/ Super-High-Definition / Extreme-High-Definition, **Hyper-HSR** — Hyper-High-Availability Seamless Redundancy, **DLT** — Distributed Ledger Technology, **3CLS** — Control, Localization, and Sensing, **CRAS** — Connected Robotic and Autonomous System, **XR/AR/VR** — Extended Reality/Augmented Reality/Virtual Reality, **BCI** — Brain-Computer Interface, **UAV** — Unmanned Aerial Vehicle

zations (SDOs) from around the world to develop new technologies specifications for the third generation (3G) of cellular networks. In July 2016, the first 3 GPP standards (part of Release 13) specifically for IoT applications were released, which is known as NB-IoT (International Telecommunication Union, 2018a, Díaz Zayas et al., 2020). Subsequently, 3GPP release 14 (initial) and release 15 (5G New Radio) were defined as the first phase of the 5G technology standards, which were approved in 2017 and June 2018, respectively. Then, the 3GPP Release 16 for second phase of the 5G standards are approved in 2018 and ended in July 2020. Recently, 3GPP release 17 are approved for enhanced support of industrial IoT in the 5G system. Starting in 2020, 5G has been widely deployed worldwide. For example, the first full 5G smartphone Samsung Galaxy S20 and the Nokia 8.3 5G with wider 5G compatibility were released in March 2020. While, the Apple iPhone 12 with 5G connectivity was released in October 2020.

In future, the 3GPP standardization for 6G is expected to be started in 2025 (Schroeder, 2021). The brief 5G, 6G, and corresponding 3GPP standards roadmap is shown in Figure 6. Further descriptions of 3GPP standard evolution, radio and system enhancements can be found on the 3GPP official website: https://www.3gpp.org/. In addition to the 5G specifications defined by 3GPP, there are two other important international committees, namely International Telecommunication Union (ITU) and International Mobile Communications (IMT-2020), which define 5G technical goals (Henry et al., 2020). Besides, for short- and medium-ranges applications, the Institute of Electrical and Electronics Engineers (IEEE) uses millimetre-wave (mmWave) unlicensed bands (45 GHz and 60 GHz) as the 5G technical standards,

Table 4. Comparison specifications, performances, and applications between current 5G and future 6G (Nguyen et al., 2021; Zhang et al., 2019b; Akyildiz et al., 2020; Saad et al., 2019; Chen & Okada, 2021; Imoize et al., 2021; You, 2022b)

Specifications, Performance, and Applications		5G	6G
1	Operating frequency	700 MHz to 95 GHz	700 MHz to 95 GHz 95 GHz to 3 THz (or above 10 THz)
2	Max. Bandwidth	1 GHz	100 GHz
3	Peak data rate	10 to 20 Gbps	1 Tbps
4	Experienced data rate	0.1 to 0.5 Gbps	10 Gbps
5	Latency	1 ms	0.1 ms
6	Reliability	99.999% ($1-10^{-5}$)	99.9999999% ($1-10^{-9}$)
7	Spectrum efficiency	30 bps/Hz	60 bps/Hz
8	Traffic density	10 Tbps/km^2	100 Tbps/km^2
9	Connectivity density	10^6 devices/km^2	10^7 devices/km^2
10	Mobility support	500 km/h	1000 km/h
11	Positioning precision	Meter level	Centimeter level
12	Receiver sensitivity	-120 dBm	< -130 dBm
13	Energy efficiency	-	1 Tb/Joule (100× over 5G)
14	Delay jitter	-	1 μs
15	Coverage	~ 70%	> 99%
16	Time buffer	Not real time	Real time
17	Autonomous vehicle	Partial	Fully
18	Haptic communication	Partial	Fully
19	Satellite integration	No	Fully
20	Extended reality (XR)	Partial	Fully
21	Artificial intelligence	Partial	Fully
22	Usage scenario/Services	eMBB, URLLC, mMTC	FeMBB, ERLLC, mURLLC, umMTC, LDHMC, ELPC (or MBRLLC, muRLLC) etc
23	Communication Technologies	i) mmWave communications	i) Sub-mmWave communications
		ii) Massive MIMO	ii) Spatial modulation (SM) MIMO
		iii) LDPC and polar codes	iii) LIS and HBF
		iv) Flexible frame structure	iv) OAM multiplexing
		v) Ultradense networks	v) Laser and VLC
		vi) Non-Orthogonal Multiple Access (NOMA)	vi) Blockchain-based spectrum sharing
		vii) Cloud/Fog/Edge computing	vii) Quantum communications and computing
		viii) SDN/NFV/Network slicing	viii) AI/Machine learning

continued on following page

Table 4. Continued

Specifications, Performance, and Applications		5G		6G	
24	Applications	i)	Virtual reality(VR) / Augmented reality (AR)/360° videos	i)	Holographic verticals and society
				ii)	Tactile/Haptic internet
		ii)	Ultra HD videos	iii)	Full-sensory digital sensing/reality
		iii)	Vehicle-to-everything (V2X)	iv)	Fully automated driving
		iv)	Smart city/factory/home	v)	Industrial internet
		v)	Telemedicine	vi)	Space travel
		vi)	Wearable devices	vii)	Deep-sea sightseeing
		vii)	Other Internet of Things (IoT)	viii)	Internet of Bio-Nano-Things

Terminology: **eMBB** — Enhanced Mobile Broadband, **mMTC** — Massive Machine Type Communications, **URLLC** — Ultra-Reliable, Low Latency Communications, **mURLLC** — Massive URLLC, **FeMBB** — Further-Enhanced Mobile Broadband, **ERLLC** — Extremely Reliable, Low Latency Communications, **umMTC** — Ultra-Massive Machine Type Communications, **LDHMC** — Long-Distance and High-Mobility Communications, **ELPC** — Extremely Low-Power Communications, **MIMO** — Multiple Input, Multiple Output, **mmWave** — Millimeter Wave, **OAM Multiplexing** — Orbital Angular Momentum Multiplexing, **LIS** — Large Intelligent Surfaces, **HBF** — Holographic Beamforming, **LDPC** — Low-Density Parity-Check Codes, **VLC** — Visible Light Communication, **NFV** — Network Function Virtualization, **SDN** — Software Defined Networking.

such as IEEE 802.11ad/aj/ay and IEEE 802.15.3c standards, to achieve data rates from 6.7 Gbps to 20 Gbps as shown in Tables 1 and 2 (Huang et al., 2018).

In fact, most industrial IoT applications do not require a high baud rate (high transfer data rate), but the main factors considered are the cost of sensors and devices, as well as battery life. For instance, agricultural remote sensing systems used for large-scale agricultural management and monitoring do not need to transmit data every second (data only needs to be sent once every half an hour), hence the access protocol of the IoT system is usually Low-Power Wide-Area Network (LPWAN), such as Sigfox, LoRaWAN, and LoWPAN, the protocol provides low power consumption and low baud rate, but long coverage distance (> 5 km). On the other hand, for the automation industry and vehicle-to-infrastructure communication, the IoT system provides intelligent, interconnected, and roboticized industrial production. Therefore, real-time data exchange is required to ensure the real-time response and precision of robot movement during factory production. In this case, Wi-Fi is usually selected for wireless interaction between IoT systems to achieve a high data transfer rate within 1 km coverage.

The relationship between baud rate, operating frequency, and coverage is shown in Figure 7. High operating frequency will provide a high baud rate, but the high operating frequency will reduce coverage. Vice versa, the high operating frequency has a low wavelength causing low penetration and high path loss. On the other hand, the relationship between energy consumption, baud rate, and coverage are illustrated in Figure 8 and Figure 9. In general, high baud rate has high energy consumption (will discuss in the next sub-section). Non-3 GPP standards (Sigfox, LoRaWAN, LoWPAN) have low energy consumption and low baud rate, but have high coverage. Here, IoT users can choose the appropriate access protocol or standard according to the required application specifications. The specification and performance comparisons between 802.15.4-based protocol (short range) and LPWAN technologies (Long range) are tabulated in the Table 7. Besides, other IoT wireless protocols are also compared in terms of coverage distance, power consumption, number of IoT device connections, bandwidth, operating frequency, and baud rate as illustrated in Table 7.

Figure 6. (a) 5G, (b) 6G, and 3 GPP standard version roadmap
Source: International Telecommunication Union (2018a), Schroeder (2021), Bhatia (2022)

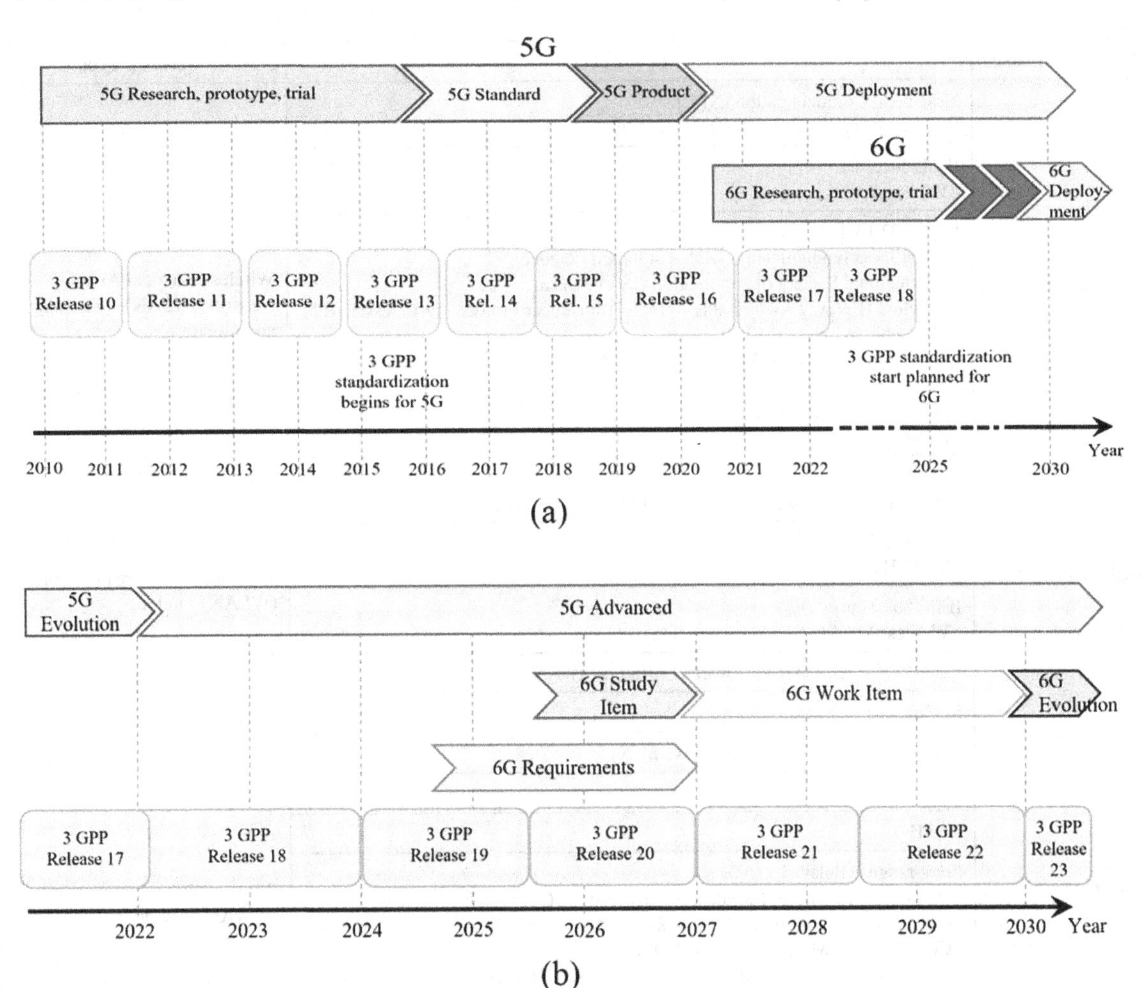

THE IMPACT OF 6G-IoT ON SOCIETY DAILY LIFE AND INDUSTRY OPERATIONS

The first 6G research items operating up to terahertz are expected to appear in 2025, starting to roll out standardization work in 2030, while the evolution of 5G will continue in parallel with early 6G research (Schroeder, 2021). In fact, in the last 40 years, microwave technology with operating frequencies exceeding 300 GHz to 3 THz has emerged and is in use to date, the so-called terahertz technology. The terahertz technology is widely utilized in the field of astronomy, medical, and security, such as space-based remote sensing and medical diagnostic imaging (Siegel, 2002, Elayan et al., 2020), due to the sub-millimeter waves that are nonionizing, and it can penetrate a wide variety of non-conducting materials. Furthermore, due to the sub-millimeter waves having a short wavelength between 0.1 mm to 1 mm, and very sensitive to subtle environmental changes. Therefore, the sub-millimeter waves are very suitable for high-sensitivity sensing and imaging applications (large bandwidth cause high resolution),

Table 5. Coverage distance range and its protocol and network type (Keysight, 2019; Qorvo & RFMW, 2020; Meneghello et al., 2019)

Coverage Distance	Protocol	Network Type
Short range (≤ 100 m)	Near-Field Communication (NFC) (use for QR codes, bar codes, and RFID tags)	Wireless Personal Area Network (WPAN) - wearable and mobile
	IEEE 802.15.1: - Bluetooth	
	IEEE 802.15.4: - Zonal Intercommunication Global-standard (Zigbee) - International Society of Automation (ISA100.11a) - Wireless Highway Addressable Remote Transducer Protocol (WirelessHART) - Microchip Wireless (MiWi) - IPv6 over Low-Power Wireless Personal Area Networks (6LoWPAN) - Thread - Subnetwork Access Protocol (SNAP) - Z-Wave	
	IEEE 802.11ad: - Wireless Gigabit Alliance (Wi-Gig or 60 GHz Wi-Fi)	
Medium range (≤ 1 km)	IEEE 802.11ah: - Wi-Fi HaLo	Wireless Local Area Network (WLAN) - indoor
	IEEE 802.11p: - Wireless Fidelity (Wi-Fi) generations	
Long range (>1 km)	Long-Range Wide-Area Network (LoRa/LoRaWAN)	Wireless Wide Area Network (WWAN) - outdoor
	SigFox	
	Wireless Smart Utility Network (Wi-SUN)	
	Ingenu	
	DASH-7	
	Long-range cellular: - 2G-Global System for Mobile Communications (2G-GSM) - General Packet Radio Service (GPRS) - Code Division Multiple Access (CDMA) - Universal Mobile Telecommunications System (UMTS) - Long Term Evolution (LTE) - Narrowband-IoT (NB-IoT) - Long Term Evolution-Machines (LTE-M) - Citizens Broadband Radio Service (CBRS) - Multi-Fire - 5G New Radio (5G NR)	

Terminology: **QR codes** — Quick Response codes, **IPv6** — Internet Protocol version 6, **RFID** — Radio Frequency Identification, **IEEE** — Institute of Electrical and Electronics Engineers

such as biosensors, atmospheric monitoring, personnel/baggage/package scanning, material thickness detection, and nanomaterial characterization devices, as shown in Figure 10.

Future 6G-IoT technology will bring mixed reality experience and immersive telepresence to life, while playing a key role in achieving global sustainable development, improving society, and increasing productivity in various industries. Various predictions and expectations for 6G-IoT use cases have been reported as scheduled in the Table 8. Obviously, every member of the future society/community will use more than 100 kinds of sensor devices and multiple wireless network platforms in their daily life.

Table 6. IEEE 802.11 protocol evolution and features (Qorvo & RFMW, 2020)

IEEE 80x.11 Protocol	Release Date	Frequency Band (GHz)	Bandwidth (MHz)	Max Throughput (bps)	Wi-Fi Generation	Distance (m)
801.11-1997	1997	2.4	22	2 M	-	-
802.11b	1999	2.4	22	11 M	Wi-Fi 1	100
802.11a	1999	5	20	54 M	Wi-Fi 2	50
802.11g	2003	2.4	20	54 M	Wi-Fi 3	100
802.11n	2009	2.4 & 5	20 & 40	600 M	Wi-Fi 4	100
802.11ad	2012	60	2160	6.7 G	Wi-Gig	10
802.11ac	2013	2.4 & 5	20, 40, 80, & 160	6.9 G	Wi-Fi 5	50
802.11af	2014	0.054 -0.79	6,7, & 8	35.6 M	White-Fi	> 3k
802.11ah	2017	0.7, 0.86, & 0.902	900	347 M	HaLow	1 k
802.11ax	2019	1 to 7	20, 40, 80, & 160	10 G	Wi-Fi 6	-
802.11ay	2019	Within 60	2160	20 G	-	10

Table 7. 802.15.4-based protocol and LPWAN technologies comparison (Oza, 2020; Keysight, 2019; Meneghello et al., 2019)

Wireless Technology	Frequency	Distance	Data Rate	Downlink	Bandwidth	Battery Life	Link Budget	Device
Wireless HART	2.4 GHz	~200 m	250 kbps	10-50 ms	3 MHz	Several years	-	30,000
ZigBee	2.4 GHz	~100 m	20 kbps - 250 kbps	10 m	600 kHz, 1.2 MHz, 2 MHz	-	-	-
ISA 100.11a	2.4 GHz	~200 m	250 kbps	~100 ms	5 MHz	Several years	-	unlimited
LoRa	169 MHz 470 MHz 868 MHz (Eur) 915 MHz (USA) 433 MHz (Asia)	5-32 km	0.3-50 kbps	-	7.8-500 kHz	>10 years	~150-157 dB	50,000
NB-IoT (LTE Cat NB2)	Cellular bands	1-10 km	< 159 kbps	1.6-10 s	180 kHz	>10 years	~ 164 dB	100,000
LTE Cat M2	Cellular bands	>11 km	4 Mbps (Down Link) 7 Mbps (Up Link)	10-15 ms	5 MHz	-	-	>>100,000
Sigfox	868 MHz 902 MHz	>50 km	100-600 kbps	-	100-600 Hz	>10 years	~146-162 dB	-
Bluetooth 5 Low Energy	2.4 GHz	< 200 m (PtP) < 1.5 km (Mesh)	1-3 Mbps	< 3ms	~ 2 MHz	-	-	32,767
WiFi	2.4 GHz 3.6 GHz 4.9 GHz 5 GHz 5.9 GHz	< 1 km	150 kbps - 78 Mbps	1-3 ms	1 – 16 MHz, ~ 22 MHz	-		-

Figure 7. Data rate versus coverage distance range at different range of operating frequencies
Source: Keskinen (2017)

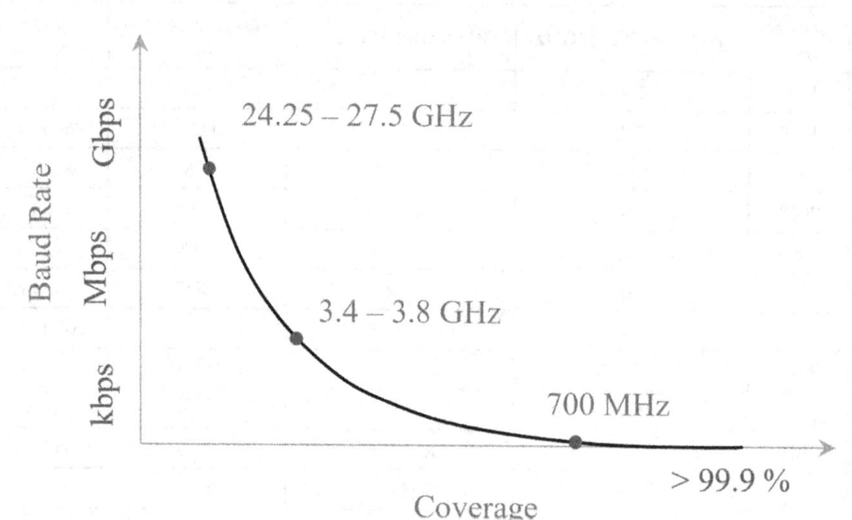

Figure 8. Energy consumption and coverage distance range are depended on the used wireless technologies

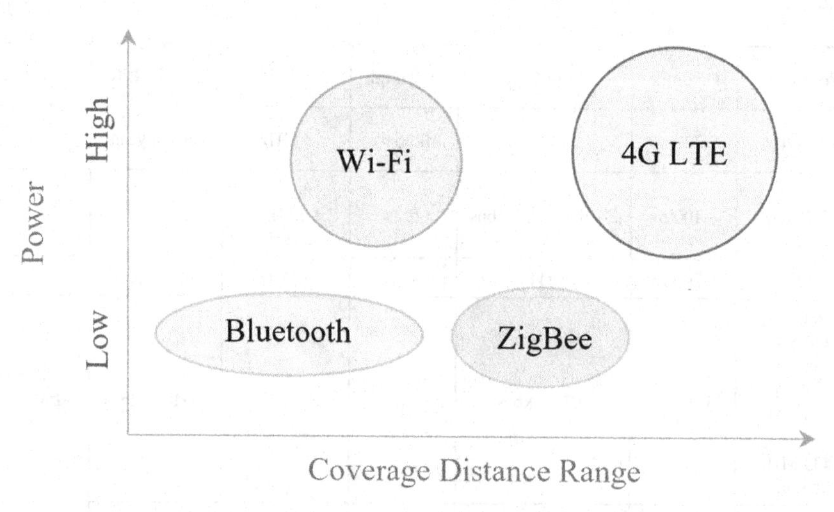

6G INTERNET OF THINGS (IoT) UNDERLYING RESEARCH

As listed in Table 4, the 6G-IoT goals to achieve data rates of 0.1–1 Tbps and spectrum efficiency of 3–60 bps/Hz, 100 GHz channel bandwidth, 1000 km/h mobility, 10^7 devices/km^2 of connectivity density, and fully automation. Hence, the coming soon 6G-IoT research may involve five main technology areas (Schroeder, 2021), namely (1) increasing operating frequencies of the internet access network to sub-millimeter waves band or terahertz, (2) improving multi-antenna techniques, (3) adopting artificial intelligence (AI) and machine learning (ML) in 6G-IoT, (4) integrating communications with more sensing capabilities, and (5) zero-energy devices.

High bandwidths up to several GHz are available at sub-THz and THz frequency regions (95 GHz to 3 THz), which provide ultra-high data rates, benefiting sensing and imaging applications, such as medical

Figure 9. Wireless technologies dependent baud rate and coverage distance range (WAN: wide area networks, PAN: personal area networks)

PAN
WAN
Wi-Fi
Cellular: 4G LTE, 5G GSM
Licensed LPWAN: LTE-M, NB-IoT
Bluetooth, Thread, ZigBee
Unlicensed LPWAN: Sigfox, LoRaWAN, LoWPAN
Baud Rate
Mbps
Kbps
bps
100 m
1 km
10 km
Coverage Distance Range

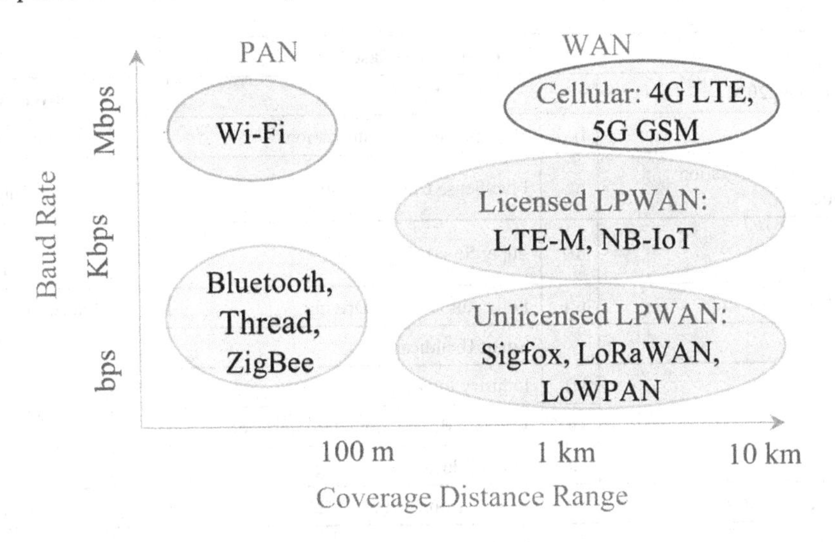

Figure 10. Terahertz applications

Astronomy
Air- and space-borne astronomy
Atmosphere monitoring
Research
Astrophysics
Materials science
Nano devices
Earth Sensing
Pollutants/hazards detection
Terahertz Applications
Telecommunication
6G Wireless communication
Biotech
Biosensors (DNA, protein)
Security & Defence
Hazardous substances and drugs detection
People, baggage and package scanning
Medical
Oncology
Dermatology
Life science
Cosmetics
Industrial
Composites QA/QC and maintenance
Pharmaceutical QC
Control of semiconductors
Thickness measurements

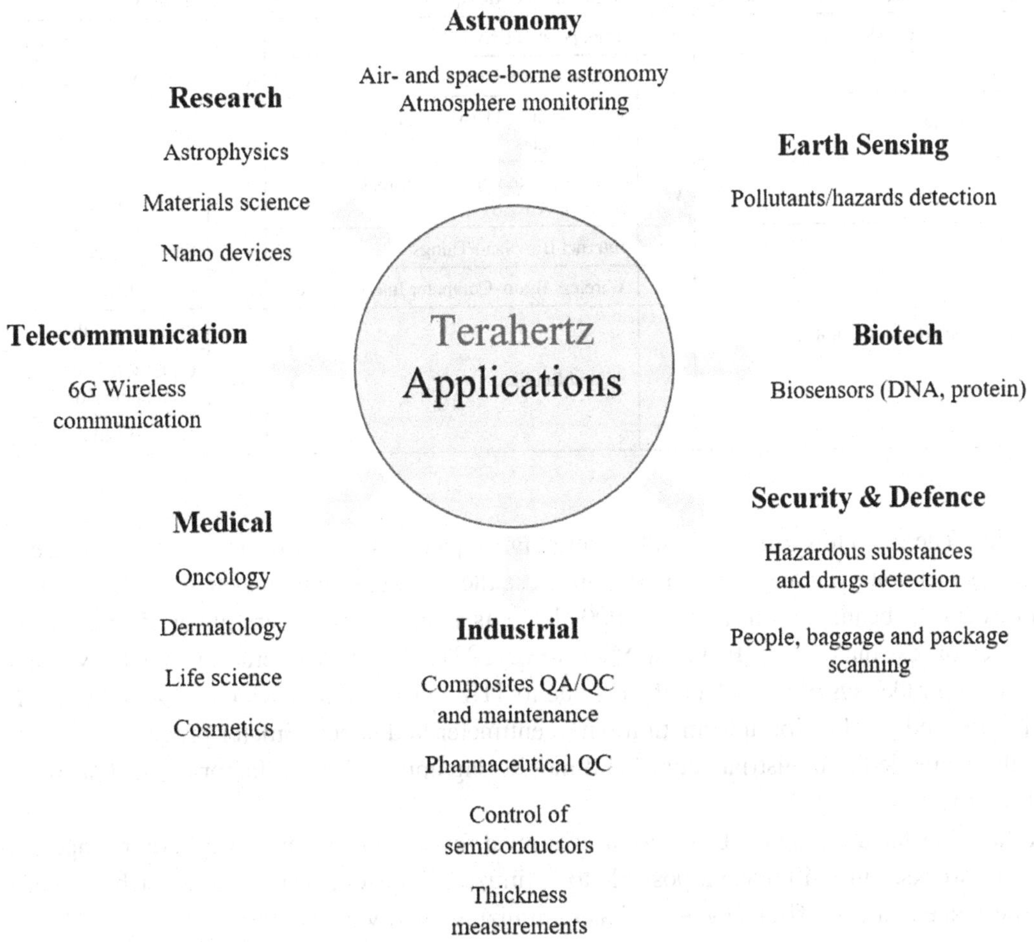

Table 8. 6G-IoT use cases (Chen, et al., 2021; Guo et al., 2021; Mahdi et al., 2021; Imoize et al., 2021 Nguyen et al., 2022; Das et al., 2022)

6G-IoT Use Cases					
Chen et al., 2021		**Guo et al., 2021**		**Nguyen et al., 2022**	
i)	Smart Home	i)	Holographic Communications	i)	Unmanned Aerial Vehicles
ii)	Satellite-Terrestrial-Integration Communication	ii)	Five-Sense Communications	ii)	Satellite Internet of Things
iii)	Smart City	iii)	Super Smart City/Home	iii)	Vehicular Internet of Things and Autonomous Diving
iv)	Underwater Communication	iv)	Fully Autonomous Driving	iv)	Internet of Healtcare Things
v)	Hospital	v)	Smart Healthcare	v)	Industrial Internet of Things
vi)	Industry	vi)	Industry Internet		
		vii)	Wireless Brain-Computer Interfaces		
		viii)	Smart Education/Training		
Mahdi et al., 2021		**Imoize et al., 2021**		**Das et al., 2022**	
i)	Holographic Vertical & Society	i)	Holographic Communication	i)	Metaverse and VR
ii)	Tactile/Haptic Internet	ii)	Tactile Internet	ii)	Smart Radio Environment
iii)	Full-Sensory Digital Sensing & Reality	iii)	Multisensory XR Applications	iii)	Smart Grid
iv)	Fully Automated Driving	iv)	Teleoperated Driving	iv)	Smart Transportation
v)	Industrial Internet	v)	Industry 4.0 & Beyond	v)	Smart Manufacturing
vi)	Space Travel	vi)	Blockchain and Distributed Ledger Technologies	vi)	Smart Education
vii)	Deep-Sea Sightseeing	vii)	Connected Robotics and Autonomous Systems (CRAS)	vii)	Smart City
viii)	Internet of Bio-Nano-Things	viii)	Internet Bio-Nano Things	viii)	Smart Farming
		ix)	Wireless Brain–Computer Interface (BCI)	ix)	Smart Banking
		x)	Digital Replica	x)	Smart Surveillance
				xi)	Underwater IoT Communication
				xii)	Smart Healthcare

diagnostics. The trend towards THz (high) operating frequencies and millimetre (short) wavelengths in 6G wireless technology, makes it possible to increase the number of antennas in a small area to provide higher directional beams. For instance, a 1000 elements antenna array operating at 250 GHz can install into an area of less than 4 cm^2 (Fairview Microwave, 2021). The THz operating frequency suitable for high-precision radar system to detect objects due to THz sensing can provide high-resolution environmental maps and localization information with centimeter and subcentimeter precision. For example, this could be applied to industrial control and monitoring applications in factories, such as robotics or virtual reality.

On the other hand, AI and ML is normally applied for self-management and control operations of massive networks. This will make it possible to optimize IoT system performance, such as maximizing user experience and cost efficiency as well as minimizing energy consumption.

6G-IoT CHALLENGES

Transmission Path Loss

As mentioned above, Terahertz (THz) frequency radio access networks from 90 GHz to 3 THz will be used for 6G communication applications by 2030 to meet increasing user demands and KPIs, as shown in the Table 9. However, long-range terrestrial radio access networks (TRAN) are not suitable for use at THz frequencies because transmitting wireless signals would deal with high transmission path losses mainly cause by atmospheric absorption and other environmental factors, such as attenuation due to atmospheric gases, clouds, and rain (You, 2022a). For instance, at sea level, $h = 0$ m, the atmospheric absorption loss is approximately 5 dB/km at 275 GHz and increases dramatically to 600 dB/km at 1 THz as shown in Figure 11.

Recently, low-altitude non-terrestrial networks (LANTN) are becoming an alternative method to solve the high loss problem for transmitted signal up to THz, since the signal attenuation decreases with the height of the sea level, h. When the height above sea level, h increases, the attenuation γ_A is expected to decrease due to the oxygen content and water vapor density, ρ will decrease as the h increases as illustrated in Figure 11. The calculated attenuation, γ_A values using MATLAB function 'gaspl' (Liebe et al., 1993) at various sea level altitudes, h are compared to the available data from ITU-R RA. 2189-1 (International Telecommunication Union, 2018b) as plotted in Figure 12. The entire calculated attenuation, γ_A is slightly less than the ITU data. It is found that the calculated γ_A needs to be multiplied by 1.2 to match the ITU data from 275 GHz to 1 THz. The infrastructures of non-terrestrial communication stations are normally handled by unmanned aerial vehicles (UAVs) (International Telecommunication

Figure 11. Specific attenuation, γ_A (dB/km) cause by atmospheric oxygen and water vapor at sea level, $h = 0$ m, atmospheric pressure, $P = 101.325$ kPa, temperature, $T = 20$ °C, and water vapor density, $\rho = 7.5$ gm^{-3}
Source: You (2022b)

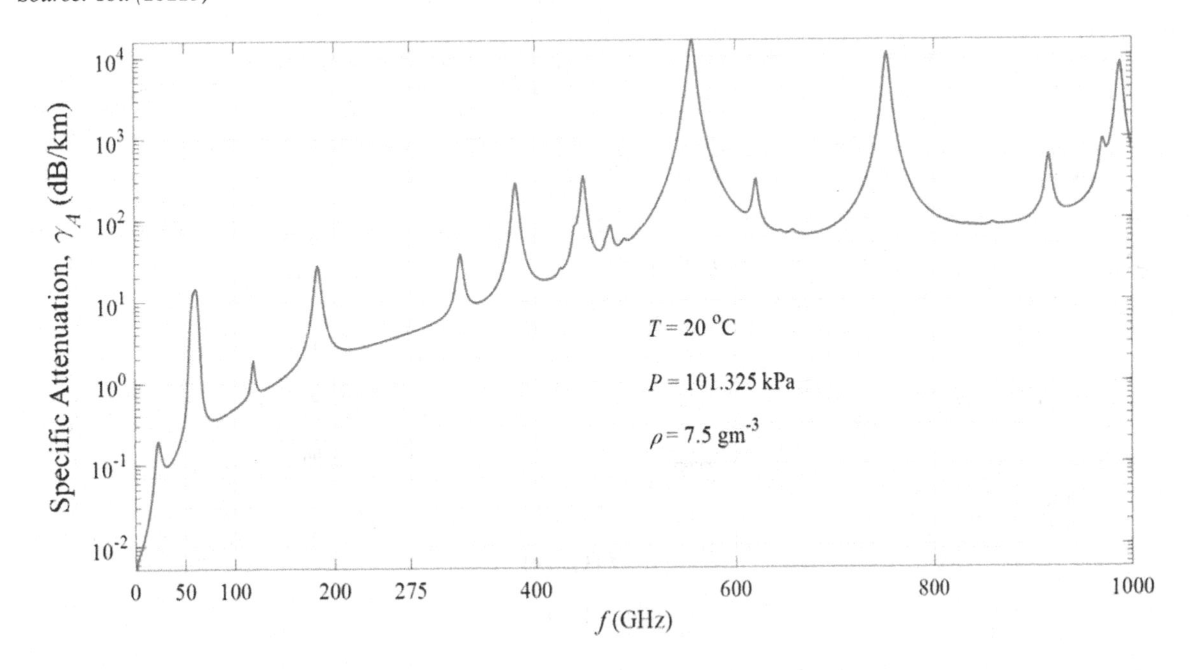

Union, 2018b; Giordani & Zorzi, 2021) which are very useful in low-altitude Internet of Things (IoT), such as remote sensing applications.

In addition to losses due to atmospheric absorption and other weather factors, free-space path loss, shadow fading, and line-of-sight (LoS) probability can also contribute to wireless signal loss. Shadow fading in wireless communication is the large-scale fluctuation of the signal envelope due to large objects (such as building) obstructing the propagation paths between the transmitter and the receiver as shown in Figure 13.

Figure 12. Comparison between data from ITU-R RA. 2189-1 and line-by-line model calculations (Liebe et al., 1993) at various height above sea levels from 275 GHz to 1 THz

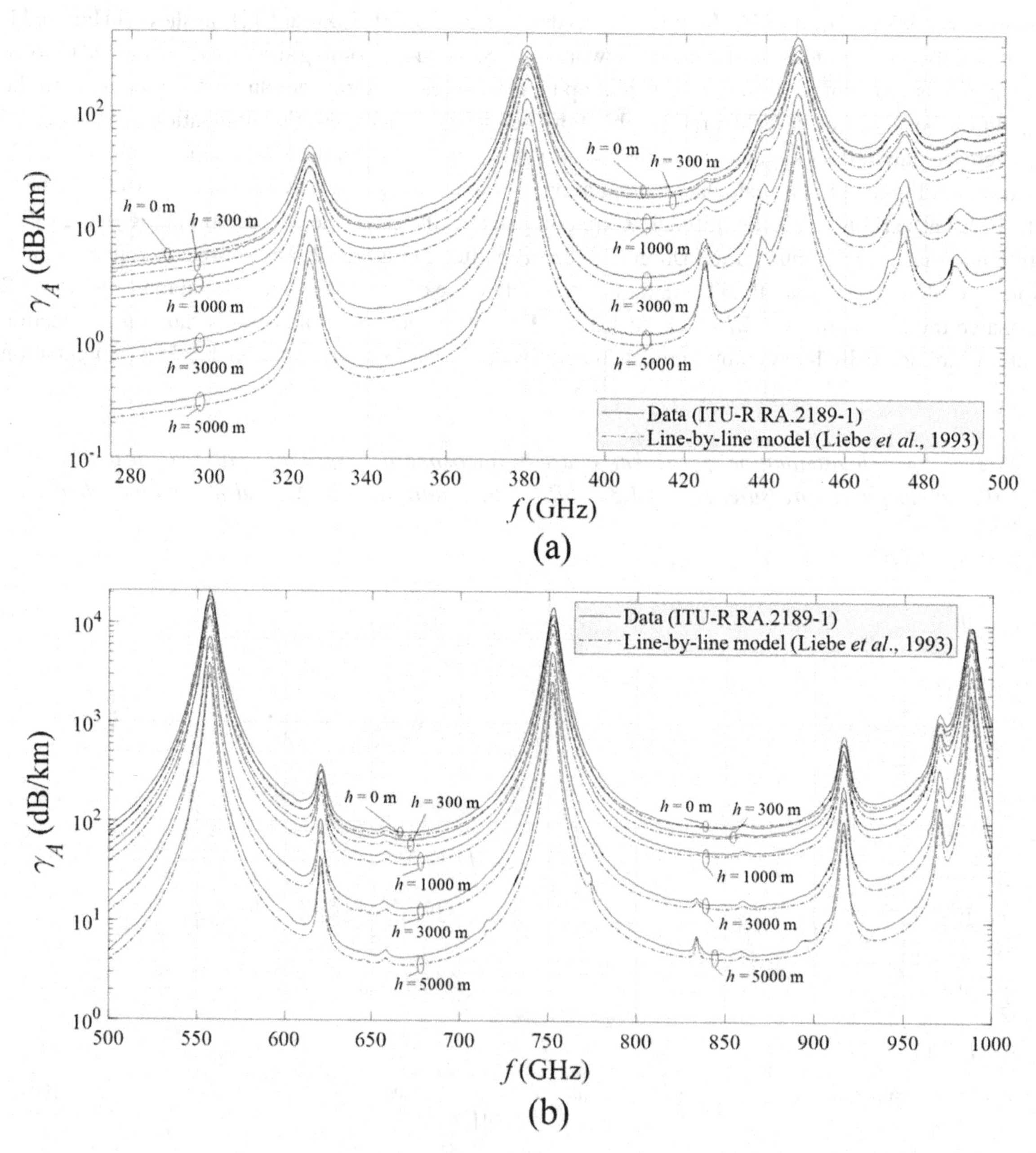

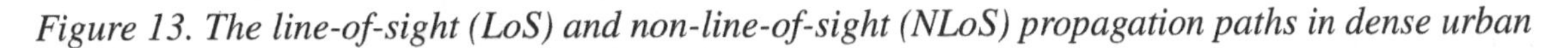

Figure 13. The line-of-sight (LoS) and non-line-of-sight (NLoS) propagation paths in dense urban

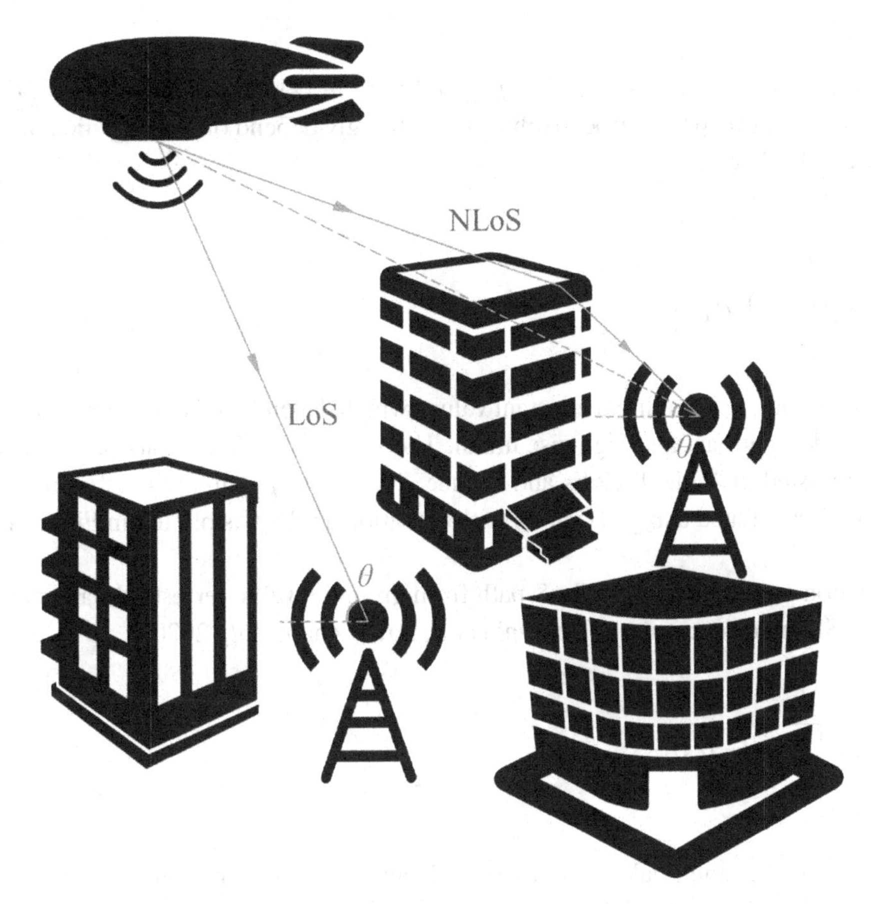

For instance, the average line-of-sight (LoS) and non-line-of-sight (NLoS) path loss, *PL* between UAVs and terrestrial base station in dense urban environment can be written as (Al-Hourani et al., 2014; Zhao et al., 2020):

$$PL = \left(p_{LoS} PL_{LoS}\right) + \left(p_{NLoS} PL_{NLoS}\right) \tag{1}$$

where PL_{LoS} and PL_{NLoS} are the path losses due to line-of-sight (LoS) and non-line-of-sight (NLoS) paths, respectively written (in unit dB) as:

$$PL_{LoS} = 20\log_{10} d + 20\log_{10} f - 147.558 + \eta_{LoS} \tag{2a}$$

$$PL_{NLoS} = 20\log_{10} d + 20\log_{10} f - 147.558 + \eta_{NLoS} \tag{2b}$$

where η_{LoS} and η_{NLoS} are the average additional loss of free-space propagation loss depending on the environment (rural, suburban, urban, dense urban, and high-rise urban). On the other hand, *d* (in unit meter) is the distance between the UAV-BS as (See Figure 14):

$$d = \sqrt{h^2 + r^2} \tag{3}$$

Symbol f is the operating frequency. The p_{LoS} and p_{NLoS} in (1) are the propagation probabilities of LoS and NLoS propagation paths respectively, which strongly depend on the elevation angle, θ (in unit degree) (Holis and Pechac, 2008):

$$p_{LoS} = \left[a - \frac{a-b}{1+\{(\theta - c)/d\}^e} \right] \times 0.01 \tag{4}$$

where a, b, c, d, and e are empirical constant values which depend on the environment, namely rural, suburban, urban, dense urban, and high-rise urban. The values of a, b, c, d, and e for various kinds of environments are listed in Table 9 (Holis and Pechac, 2008). The p_{LoS} in (4) satisfies the condition 0 £ p_{LoS} £ 1. The pLoS calculated using (4) varies with elevation angle, θ is plotted in Figure 15 (Holis and Pechac, 2008).

The propagation probability, p_{LoS} of LoS path from ground level to terrestrial base station (BS) can be simplified as Sigmoid function (Al-Hourani *et al.*, 2014; Zhao *et al.*, 2020):

$$p_{LoS} = \frac{1}{1+\alpha \exp\{-\beta(\theta - \alpha)\}} \tag{5a}$$

$$p_{NLoS} = 1 - p_{LoS} \tag{5b}$$

where α and β are the empirical constant values depending on the environment.

Figure 14. (a) UAV-BS coverage model. (b) The coverage area depends on the altitude.

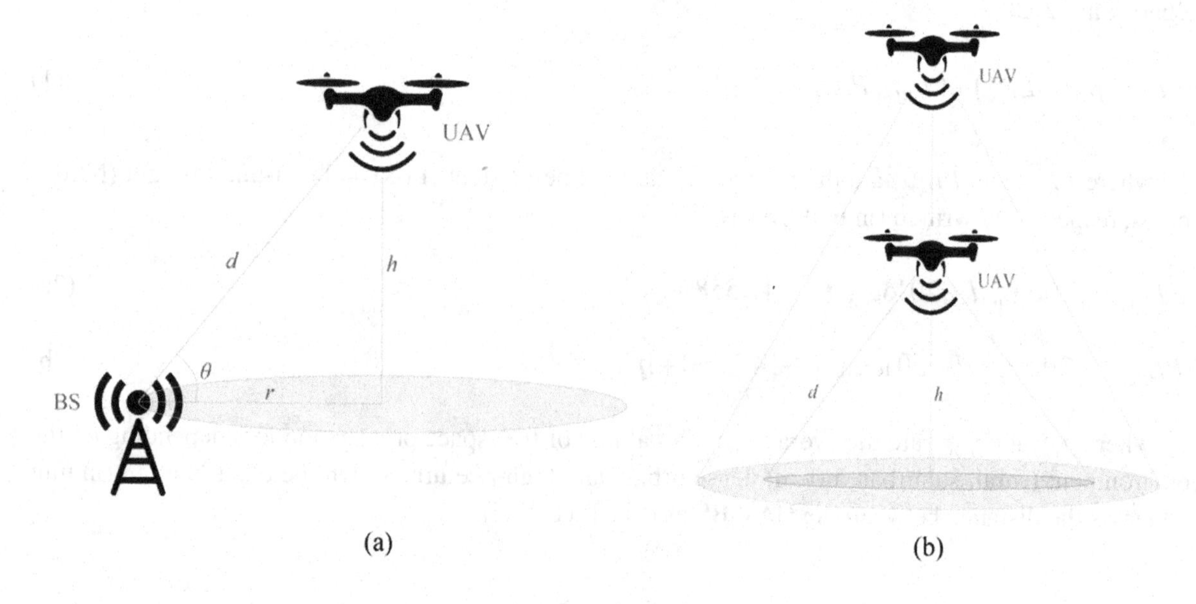

Table 9. Coefficient values of a, b, c, d, and e in (4) (Holis & Pechac, 2008)

Area	*a*	*b*	*c*	*d*	*e*
Suburban	101.6	0	0	3.25	1.241
Urban	120.0	0	0	24.30	1.229
Dense urban	187.3	0	0	82.10	1.478
High-rise urban	352.0	-1.37	-53	173.80	4.670

Figure 15. Variation in propagation probability, pLoS of LoS with elevation angle, θ at various kinds of environments
Source: Holis and Pechac (2008)

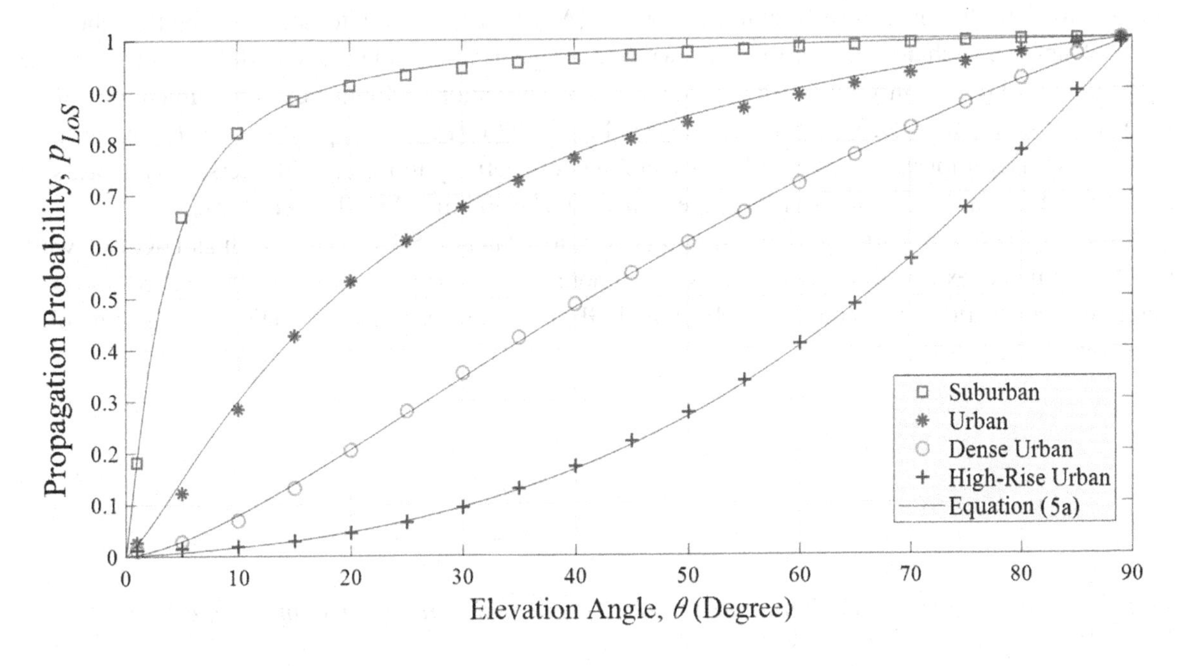

Wireless Network Coverage

Based on 6G-IoT KPI, the possible geo-location update data rate requirement in the future maybe 0.1 Tbps (100 Gbps) and near 100% geographical coverage (land, sea, and sky), as well as sub-centimeter geo-location accuracy (You et al., 2021). Hence, for fully automated driverless vehicle, operating frequency up to sub-millimeter wave is essential to be adopted, in which enter the era of 6G wireless technology.

Beside solving wireless signal loss, the non-terrestrial networks, such as unmanned aerial vehicle (UAV) access networks, are also used to solve the wireless network coverage for the transmitted signal of IoT devices up to THz, since the network access coverage increases with the height, h of the sea level. The infrastructures of non-terrestrial communication stations are normally handled by unmanned aerial vehicles, such as drones, vulture, and blimps. The aerial radio access networks can be divided into three types of platforms according to the height of the access network above sea level, h, namely low-altitude platforms (LAPs; h = 100 m to 10 km) and high-altitude platforms (HAPs; h = 10 km to 50 km), and spaceborne platforms [low earth orbits (LEO): h = 50 km to 1500 km, medium earth orbits

(MEO): h = 7000 km to 25000 km, and geostationary earth orbits (GEO): h = 35786 km] for satellite networks (Rinaldi et al., 2020, Dao et al., 2021), respectively. The integration of terrestrial and aerial access networks infrastructure is shown in Figure 16. The higher the UAV is off the terrestrial base station or ground, the greater the coverage area. Therefore, HAPs networks have wider coverage and longer endurance compared to LAPs.

Satellite networks connections complement terrestrial networks especially in remote areas for smart agricultural applications, asset tracking [such as Global Positioning System (GPS)], maritime and intermodal transportation, oil and gas industry exploration (Johan, 2021). Therefore, the installed base of satellite IoT connections will increase exponentially to 15.7 million units by 2025 as shown in Figure 17 (Johan, 2021). On the LAPs, drones are mainly used to supplement terrestrial coverage by providing connections to hotspots and scenes with weak terrestrial signals. On the other hand, airplanes, balloons, and airships are the main infrastructure of HAPs. HAPs seem like an alternative to satellites because of its advantages such as rapid deployment, wide coverage, low upgrade costs, high flexibility, and low propagation delay. Recently, the use of HAPs stations as International Mobile Telecommunications (IMT) base stations, so-called HIBS for mobile service below 2.7 GHz was proposed by IMT. The HIBS is particularly useful for providing low-latency mobile connectivity to unserved/underserved areas, such as rural and remote areas, and over large areas around 31,500 km^2 (Weissberger, 2021).

On the spaceborne platforms, LEO supports very low-latency 5G services, such as URLLC, while GEO implements extremely high data rates to promote eMBB services. Table 10 summarizes the comparative specifications between LEO, MEO, and GEO satellite systems and their coverage capability are shown in Figure 18.

Figure 16. 6G access networks infrastructure covers space, air, ground, and underwater/sea
Source: Dao et al. (2021) and You (2022b)

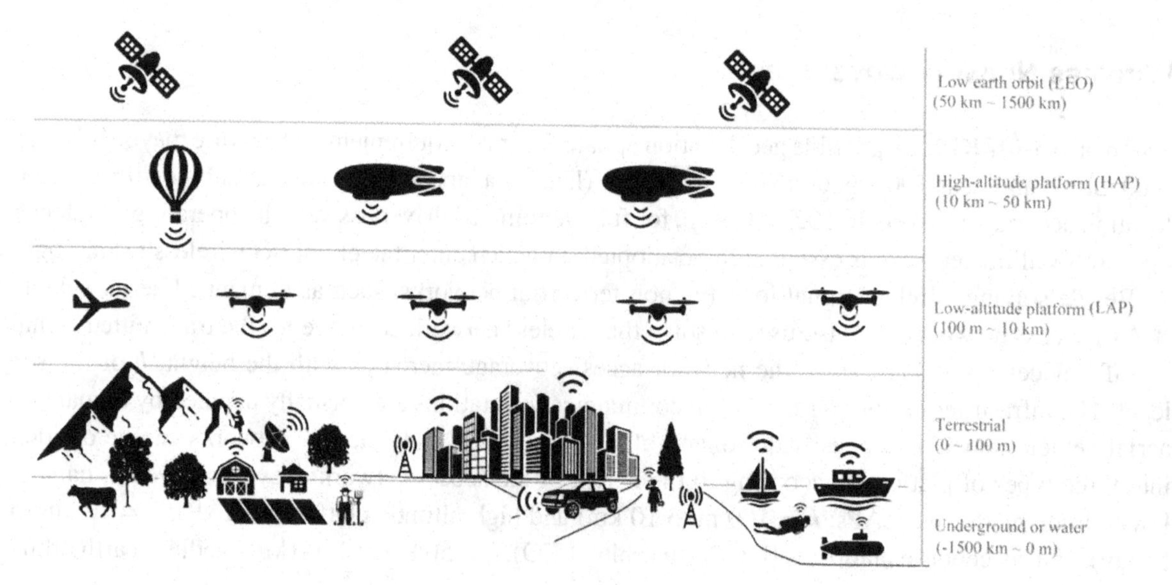

Figure 17. Global satellite IoT subscribers
Source: Johan (2021)

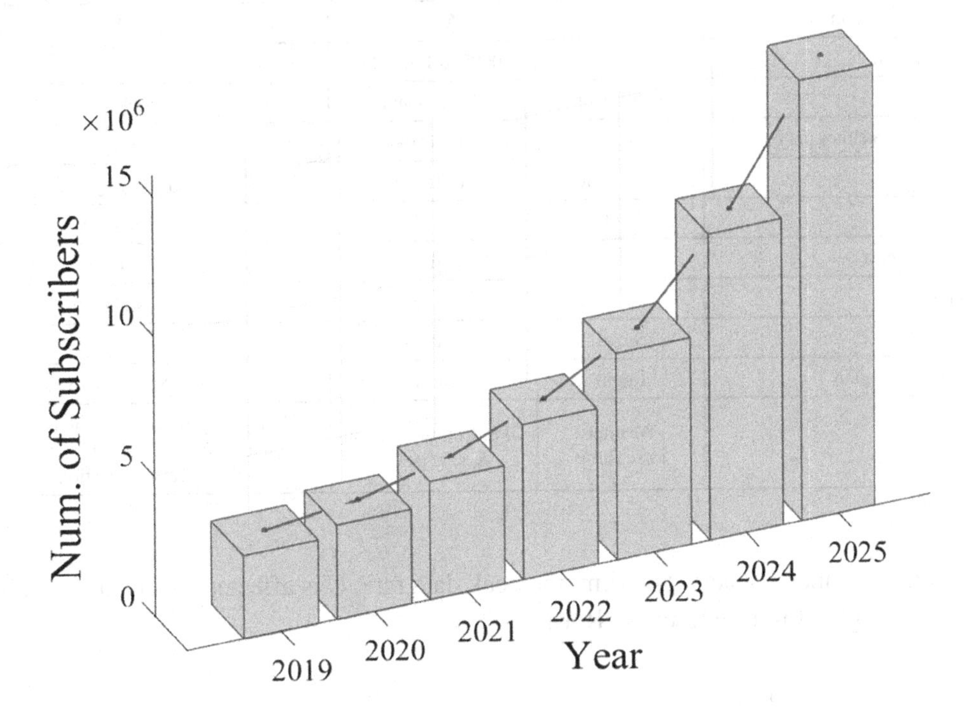

Figure 18. Orbit altitudes and coverage areas by LEO, MEO, and GEO, respectively

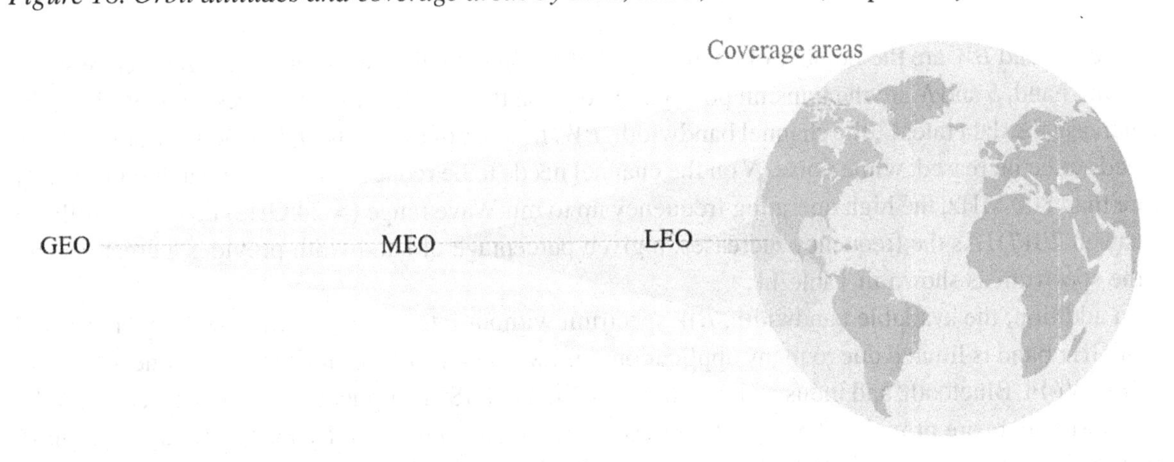

Transfer Data Rate and Latency

Although 5G communication can provide a peak data rate, *C* of 20 Gbps, however, when the capacity needs to be shared with multiple users and the large number of IoT devices, the performance of data rate will be significantly reduced (may be decreased by 80% above) (typical 0.1 Gbps of experienced data rate). Therefore, the higher target peak data rate, *C* that exceeds the expected value of 20 Gbps needs to be considered and the data rate limit needs to be increased based on the continuous increase in the number of connected IoT devices.

Table 10. Comparison specifications between LEO, MEO, and GEO satellite systems (INFOLYSiS, 2017)

Specification	LEO	MEO	GEO
Height, h from ground	700 to 1400 km	10000 to 15000 km	35786 km
Orbital period, t	10 to 40 minutes	2 to 8 hours	24 hours
Number of required satellites per operator	40+	10 to 15	3 to 4
Satellite life	3 to 7 years	10 to 15 years	10 to 15 years
Space segment cost	High	Low	Medium
Terrestrial gateway cost	High	Medium	Low
Propagation loss	Least	High	Highest
Coverage	Small	Medium	Large
Elevation angle, θ variation	Rapid	Slow	No variation
Main application	Weather forecasting	GPS, Communication & navigation	Telephony, data/TV distribution, mobile communication, broadcasting, low-speed data communication.

Based on the Shannon capacity theorem, the peak data rate, C is affected by channel bandwidth, B and the received signal-to-noise ratio, SNR as:

$$C = (n \times BW)\log_2\left(1 + \frac{S}{N}\right) \tag{6}$$

where n and BW are the number of channels and the channel bandwidth (in Hertz), respectively. On the other hand, S and N are the transmit power and noise on the channel (in Watts), respectively. Clearly, to increase the data rate, C, the channel bandwidth, BW, number of channels, n, and power transmission, S needs to be increased, while noise, N on the channel needs to be reduced. To obtain bandwidth, BW of more than 800 MHz, the high operating frequency up to mmWave range (> 24 GHz) need to be utilized (Huawei, 2017). As the frequency increases, a given percentage of bandwidth provides a greater share of the spectrum as shown in Table 11.

In addition, the available bandwidth, BW spectrum without interference in the low frequencies and sub-6GHz band is limited due to many applications and wireless protocols falls in that frequency range, such as Wi-Fi, Bluetooth, and Industrial, Scientific, & Medical (ISM) applications. For instance, recently, several suppliers are provided 2.5 Gbps transceivers in a single box using frequency band above sub-6 GHz (6 GHz to 42 GHz) for radio access network (Madarasz, 2019). In order to improve the received signal-to-noise ratio, S/N (improve latency), a high transmit power, S needs to be generated. Therefore, the usual high data rate wireless protocol is high energy consumption. Besides, high-quality, high-sensitivity, and low-noise receivers also contribute greatly to data rate performance.

As the connectivity of 6G IoT devices increases and the IoT devices are required to operate for increasingly complex tasks, such as artificial intelligence (AI) and machine learning (ML), thus the peak data rate, C is required to consider the different bandwidth, BW, transmit power, S, and interference, I of each application. Hence, the conventional Shannon capacity theorem is modified as (Wang *et al.*, 2020):

$$C \approx \sum_{m=1}^{M}\sum_{n=1}^{N} BW_{mn}\log_2\left(1+\frac{S_{mn}}{N_{mn}+I_{mn}}\right) \tag{7}$$

where M (³ 5) is the total number of global network coverage, such as satellite communication network (space), UAV communication network (air), terrestrial communication (ground), maritime and underwater communication networks (sea). On the other hand, N is the total channel of the 6G system, such as ultra-massive MIMO, OAM, HST/V2V/D2D, industry IoT, VR/AR/MR, wearable displays, mobile robots, and UAV. The added interference, I term in (7) may be interference randomization, interference coordination, interference cancellation, and interference alignment.

Figure 19 shows the bandwidth, BW required to achieve a specific maximum data transfer rate, C with the received signal quality of SNR = 50 dBm and 60 dBm, respectively. Based on (6) with $n = 1$, up to 100 GHz of usable BW has to be available at signal-to-noise ratio, S/N = 60 dBm in order to reach C = 1 Tbps. If the available BW is less than 100 GHz, then the S/N request has to be increased (> 60 dBm). Hence, recently, the Federal Communications Commission (FCC) proposed that the Spectrum Horizons License is located at operating frequency range of 116 to 123 GHz (BW = 7 GHz), 174.8 to 182 GHz (BW = 7.2 GHz), 185 to 190 GHz (BW = 5 GHz), and 244 to 246 GHz (BW = 2 GHz) (Fairview Microwave, 2021). Besides, Table 12 lists potentially usable frequency bands for 6G wireless technologies from 275 GHz to 3 THz which enable providing greater than 8 GHz to 110 GHz contiguous frequency bands (You, 2022a).

Security and Protect Privacy

In 2015, a survey conducted by Hewlett-Packard (HP) (Fortify, 2015) found that 70% of IoT devices are vulnerable to various attacks when connected to the Internet. By the end of 2020, up to 25% of enterprises

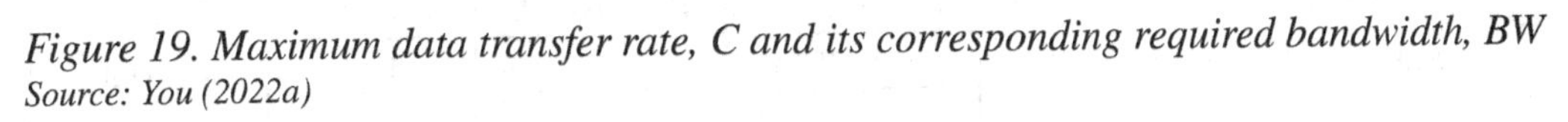

Figure 19. Maximum data transfer rate, C and its corresponding required bandwidth, BW
Source: You (2022a)

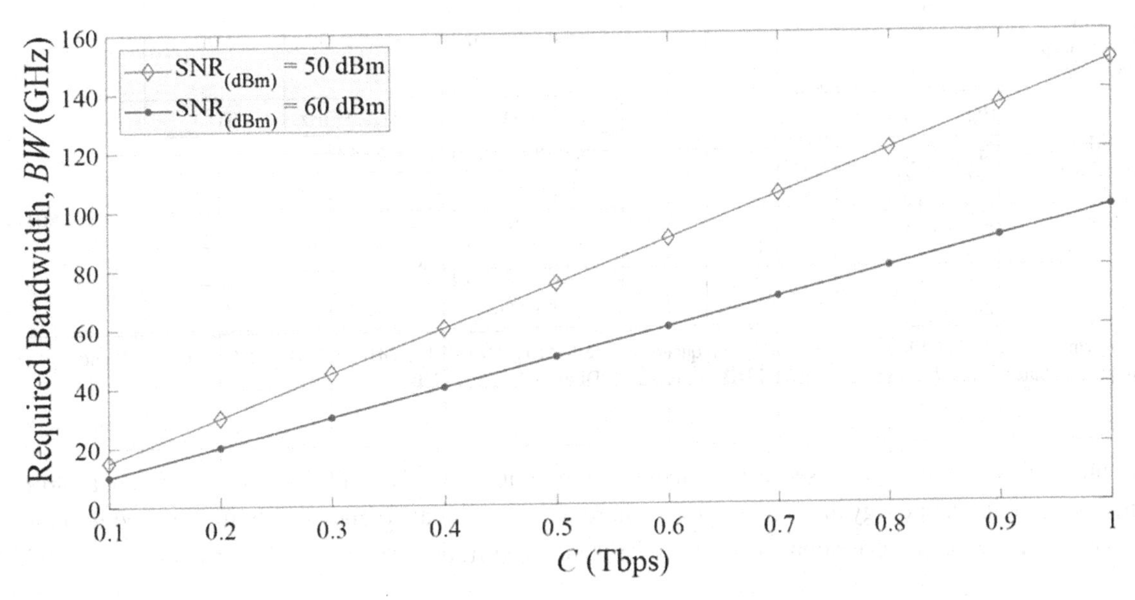

Table 11. Spectrum requirements for various 5G usage (Huawei, 2017)

5G Services	Frequencies Group	5G NR	Uplink	Downlink	Duplex Mode	Bandwidth, *BW*
eMBB, uRLLC, mMTC Wide-area and deep indoor coverage.	**Low frequencies (< 2.5 GHz)** 600 MHz, 700 MHz, 800 MHz, 900 MHz. 1.427–1.518 GHz, 1.695–1.785 GHz, 1.805–2.205 GHz, 2.30–2.69 GHz.	FR 1	663–698 MHz	617–652 MHz	FDD	>20 MHz
			699–716 MHz	729–746 MHz	FDD	
			703–748 MHz	758–803 MHz	FDD	
			824–849 MHz	869–894 MHz	FDD	
			832–862 MHz	791–821 MHz	FDD	
			880–915 MHz	925–960 MHz	FDD	
			1.427–1.432 GHz	1.427–1.432 GHz	TDD	
			1.427–1.470 GHz	1.475–1.518 GHz	FDD	
			1.432–1.517 GHz	1.432–1.517 GHz	TDD	
			1.695–1.710 GHz	1.995–2.020 GHz	FDD	
			1.710–1.785 GHz	1.805–1.880 GHz	FDD	
			1.710–1.780 GHz	2.110–2.200 GHz	FDD	
			1.850–1.910 GHz	1.930–1.990 GHz	FDD	
			1.850–1.915 GHz	1.930–1.995 GHz	FDD	
			1.880–1.920 GHz	1.880–1.920 GHz	TDD	
			1.920–1.980 GHz	2.110–2.170 GHz	FDD	
			2.010–2.025 GHz	2.010–2.025 GHz	TDD	
			2.300–2.400 GHz	2.300–2.400 GHz	TDD	
			2.496–2.690 GHz	2.496–2.690 GHz	TDD	
			2.500–2.570 GHz	2.620–2.690 GHz	FDD	
			2.570–2.620 GHz	2.570–2.620 GHz	TDD	
eMBB, uRLLC, mMTC Wide-area and best compromise between capacity, but no deep coverage.	**Medium frequencies (Sub-6 GHz)** 3.3 – 4.2 GHz 4.4 – 5.0 GHz 5.15 – 5.925 GHz	FR 1	3.30–3.80 GHz	3.30–3.80 GHz	TDD	> 100 MHz
			3.30–4.20 GHz	3.30–4.20 GHz	TDD	
			4.40– 5.00 GHz	4.40–5.00 GHz	TDD	
			5.15– 5.925 GHz	5.15– 5.925 GHz	TDD	
			5.855–5.925 GHz	5.855–5.925 GHz	TDD	
eMBB Specific use cases and extremely high data rates.	**High frequencies (> 6 GHz)** 24.25–27.5 GHz 27.5–29.5 GHz 31.8–33.4 GHz 37–40.5 GHz 40.5–42.5 GHz 42.5–43.5 GHz 45.5–47 GHz 47–47.2 GHz 47.2–50.2 GHz 50.4–52.6 GHz 66–71 GHz 71–76 GHz 81–86 GHz	FR 2	24.25–27.5 GHz	24.25–27.5 GHz	TDD	> 800 MHz
			26.5–29.5 GHz	26.5–29.5 GHz	TDD	
			27.5–28.35 GHz	27.5–28.35 GHz	TDD	
			37–40 GHz	37–40 GHz	TDD	
			39.5–43.5 GHz	39.5–43.5 GHz		
			47.2–48.2 GHz	47.2–48.2 GHz		

* Terminology: **5G NR FR 1** — 5G New Radio Frequency Range 1 (410 MHz to 7.125 GHz), **5G NR FR 2** — 5G New Radio Frequency Range 2 (24.25 GHz to 52.6 GHz), **FDD** — Frequency Division Duplex, **TDD** — Time Division Duplex.

are attacked due to compromised IoT devices. Hence, more recently, IoT, software-defined networking, and blockchain technology are integrated to inherently increase efficiency (without the need for central control and management), enhance security, and increase transparency for all parties involved, while

Table 12. Potential 6G wireless bands from 275 GHz to 3 THz with usable bandwidth, BW of 8 GHz to 110 GHz (You, 2022a)

f (GHz)	*BW* (GHz)	*f* (GHz)	*BW* (GHz)	*f* (GHz)	*BW* (GHz)
275 – 310	35	925 – 955	30	2090 – 2130	40
330 – 365	35	1005 – 1060	55	2285 – 2316	31
395 – 425	30	1259 – 1277	18	2322 – 2332	10
426 – 435	9	1279 – 1295	16	2490 – 2560	70
456 – 470	14	1298 – 1312	14	2570 – 2600	30
478 – 486	8	1330 – 1370	40	2700 – 2735	35
490 – 520	30	1450 – 1535	85	2810 – 2865	55
593 – 618	25	1546 – 1570	24	2895 – 2945	50
628 – 720	92	1816 – 1845	29		
790 – 900	110	1950 – 2010	60		

enabling secure machine-to-machine transactions (Nguyen et al., 2020; Saad et al., 2020; Pajooh et al., 2022; Shammar et al., 2021). The importance of IoT security depends on the kind of IoT application as summarized in the Table 13. Besides, the security protocol of IoT also depends on the type of wireless access technology used by the IoT devices as listed in the Table 14.

In fact, the concept of blockchain was first proposed by Satoshi Nakamoto in 2008, and it is used to store financial transactions and related protocols to ensure the validity of blockchain in the network (Shammar et al., 2021). The blockchain is built up of sequential blocks that can store various types of transactions. For instance, Watson IoT Platform can choose the data to be managed, analyzed, customized, and shared among permissioned customers and partners (IBM, 2023). However, most existing blockchain schemes are unable to meet the specific requirements of all IoT (Liu et al., 2019). The reason is that recent IoT environments are resource-constrained with limited capabilities in terms of computation, energy (limited battery life), and storage, which discourages the use of blockchain, which has high computational complexity, limited scalability, high bandwidth overhead, and latency (Abdulkader et al., 2019).

Energy Efficient and Reliable Devices/Services

The evolution of 5G wireless access network is transferred from the macrocell environment to the small cell coverage area as shown in Figure 20. In fact, the small cell is a miniature version of the traditional macrocell, which has all the same characteristics and features as the traditional macrocell (Tuan, 2020). However, the small cell feature is suitable for 5G deployments that promise ultra-high data rates, a million devices per square mile, and millisecond latency. The small cell hardware units are designed to reduce complexity, hence the hardware implementation is faster, easier, and low-power consumption (extends battery life).

For indoor cases, the performance of the wireless access signal will degrade interior of the building, especially in large buildings having multiple rooms, due to high loss building materials, such as low emissivity glass, metal, and concrete can degrade the wireless signals. Therefore, the distributed antenna system (DAS) is proposed to solve the indoor signals degradation issues by distributing the available external cell signals over the system of small antennas installed around the building in order to disperse

Table 13. IoT applications security requirements (Iqbal et al.,2020)

IoT Applications	Important Security Requirements					
	Availability	Confidentiality	Integrity	Non-Repudiation	Privacy	Authentication
Smart Grids	Yes	Yes	Yes	Yes	Yes	Trivial
Healthcare	Trivial	Yes	Yes	Trivial	Yes	Yes
Transportation Systems	Yes	Trivial	Trivial	Yes	Yes	Yes
Smart City	Yes	Yes	Yes	Trivial	Trivial	Yes
Smart Manufacturing	Yes	Yes	Yes	Trivial	Trivial	Yes
Smart Home	Yes	Yes	Yes	Trivial	Trivial	Yes
Smart Wearables	Trivial	Yes	Yes	Trivial	Yes	Yes
Smart Farming	Yes	Trivial	Trivial	Trivial	Trivial	Yes
Smart Supply Chain	Yes	Yes	Yes	Trivial	Yes	Yes
Smart Security Systems	Yes	Yes	Yes	Trivial	Yes	Yes

Table 14. IoT signal processing devices, techniques, and security algorithm/protocol (Pattnaik et al., 2022)

Wireless Technology	Modulation Scheme	Multiplexing/MAC Scheme	Security Algorithm
NFC	ASK, FSK, BPSK	Electromagnetic coupling, TDMA	Encryption Cryptographic, Secure Channel, Key Agreements
Bluetooth	GFSK, DQPSK, 8DPSK	TDD, FHSS	E0, E1, E21, E22, E3, 56-128 bit
BLE	GFSK, FHSS Star	FHSS	AES-128
ANT	GFSK	TDMA	AES-128 64 bit
Zigbee	BPSK, O-QPSK	DSSS, CSMA/CA TDMA+CSMA/CA	AES-128
Zwave	FSK, GFSK	FHSS, CSMA/CA	AES-128
WiFi	BPSK, QPSK, COFDM, CCK, M-QAM	CSMA/CA+PCF	CCMP 128
LoRaWAN	LoRa CSS, GFSK	Pure-ALOHA	AES-128 encryption
NB-IoT	QPSK, BPSK, GFSK, BPSK	SC-FDMA (Uplink) OFDMA (Downlink)	3GPP 128-256 bit
Sigfox	BPSK (Uplink), DBPSK (Uplink), GFSK (Downlink)	R-FDMA	AES-128 encryption

* Terminology: **NFC** — Near-Field Communication, **ASK** — Amplitude-Shift Keying, **FSK** — Frequency-Shift Keying, **BPSK** — Binary Phase-Shift Keying, **GFSK** — Gaussian Frequency-Shift Keying, **QPSK** — Quadrature Phase-Shift Keying, **M-QAM** — M-ary Quadrature Amplitude Modulation, **FHSS** — Frequency-Hopping Spread Spectrum, **DSSS** — Direct-Sequence Spread Spectrum **LoRa CCS** — Long Range Chirp Spread Spectrum, **CCK** — Complementary Code Keying, **OFDMA** — Orthogonal Frequency-Division Multiple Access, **COFDM** — Coded Orthogonal Frequency-Division Multiplexed, **CSMA** — Carrier Sense Multiple Access, **FDMA** — Frequency Division Multiple Access, **TDMA** — Time Division Multiple Access, **TDD** — Time Division Duplex, **PCF** — Point Coordination Function. **AES-128** — Advanced Encryption Standard-128 bit. **CCMP-128** — Counter Mode CBC-MAC Protocol-128 bit. **MAC** — Medium Access Control.

and amplify the cell signal throughout the building to achieve perfect coverage. In addition, the small cells and distributed antenna system (DAS) solutions can support multiple standards, such as the 3G/4G cellular and implement carrier aggregation with the LTE Advanced (LTE-A) systems (Tuan, 2020). There are three types of small cells, namely femtocells, picocells, and microcells as listed in Table 15.

Table 15. Macrocells and small cell types of the wireless network

Cell Type	Coverage Radius (m)	Indoor/ Outdoor	Transmit Power (dBm)	Number of Users	Cost	Applications
Femtocells	10 to 50	Indoor	20	4 to 16	Low	Residential, home, and small offices.
Picocells	100 to 250	Indoor/ Outdoor	24	32 to 64	Low	Offices, hospitals, shopping complexes, train stations, schools, universities, and in-aircraft.
Microcells	250 to 2500	Indoor/ Outdoor	33 to 37	100 to 2000	Medium	Mall, hotels, stations, transportation hub, and urban.
Macrocells	5000	Outdoor	> 40	>2000	High	Suburban

The efficiency of the 5G access network has been further improved, and it has begun to focus on the antenna design that will be installed on the base station, so-called multi-user MIMO (MU-MIMO) systems or large-scale antenna systems (LSAS). The general MU-MIMO is the base station (BS) with multiple antennas simultaneously serves a set of single-antenna users and the multiplexing gain can be shared by all users over the same frequency band. The base station uses antenna array ($N > 100$ antenna elements) as a beamforming antenna to simultaneously implement oriented signal transmission and reception (full-duplex). In order to change the direction of the array, the beamformer adjusts the phase shifts and amplitude of the signal at each antenna element. The change of phase shifts alone would be sufficient to achieve beam steering in different directions. In addition, the ability to change the amplitude enables optimization of the side lobe suppression.

Therefore, the beamforming of each antenna can increase the received power level of the user's device, mitigates interference to other users, and improve overall system efficiency. Recently, the problem of bandwidth and energy consumption is partly solved by massive MIMO (a large number of antenna elements, N). For instance, the 256 elements transmit antenna array normally having the gain of 28 dB. If 6.2 mW of transmit power is supplied for each antenna element, thus the total transmitted power can be achieved by 32 dBm [$10\log_{10}(256\times6.2\ \text{mW}/1\text{mW})$]. Finally, the combination of 32 dBm of transmitted power and 28 dB of antenna gain meets the 60 dBm of effective isotropic radiated power (EIRP) target (Admin, 2020).

For 3GPP, the 5G NR standard distinguishes between two ranges for carrier frequencies, namely Frequency Range 1 (Low frequencies and sub-6 GHz) and Frequency Range 2 (mmWave band from 24.25 GHz to 52.6 GHz). It should be noted that the sub-6 GHz band is extended operating frequencies under 5.925 GHz to 7.125 GHz. Although the coverage of sub-6 GHz is larger than the mmWave frequency band, operating frequencies below 6 GHz require a larger antenna aperture size, which leads to antenna weight, surface roughness, and antenna manufacturing costs. Hence, nowadays, hybrid beamforming (analog + digital) is most commonly used in mmWave bands to increase bandwidth usage (> 800 MHz), high data rates (20 Gbps download rates), and wireless network capacity (10^6 devices/km^2). For instance, China is planning and preparing 5G mmWave deployment for the 2022 Winter Olympics (Admin, 2021).

In future, 6G-IoT has to support massive numbers of IoT devices through more diverse network channels than previous systems, thus it is expected to face various security and privacy difficulties. Ultra-massive connection network costs are high, and hardware-based networking fails to accommodate them. Hence, software-defined networking (SDN) is employed to handle such a large number of connected devices and massive data operations. On the other hand, the progress of the 6G-IoT research will be directly affected by the development of the sub-millimeter wave monolithic microwave integrated circuits (MMICs)

Figure 20. The evolution of 5G wireless networks
Source: Tuan (2020)

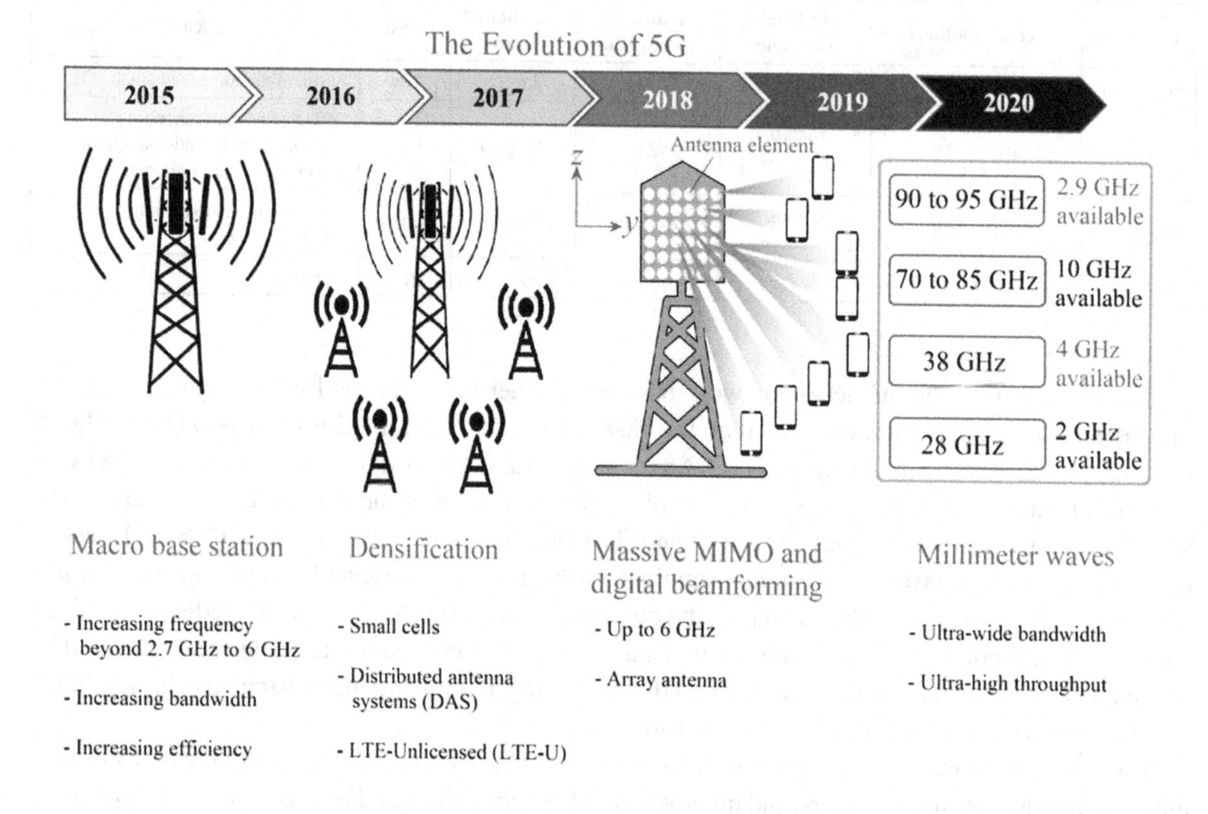

industry. In addition to high operating frequencies (> 300 GHz), MMICs used in 6G-IoT also require to have good thermal resistance, low DC power consumption, high sensitivity, high output RF power, high thermal stability, nanoscale integrated circuits, broadband, and less expensive. The specification and properties of the current MMICs technologies is shown in Figure 21.

The greater semiconductor challenge for MMICs above 100 GHz is achieving sufficient gain, output power, and efficiency. For recent complementary metal-oxide-semiconductor (CMOS) and silicon germanium (SiGe) amplifiers, the saturated output power up to 170 GHz does not exceed 15 dBm, with an efficiency below 10% which cannot overcome the higher path loss, especially using regular modulation schemes, such as 64-QAM (Collaert & Peeters, 2022). To achieve data rate $C = 100$ Gbps, modulation schemes will need spectral efficiency, *C/BW* greater than 14 bps/Hz, which is beyond what's available today (Fairview Microwave, 2021). Indium phosphide (InP) amplifier provides better performance than CMOS and SiGe with output power greater than 20 dBm (typical) with efficiencies more than 20% up (max 30%) at 170 GHz and 220 GHz. However, the cost of the InP material is high and wafer/chip handling is difficult. On the other hand, Gallium Nitride (GaN) has high thermal conductivity and is easy to dissipate heat, so it can handle and deliver high output power and is suitable for use in harsh environments (Eichler & Ziegler, 2022). Research to reduce the size of GaN transistors may enable this amplifier to be used from 100 GHz to 300 GHz (Waliwander, 2022). Gallium Arsenide (GaAs) semiconductor exhibits good broadband noise up to THz frequencies. Therefore, GaAs Schottky diodes can be used to create ultra-low noise receivers up to 4.7 THz (Eichler & Ziegler, 2022). It should be noted that GaAs, GaN, and InP are III-V semiconductors.

Figure 21. Coverage distance versus operating frequency of various IC technology, including its specification of data rate (Gbps), fabrication technology scaling (nm), modulation technique, and antenna gain (dBi)
Source: Strinati et al. (2019) and Waliwander (2022)
** Terminology: ASK—Amplitude-Shift Keying, PSK—Phase-Shift Keying, QAM—Quadrature Amplitude Modulation, QPSK—Quadrature Phase-Shift Keying, CMOS—Metal-Oxide-Semiconductor, InP—Indium Phosphide, mHEMTs—Metamorphic High Electron Mobility Transistors, GaAs—Gallium Arsenide, OOK—On-Off Keying.*

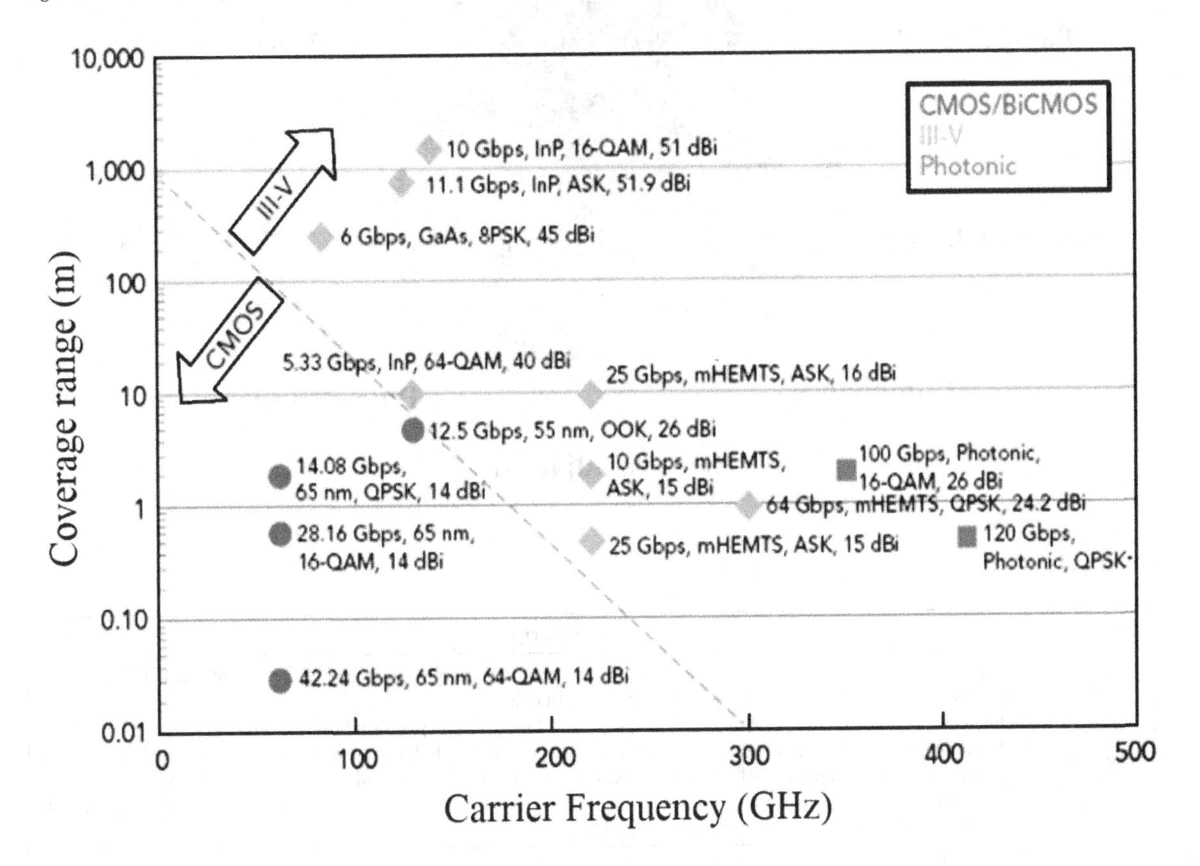

INTERNET OF THINGS (IoT) BUSINESS MODELS

By 2030, IoT devices are expected to be more than 80 billion and be able to analyze user data rapidly and make informed decisions in an autonomous manner. Hence, IoT technology is rapidly becoming an essential part of businesses operating in different business models. In general, IoT business model is a business model that delivers value to customers through IoT solutions, including complex third-party integration, mobility and automation, data collection, processing and analysis, cloud-neutral architecture, and more. Recently, there are nine IoT business models available, namely platform bussiness, subscription, asset sharing, asset tracking, data-driven, service-adjacent, compliance, outcome-oriented, and pay-per-use, as shown in Figure 22. The brief description of each business model is tabulated in Table 16.

Figure 22. Nine IoT business models

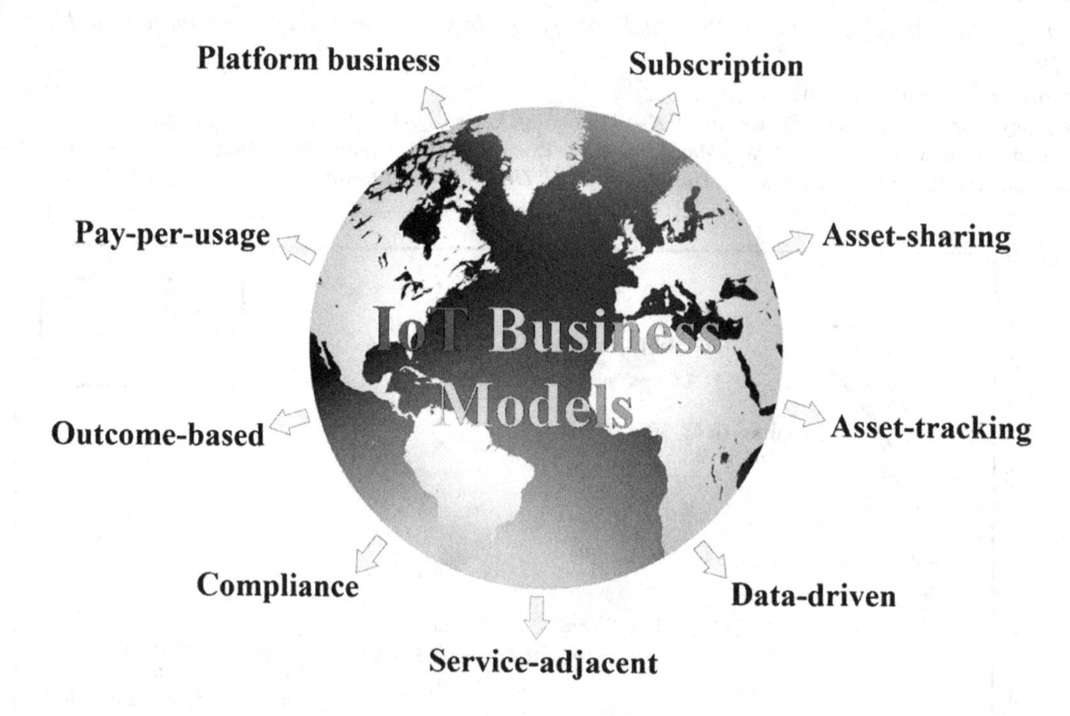

Table 16. Brief description of IoT business models

	Model	Description
1	Platform Business	- A platform-based model brings producers and consumers together to eliminate the possibility of conflict in favorable markets, and make money on the transaction or the booth rental. - Interconnection and interoperability are the keys in developing an IoT platform business model. - For example, for smart homes, every IoT system is separate, such as lighting bulbs, music streaming speakers, and smart plugs (manufactured by different companies). Hence, the user is requested to activate each IoT system using different mobile apps. A platform like Echo + Alexa makes this easy as the user can combine commands only through the voice instead of opening three different mobile apps.
2	Subscription	- Using the always-on connectivity of IoT devices to develop a recurring revenue business or subscription model. - Enables manufacturer to deliver continuous value to customers for a regular fee (Instead of having a one-time sale) - As IoT devices gather more data about customer experiences over time, manufacturers can learn about their customers and then improve and deliver valuable features tailored to their specific needs. - Subscription model is the IoT in the "as a Service" business model. - For instance, customer involves regular payments for products such as media, online movie and music services, software platforms, hosting providers, HR portals, analytics platforms, and more.
3	Asset-Sharing	- This IoT business model is based on selling your extra capacity back to the market in order to maximize the use of the IoT product by multiple clients. - Each customer pays a reduced price as well as the manufacturer can get to market faster than having one customer pay for their entire product. - For example, IoT-based batteries provide energy to commercial buildings, and in case of extra capacity, the remain energy is sold back to the electric grid. The batteries are a shared asset between the building and the electric grid.
4	Asset-Tracking	- Connected IoT devices in the supply chain help businesses identify, control, and track assets in real-time. - For example, using a custom IoT solution to identify water leaks in a building, a technician can point a tablet (with the app installed) at the ground and walk along a water pipe to collect data that is automatically pushed to the cloud infrastructure and aggregated into detailed information data report. The app uses thermal sensors to detect the ground temperature and analyzes temperature deviations to find leaking areas.

continued on following page

Table 16. Continued

	Model	Description
5	Data Driven	- Data-driven business models are powered by data generated by IoT devices. The IoT devices provide value to customers and collect data that can be used in other products or sold to third parties. - The IoT engine system collects a huge amount of data (Big data) from all IoT users/customers, then the valuable data is provided to advertisers and other third-party companies that use the data to promote their products and services. - For instance, LinkedIn or Facebook are tools that collect data from users and provide it to advertisers. Then, online shopping advertisers can follow up and provide product suggestions to users/consumers based on their previous browsing or purchase records.
6	Service-Adjacent	- A service that enhances the use of the IoT device but doesn't necessarily sell the IoT device itself. The IoT device is the enabler of your service. - For instance, use an IoT device to monitor a network or system, predict maintenance timelines of the network or system and sell a maintenance contract to customers.
7	Compliance	- IoT is used to remotely keep a check on and monitor essential compliance metrics, including varying safety, economic, and environmental regulations. - To reduce cost and makes the process much more responsive. - Safety in processes can be maintained as IoT-based devices provide updated and constant information in real time. - For instance, the IBM platform combines weather, satellite and IoT data to help utilities track assets in real-time and offers up-to-date information to help with work planning, budget allocation, hazard reporting, and regulatory reporting.
8	Outcome-Based	- Customers pay for the outcome the IoT product provides, as opposed to the IoT product itself. - This business model focuses more on what customers get from an IoT device (desired information or service) rather than the sale of the IoT device itself. - For instance, a self-monitoring IoT device can automatically reorder replacement parts or create a service request, and the customer then pays for the replacement part or service, not the self-monitoring IoT device.
9	Pay-per-Usage	- Installing sensors on IoT hardware devices can monitor how often customers use IoT devices. Manufacturers of IoT devices can bill customers based on the time they use the IoT devices. o Use the data produced by the IoT device to track customer/usage. o For example, a mobile phone user/customer can choose the most profitable and convenient network service package (such as internet speed) from the operator.

CONCLUSION

This chapter attempts to summarize a lot of facts and list them in the form of tables and charts so that readers can compare recent 5G-IoT and future 6G-IoT, as well as understand their standard protocols, performance, specifications, and applications more easily and quickly.

Despite that, this chapter also explores the opportunities brought by 5G-IoT and 6G-IoT to significantly change in future human social activities, industry operations (industry 4.0), and business models (wireless network-based business). For instance, in future, majority people will deal with hundreds kind of sensors and various wireless networking platforms directly or indirectly in their daily life. Possible types of sensors are RFID cards for home or office security doors, touch screen sensors for mobile phones/laptops, voice recognition sensors for smart home systems, face recognition sensors in market payment systems, and various sensors in self-driving cars. On the other hand, the types of wireless networking platforms are Bluetooth-based headsets and computer mice, while WiFi platforms are used for online meetings and online teaching/learning, as well as online billing/banking transactions. In general, most functions and operations in the sensory devices and network platforms are supported by artificial intelligence (AI) and machine learning (ML).

However, there are certain challenges for recent and future IoT technologies. OpenAI has recently released an AI-based app called ChartGPT, which can interact in a conversational way. In addition, the app is able to instead humans to write essays/theses and able to answer technical questions. Maybe in the future, this app will change the way of teaching and learning in academia. Another example is the evolution of auto insurance policies for future self-driving cars. Do self-driving car owners have to buy auto insurance or the car manufacturer? In the event of an accident, who is responsible, the car owner, the automaker, or the wireless network company?

Indeed, the development of IoT is strongly affected by the development of monolithic microwave integrated circuits (MMIC). At this point, most MMIC that are capable of operating at THz are still in the research stage to overcome the existing shortcomings of the THz MMIC, such as gain, output power, efficiency, and thermal issues. The future MMIC not only requires good performance and operating at THz frequencies, but also needs to be applied in harsh environments, such as MMIC used for an automotive engine. After all, the lack of MMIC in the current market is due to most MMIC production factories previously being locked down due to the COVID-19 pandemic. Furthermore, the recent technological competition between China and the United States will more or less affect the global MMIC production supply chain.

ACKNOWLEDGMENT

This research received no specific grant from any funding agency in the public, commercial, or not-for-profit sectors.

REFERENCES

Abdulkader, O., Bamhdi, A. M., Thayananthan, V., Elbouraey, F., & Al-Ghamdi, B. (2019). A lightweight blockchain based cybersecurity for IoT environments. *Proc. 6th IEEE Int. Conf. Cyber Secur. Cloud Comput. (CSCloud)/5th IEEE Int. Conf. Edge Comput. Scalable Cloud (Edge-Com)*, 139–144. 10.1109/CSCloud/EdgeCom.2019.000-5

Admin. (2020). Beamforming in 5G mmWave radios. *5G mmWave*. https://www.5gmmwave.com/5g-mmwave/beamforming-in-5g-mmwave-radios/

Admin. (2021). Economics of 5G mmWave:is it cost effective? *5G mmWave*. https://www.5gmmwave.com/5g-mmwave/economics-of-5g-mmwave-is-it-cost-effective/

Agiwal, M., Roy, A., & Saxena, N. (2016). Next generation 5G wireless networks: A comprehensive survey. *IEEE Communications Surveys and Tutorials*, *18*(3), 1617–1655. doi:10.1109/COMST.2016.2532458

Akhtar, M. W., Hassan, S. A., Ghaffar, R., Jung, H. J., Garg, S., & Hossain, M. S. (2020). The shift to 6G communications: Vision and requirements. *Human-centric Computing and Information Sciences*, *10*(53), 1–27. doi:10.118613673-020-00258-2

Akyildiz, I. F., Kak, A., & Nie, S. (2020). 6G and beyond: The future of wireless communications systems. *IEEE Access : Practical Innovations, Open Solutions*, *8*, 44983–44998. doi:10.1109/ACCESS.2020.3010896

Al-Hourani, A., Kandeepan, S., & Lardner, S. (2014). Optimal LAP altitude for maximum coverage. *IEEE Wireless Communications Letters*, *3*(6), 569–572. doi:10.1109/LWC.2014.2342736

Alsharif, M. H., Kelechi, A. H., Albreem, M. A., Chaudhry, S. A., Zia, M. S., & Kim, S. (2020). Sixth generation (6G) wireless networks: Vision, research activities, challenges and potential solutions. *Symmetry*, *12*(4), 1–27. doi:10.3390ym12040676

Barakat, B., Taha, A., Samson, R., Steponenaite, A., Ansari, S., Langdon, P. M., Wassell, I. J., Abbasi, Q. H., Imran, M. A., & Keates, S. (2021). 6G opportunities arising from internet of things use cases: A review paper. *Future Internet*, *13*(6), 1–29. doi:10.3390/fi13060159

Bhatia, B. B. (2022). 6G Spectrum studies for ITU. *ITU-APT Foundation of India*. https://www.itu-apt.org/system/static/uploads/mediap/6G%20Spectrum%20Studies%20for%20ITU%20(1).pdf

Borgia, E. (2014). The internet of things vision: Key features, applications and open issues. *Computer Communications*, *54*(12), 1–31. doi:10.1016/j.comcom.2014.09.008

Carey, L. D., Niu, J. G., Yang, P., Kankiewicz, J. A., Larson, V. E., & Haar, T. H. V. (2008). The vertical profile of liquid and ice water content in midlatitude mixed-phase altocumulus clouds. *Journal of Applied Meteorology and Climatology*, *47*(9), 2487–2495. doi:10.1175/2008JAMC1885.1

Chen, N., & Okada, M. (2021). Toward 6G internet of things and the convergence with RoF system. *IEEE Internet of Things Journal*, *8*(11), 8719–8733. doi:10.1109/JIOT.2020.3047613

Chettri, L., & Bera, R. (2020). A comprehensive survey on internet of things (IoT) toward 5G wireless systems. *IEEE Internet of Things Journal*, *7*(1), 16–32. doi:10.1109/JIOT.2019.2948888

Collaert, N., & Peeters, M. (2022). InP+CMOS heterogeneous integration for the next generation of wireless. *Microwave Journal*, *65*(11), 50–56.

Dang, S. P., Amin, O., Shihada, B., & Alouini, M. S. (2020). What should 6G be? *Nature Electronics*, *3*(1), 20–29. doi:10.103841928-019-0355-6

Dao, N. N., Pham, Q. V., Tu, N. H., Thanh, T. T., Bao, V. N. Q., Lakew, D. S., & Cho, S. (2021). Survey on aerial radio access networks: Toward a comprehensive 6G access infrastructure. *IEEE Communications Surveys and Tutorials*, *23*(2), 1193–1225. doi:10.1109/COMST.2021.3059644

Das, S. K., Mudi, R., Rahman, M. S., & Fapojuwo, A. O. (2022). (Preprint). Distributed learning for 6G–IoT networks: A comprehensive survey. *TechRxiv*. Advance online publication. doi:10.36227/techrxiv.20069051.v3

De Alwis, C. (2021). Survey on 6G frontiers: Trends, applications, requirements, technologies and future research. *IEEE Open Journal of the Communications Society*, *2*, 836–886. doi:10.1109/OJCOMS.2021.3071496

DeTomasi, S. (2019). Navigating the 5G NR standards. In P. Hindle (Ed.), *Advanced 5G Over the Air Testing (OTA) Challenges and Solutions* (pp. 10–13). Microwave Journal.

Díaz Zayas, A., Rivas Tocado, F. J., & Rodríguez, P. (2020). Evolution and testing of NB-IoT solutions. *Applied Sciences, 10*(21), 1–17.

Eichler, T., & Ziegler, R. (2022). *Fundamentals of THz Technology for 6G*. Rohde & Schwarz.

Elayan, H., Amin, O., Shihada, B., Shubair, R. M., & Slim Alouini, M. (2020). Terahertz band: The last piece of RF spectrum puzzle for communication systems. *IEEE Open Journal of the Communications Society, 1*, 1–32. doi:10.1109/OJCOMS.2019.2953633

Fortify, H. (2015). *Internet of Things security study: Smartwatches*. Hewlett-Packard.

Frustaci, M., Pace, P., Aloi, G., & Fortino, G. (2018). Evaluating critical security issues of the IoT world: Present and future challenges. *IEEE Internet of Things Journal, 5*(4), 2483–2495. doi:10.1109/JIOT.2017.2767291

Giordani, M., & Zorzi, M. (2021). Non-terrestrial networks in the 6G era: Challenges and opportunities. *IEEE Network, 35*(2), 244–251. doi:10.1109/MNET.011.2000493

Guo, F. X., Richard Yu, F., Zhang, H. L., Ji, H., & Victor Leung, C. M. (2021). Enabling massive IoT toward 6G: A comprehensive survey. *IEEE Internet of Things Journal, 8*(15), 11891–11915. doi:10.1109/JIOT.2021.3063686

Han, C., Wu, Y. Z., Chen, Z., & Wang, X. D. (2019). Terahertz communication (TeraCom): challenges and impact on 6G wireless systems. arXiv preprint, arXiv:1912.06040. doi:10.1109/ACCESS.2020.2977406

Henry, S., Alsohaily, A., & Sousa, E. S. (2020). 3GPP 5G new radio system with the ITU IMT-2020 requirements. *IEEE Access : Practical Innovations, Open Solutions, 8*, 42828–42840. doi:10.1109/ACCESS.2020.2977406

Himmler, A., & Alexander, R. N. (2021). *How 4D Imaging Radar Sensors Can Be Validated*. Autocar Professional. Available online: https://www.autocarpro.in/opinion-blogs/how-4d-imaging-radar-sensors-can-be-validated-79866

Holis, J., & Pechac, P. (2008). Elevation dependent shadowing model for mobile communications via high altitude platforms in built-up areas. *IEEE Transactions on Antennas and Propagation, 56*(4), 1078–1084. doi:10.1109/TAP.2008.919209

Hosseinzadeh, M., Hemmati, A., & Rahmani, A. M. (2022). 6G-enabled internet of things: Vision, techniques, and open issues. *Computer Modeling in Engineering & Sciences, 133*(3), 509–556. doi:10.32604/cmes.2022.021094

Huang, J., Liu, Y., Wang, C. X., Sun, J., & Xiao, H. L. (2018). 5G millimetre-wave channel sounders, measurements, and models: Recent developments and future challenges. *IEEE Communications Magazine, 57*(1), 138–145. doi:10.1109/MCOM.2018.1701263

Huang, T. Y., Yang, W., Wu, J., Ma, J., Zhang, X., & Zhang, D. (2019). A survey on green 6G network: Architecture and technologies. *IEEE Access : Practical Innovations, Open Solutions*, *7*, 175758–175768. doi:10.1109/ACCESS.2019.2957648

Huawei. (2017). *5G Spectrum Public Policy Position*. Huawei Technologies Co., Ltd.

IBM. (2023). *How does IoT work with blockchain?* IBM. https://www.ibm.com/my-en/topics/blockchain-iot

Imoize, A. L., Adedeji, O., Tandiya, N., & Shetty, S. (2021). 6G enabled smart infrastructure for sustainable society: Opportunities, challenges, and research roadmap. *Sensors (Basel)*, *21*(5), 1–58. doi:10.339021051709 PMID:33801302

INFOLYSiS. (2017). A Comparative CAPEX Techno-Economic Analysis of NFV Applicability on the Ground Segment of GEO/MEO/LEO Satellite Systems. INFOLYSIS P.C.

International Telecommunication Union. (2005). *Specific Attenuation Model for Rain for Use in Prediction Methods*. Recommendation ITU-R P. 838-3.

International Telecommunication Union. (2012). *Series Y: Global Information Infrastructure, Internet Protocol Aspects and Next-Generation Networks*. Recommendation ITU-T Y.2060.

International Telecommunication Union. (2017a). *Characteristics of Precipitation for Propagation Modelling*. Recommendation ITU-R P. 837-7.

International Telecommunication Union. (2017b). *Reference Standard Atmospheres*. ITU-R P.

International Telecommunication Union. (2018a). *IoT Standards Part II: 3GPP Standards. Training on Planning Internet of Things (IoTs) Networks.* ITU Report.

International Telecommunication Union. (2018b). *Sharing between the Radio Astronomy Service and Active Services in the Frequency Range 275-3000 GHz*. Recommendation ITU-R Report RA.2189-1.

International Telecommunication Union. (2019a). *Attenuation Due to Clouds and Fog*. Recommendation ITU-R P. 840-8.

International Telecommunication Union. (2019b). *The Radio Refractive Index: Its Formula and Refractivity Data*. ITU-R P.

International Telecommunication Union. (2019c). *Attenuation by Atmospheric Gases and Related Effects*. ITU-R P.

International Telecommunication Union. (2021). *Propagation Data and Prediction Methods Required for the Design of Terrestrial Line-of-Sight Systems*. ITU-R P.

International Telecommunication Union. (2022). *Future Technology Trends of Terrestrial International Mobile Telecommunications Systems Towards 2030 and Beyond*. Report ITU-R M.2516-0.

IoT for all. (2021). What is the Internet of Things, or IoT? A simple explanation. *IoT For All*. https://www.iotforall.com/what-is-internet-of-things

Iqbal, W., Abbas, H., Daneshmand, M., Rauf, B., & Bangash, Y. A. (2020). An in-depth analysis of IoT security requirements, challenges, and their countermeasures via software-defined security. *IEEE Internet of Things Journal*, *7*(10), 10250–10276. doi:10.1109/JIOT.2020.2997651

Ji, B. F., Wang, Y., Song, K., Li, C., Wen, H., Menon, V. G., & Mumtaz, S. (2021). A survey of computational intelligence for 6G: Key technologies, applications and trends. *IEEE Transactions on Industrial Informatics*, *17*(10), 7145–7154. doi:10.1109/TII.2021.3052531

Jiang, W., Han, B., Habibi, M. A., & Schotten, H. D. (2021). The road towards 6G: A comprehensive survey. *IEEE Open Journal of the Communications Society*, *2*, 334–366. doi:10.1109/OJCOMS.2021.3057679

Johan, F. (2021). *The Satellite IoT Communications Market*. Berg Insight.

Keskinen, M. (2017). Mobile Network Evolution. University of Oulu.

Kim, J. H. (2021). 6G and Internet of Things: A survey. *Journal of Management Analytics*, *8*(2), 316–332. doi:10.1080/23270012.2021.1882350

Lee, Y. L., Qin, D. H., Wang, L. C., & Sim, G. H. (2020). 6G massive radio access networks: Key applications, requirements and challenges. *IEEE Open Journal of Vehicular Technology*, *2*, 54–66. doi:10.1109/OJVT.2020.3044569

Letaief, K. B., Chen, W., Shi, Y. M., Zhang, J., & Angela Zhang, Y. J. (2019). The roadmap to 6G: AI empowered wireless networks. *IEEE Communications Magazine*, *57*(8), 84–90. doi:10.1109/MCOM.2019.1900271

Li, Q., Zhu, Q., Zheng, J. S., Liao, K. H., & Yang, G. S. (2014). Soil moisture response to rainfall in forestland and vegetable plot in Taihu Lake Basin, China. *Chinese Geographical Science*, *25*(4), 1–12.

Liebe, H., Hufford, G., & Cotton, M. (1993). Propagation modeling of moist air and suspended water/ice particles at frequencies below 1000 GHz. *Proceedings of AGARD*, 3-1–3-11.

Lin, J., Yu, W., Zhang, N., Yang, X., Zhang, H., & Zhao, W. (2017). A survey on internet of things: Architecture, enabling technologies, security, privacy, and applications. *IEEE Internet of Things Journal*, *4*(5), 1125–1142. doi:10.1109/JIOT.2017.2683200

Liu, H., Shen, F., Liu, Z., Long, Y., Liu, Z., Sun, S., Tang, S., & Gu, D. (2019). A secure and practical blockchain scheme for IoT. *Proc. 18th IEEE Int. Conf. Trust, Secur. Privacy Comput. Commun./13th IEEE Int. Conf. Big Data Sci. Eng. (TrustCom / BigDataSE)*, 538–545. 10.1109/TrustCom/BigDataSE.2019.00078

Madarasz, T. (2019). Microwave will drive the development of 5G. In P. Hindle (Ed.), *Design Ideas and Tradeoffs for 5G Infrastructure* (pp. 10–13). Microwave Journal.

Mahdi, M. N., Ahmad, A. R., Qassim, Q. S., Natiq, H., Subhi, M. A., & Mahmoud, M. (2021). From 5G to 6G technology: meets energy, internet-of-things and machine learning: a survey. *Applied Sciences (Basel, Switzerland)*, *11*(17), 8117. doi:10.3390/app11178117

Meneghello, F., Calore, M., Zucchetto, D., Polese, M., & Zanella, A. (2019). IoT: Internet of threats? a survey of practical security vulnerabilities in real IoT devices. *IEEE Internet of Things Journal*, *6*(5), 8182–8201. doi:10.1109/JIOT.2019.2935189

Michailidis, E. T., Potirakis, S. M., & Kanatas, A. G. (2020). AI-inspired non-terrestrial networks for IIoT: Review on enabling technologies and applications. *IoT*, *1*(1), 21–48. doi:10.3390/iot1010003

Microwave, F. (2021). 6G fantastic, yes. Fantasy? Not so much. *Microwaves & RF*, *60*(1), 10–13.

Nguyen, Ding, Pubudu, Pathirana, Seneviratne, Li, Niyato, Dobre, & Poor. (2022). 6G internet of things: a comprehensive survey. *IEEE Internet of Things Journal*, *9*(1), 836–886.

Oza, T. (2020). Understanding the underlying sensor and wireless technologies in IIoT apps. *Microwaves & RF*, *9*, 18–22.

Padhi, P. K., & Charrua-Santos, F. (2021). 6G enabled industrial internet of everything: Towards a theoretical framework. *Applied System Innovation*, *4*(1), 11. doi:10.3390/asi4010011

Pajooh, H. H., Demidenko, S., Aslam, S., & Harris, M. (2022). Blockchain and 6G-enabled IoT. *Inventions (Basel, Switzerland)*, *7*(4), 109. doi:10.3390/inventions7040109

Palattella, M. R., Dohler, M., Grieco, A., Rizzo, G., Torsner, J., Engel, T., & Ladid, L. (2016). Internet of things in the 5G era: Enablers, architecture, and business models. *IEEE Journal on Selected Areas in Communications*, *34*(3), 510–527. doi:10.1109/JSAC.2016.2525418

Pattnaik, S. K., Samal, S. R., Bandopadhaya, S., Swain, K., Choudhury, S., Das, J. K., Mihovska, A., & Poulkov, V. (2022). Future wireless communication technology towards 6G IoT: An application-based analysis of IoT in real-time location monitoring of employees inside underground mines by using BLE. *Sensors (Basel)*, *22*(9), 3438. doi:10.339022093438 PMID:35591138

Qadir, Z., Le, K. N., Saeed, N., & Munawar, H. S. (2022). Towards 6G internet of things: Recent advances, use cases, and open challenges. *ICT Express*. Advance online publication. doi:10.1016/j.icte.2022.06.006

Qorvo & RFMW. (2020). *Wi-Fi Convergence with IoT and 5G*. Microwave Journal.

Riazul Islam, S. M., & Kwak, D. (2015). The Internet of Things for health care: A comprehensive survey. *IEEE Access : Practical Innovations, Open Solutions*, *3*, 678–708. doi:10.1109/ACCESS.2015.2437951

Rinaldi, F., Maattanen, H.-L., Torsner, J., Pizzi, S., Andreev, S., Iera, A., Koucheryavy, Y., & Araniti, G. (2020). Non-terrestrial networks in 5G & beyond: A survey. *IEEE Access : Practical Innovations, Open Solutions*, *8*, 165178–165200. doi:10.1109/ACCESS.2020.3022981

Saad, W., Bennis, M., & Chen, M. Z. (2019). A vision of 6G wireless systems: Applications, trends, technologies, and open research problems. *IEEE Network*, *34*(3), 134–142. doi:10.1109/MNET.001.1900287

Said, O., Al-Makhadmeh, Z., & Tolba, A. A. (2020). EMS: An energy management scheme for green IoT environments. *IEEE Access : Practical Innovations, Open Solutions*, *8*, 44983–44998. doi:10.1109/ACCESS.2020.2976641

Schroeder, C. (2021). Early indications of 6G. *Microwave Journal*, *64*, 5–9.

Sekaran, R., Patan, R., Raveendran, A., Al-Turjman, F., Ramachandran, M., & Mostarda, L. (2020). Survival study on blockchain based 6G-enabled mobile edge computation for IoT automation. *IEEE Access : Practical Innovations, Open Solutions*, *8*, 143453–143463. doi:10.1109/ACCESS.2020.3013946

Shammar, E. A., Zahary, A. T., & Al-Shargabi, A. A. (2021). A survey of IoT and blockchain integration: Security perspective. *IEEE Access : Practical Innovations, Open Solutions*, *9*, 156114–156150. doi:10.1109/ACCESS.2021.3129697

Siegel, P. H. (2002). Terahertz technology. *IEEE Transactions on Microwave Theory and Techniques*, *50*(3), 910–928. doi:10.1109/22.989974

Siles, G. A., Riera, J. M., & García-del-Pino, P. (2015). Atmospheric attenuation in wireless communication systems at millimeter and THz frequencies. *IEEE Antennas & Propagation Magazine*, *57*(1), 48–61. doi:10.1109/MAP.2015.2401796

Spyridis, Y., Lagkas, T., Sarigiannidis, P., Argyriou, V., Sarigiannidis, A., Eleftherakis, G., & Zhang, J. (2021). Towards 6G IoT: Tracing mobile sensor nodes with deep learning clustering in UAV networks. *Sensors (Basel)*, *21*(11), 1–16. doi:10.339021113936 PMID:34200449

Strinati, E. C., Barbarossa, S., Gonzalez-Jimenez, J. L., Kténas, D., Cassiau, N., Maret, L., & Dehos, C. (2019). 6G: The next frontier –from holographic messaging to artificial intelligence using subterahertz and visible light communication. *IEEE Vehicular Technology Magazine*, *14*(3), 42–50. doi:10.1109/MVT.2019.2921162

Technologies, K. (2018). *Smart Cows and Tips for Designing Mission-Critical IoT Products*. Keysight Technologies.

Technologies, K. (2019). *How to Ensure IoT Devices Work in Their Intended Environment (Locate and Identify Interference)*. Keysight Technologies.

Technologies, K. (2020a). *Unleash the Power of IoT*. Keysight Technologies.

Technologies, K. (2020b). *The 5C's of IoT*. Keysight Technologies.

Tuan, N. Y. (2020). Small cells help keep 5G connected. In B. Wong (Ed.), *Focus on: 5G technology and challenges* (pp. 2–4). Microwaves & RF.

Waliwander, T. (2022). THz – To be or not to be in 6G. *Microwave Journal*, *65*(5), 54–66.

Wang, C. X., Huang, J., Wang, H. M., Gao, X. Q., You, X. H., & Hao, Y. (2020). 6G wireless channel measurements and models. *IEEE Vehicular Technology Magazine*, *15*(4), 22–32. doi:10.1109/MVT.2020.3018436

Weissberger, A. (2021). ITU-R future report: high altitude platform stations as IMT base stations (HIBS). *IEEE Communication Society Technology Blog*. https://techblog.comsoc.org/2021/02/17/itu-r-future-report-high-altitude-platform-stations-as-imt-base-stations-hibs/

Wikipedia. (2021). Internet of things. In *Wikipedia, the free encyclopedia*. https://en.wikipedia.org/wiki/Internet_of_things

You, K. Y. (2022a). Propagation channel modelling for low-altitude platform non-terrestrial networks from 275 GHz to 3 THz. *International Journal Wireless and Microwave Technologies*, *3*(3), 1–17. doi:10.5815/ijwmt.2022.03.01

You, K. Y. (2022b). Survey on 5G and future 6G access networks for IoT applications. *International Journal Wireless and Microwave Technologies*, *4*(4), 26–47. doi:10.5815/ijwmt.2022.04.03

You, X. H. (2021). Towards 6G wireless communication networks: vision, enabling technologies, and new paradigm shifts. *Science China Information Sciences, 64*, 110301:1–110301:74.

Zhang, L., Liang, Y. C., & Niyato, D. (2019a). 6G visions: Mobile ultra-broadband, super internet-of-things, and artificial intelligence. *China Communications*, *16*(8), 1–14. doi:10.23919/JCC.2019.08.001

Zhang, Z. Q., Xiao, Y., Ma, Z., Xiao, M., Ding, Z., Lei, X., Karagiannidis, G. K., & Fan, P. (2019b). 6G wireless networks vision, requirements, architecture, and key technologies. *IEEE Vehicular Technology Magazine*, *14*(3), 28–41. doi:10.1109/MVT.2019.2921208

Zhao, T. F., Wang, H., & Ma, Q. W. (2020). The coverage method of unmanned aerial vehicle mounted base station sensor network based on relative distance. *International Journal of Distributed Sensor Networks*, *16*(5), 1–12. doi:10.1177/1550147720920220

Zong, B. Q., Fan, C., Wang, X., Duan, X., Wang, B., & Wang, J. (2019). 6G technologies: Key drivers, core requirements, system architectures, and enabling technologies. *IEEE Vehicular Technology Magazine*, *14*(3), 18–27. doi:10.1109/MVT.2019.2921398

ADDITIONAL READING

Chen, S., Yin, Y., Wang, Z., & Gui, F. (2020). Low-altitude protection technology of anti-UAVs based on multisource detection information fusion. *International Journal of Advanced Robotic Systems*, *17*(5), 1–12. doi:10.1177/1729881420962907

Chen, Z., Ma, X. Y., Zhang, B., Zhang, Y. X., Niu, Z. Q., Kuang, N. Y., Chen, W. J., Li, L. X., & Li, S. Q. (2019). A survey on Terahertz communications. *China Communications*, *16*(2), 1–35.

Hirata, A., & Yaita, M. (2015). Ultrafast terahertz wireless communications technologies. *IEEE Transactions on Terahertz Science and Technology*, *5*(6), 1128–1132.

Lu, L., Geoffrey Li, Y., Swindlehurst, A. L., Ashikhmin, A., & Zhang, R. (2014). An overview of massive MIMO: Benefits and challenges. *IEEE Journal of Selected Topics in Signal Processing*, *8*(5), 742–758. doi:10.1109/JSTSP.2014.2317671

Miller, L., & Cavazos, J. (2022). *5G & Beyond for Dummies*. John Wiley & Sons.

Promwongsa, N., Ebrahimzadeh, A., Naboulsi, D., Kianpisheh, S., Belqasmi, F., Glitho, R., Crespi, N., & Alfandi, O. (2021). A comprehensive survey of the tactile internet: State-of-the-art and research directions. *IEEE Communications Surveys and Tutorials*, *23*(1), 472–523. doi:10.1109/COMST.2020.3025995

Sambo, Y. A., Shakir, M. Z., Qaraqe, K. A., Serpedin, E., & Imran, M. A. (2014). Expanding cellular coverage via cell-edge deployment in heterogeneous networks: Spectral efficiency and backhaul power consumption perspectives. *IEEE Communications Magazine*, *52*(6), 140–149. doi:10.1109/MCOM.2014.6829956

Song, H. J., & Nagatsuma, T. (2011). Present and future of terahertz communications. *IEEE Transactions on Terahertz Science and Technology*, *1*(1), 256–263. doi:10.1109/TTHZ.2011.2159552

Yang, P., Xiao, Y., Xiao, M., & Li, S. Q. (2019). 6G Wireless communications: Vision and potential techniques. *IEEE Network*, *33*(4), 70–75. doi:10.1109/MNET.2019.1800418

You, K. Y. (2018). Introductory chapter: RF/microwave applications. In K. Y. You (Ed.), *Emerging microwave technologies in industrial, agricultural, medical and food processing* (pp. 1–7). IntechOpen. doi:10.5772/intechopen.73574

KEY TERMS AND DEFINITIONS

3rd Generation Partnership Project (3GPP): European Telecommunications Standards Institute (ETSI) formed in December 1998, partnered with other standard development organizations (SDOs) from around the world to develop new technologies specifications for the third generation (3G) of cellular networks.

Aerial Radio Access Networks (ARAN): A type of network infrastructure used mobile networks that consist of airborne radio base stations (such as unmanned aerial vehicles, drones, and satellites) with antennas that cover a specific region according to its capacity, design, and propagation.

Antenna: A transducer designed to transmit or receive electromagnetic waves.

Artificial Intelligence (AI): A simulation of human intelligence processes in machines using computer systems.

Bluetooth: A short-range wireless technology based on the IEEE 802.15.1 standard, used for exchanging data between fixed and mobile devices over short distance (within 10 m) using ISM band from 2.402 GHz to 2.4835 GHz.

Fifth Generation Wireless Technology (5G): Digital cellular mobile communication networks that began wide deployment in 2019.

Internet of Things (IoT): A system of interrelated computing devices, mechanical, and digital machines provided with unique identifiers (UIDs) and the ability to transfer data over a network without requiring human-to-human or human-to-computer interaction.

Machine learning (ML): A subfield of artificial intelligence (AI) that enables machine systems to automatically learn, self-improvement from past experience, and make decisions without being explicitly programmed.

Microwave (MW): A form of electromagnetic radiation with wavelengths ranging from 1 m to 1 mm, which is corresponding to operating frequencies ranging from 300 MHz to 300 GHz.

Millimeter Wave (mmWave): A form of electromagnetic radiation with wavelengths ranging from 10 mm to 1 mm, which is corresponding to operating frequencies ranging from 30 GHz to 300 GHz (within UHF and EHF bands of microwave).

Radio Frequency (RF): A part of the electromagnetic spectrum from about 3 kHz to 300 MHz, which is corresponding to wavelengths ranging from 100 km to 1 m.

Sixth Generation Wireless Technology (6G): A successor to 5G cellular technology, in which it is expected to be extended the operating frequency up to 3 THz (or above) with data rates of 0.1–1 Tbps, spectrum efficiency of 3–60 bps/Hz, 100 GHz channel bandwidth, and 1000 km/h mobility.

Submillimeter Wave (Sub-mmWave): A form of electromagnetic radiation with wavelengths ranging from 1 mm to 0.01 mm, which is corresponding to operating frequencies ranging from 0.3 THz to 30 THz.

Terahertz Wave (THz Wave): Also known as submillimeter wave.

Terrestrial Radio Access Networks (TRAN): A type of network infrastructure used for mobile networks that consist of terrestrial radio base stations with large antennas that cover a specific region according to its capacity, design, and propagation.

Wireless Fidelity (WiFi): A family of wireless network protocols based on the IEEE 802.11 standards, which are used to connect computers, tablets, smartphones, and other devices to the internet using radio waves.

LIST OF ABBREVIATIONS

2G-GSM: 2G-Global System for Mobile Communications
3CLS: Control, Localization, and Sensing
3D-InteCom: Three-Dimensional Integrated Communications
3GPP: 3rd Generation Partnership Project
5G: Fifth Generation Wireless Technology
5G NR: 5G New Radio
6G: Sixth Generation Wireless Technology
6LoWPAN: IPv6 over Low-Power Wireless Personal Area Networks
AES-128: Advanced Encryption Standard-128 bit
AI: Artificial Intelligence
AR: Augmented Reality
ARAN: Aerial Radio Access Networks
ASK: Amplitude-Shirt Keying
BCI: Brain-Computer Interface
BigCom: Big Communications
BPSK: Binary Phase-Shift Keying
BS: Base Station
BW: Bandwidth
CAeC: Contextually Agile eMBB Communications
CBRS: Citizens Broadband Radio Service
CCK: Complementary Code Keying
CCMP-128: Counter Mode CBC-MAC Protocol-128 bit
CDMA: Code Division Multiple Access
CMOS: Complementary Metal Oxide Semiconductor
CNC: Computer Numerical Control
COC: Computation Oriented Communications
COFDM: Coded Orthogonal Frequency-Division Multiplexed
CRAS: Connected Robotic and Autonomous System
CSMA: Carrier Sense Multiple Access
D2D: Devices-to-Device
DAS: Distributed Antenna Systems
DLT: Distributed Ledger Technology
DSSS: Direct-Sequence Spread Spectrum
ECC: European Electronic Communications Committee
EDuRLLC: Event Defined uRLLC
EIRP: Effective Isotropic Radiated Power
ELPC: Extremely Low-Power Communications
eMBB: Enhanced Mobile Broadband
eMBB-Plus: Enhanced Mobile Broadband Plus
eURLLC: Extremely Ultra-Reliable, Low Latency Communications
FCC: Federal Communications Commission

FDD: Frequency Division Duplex
FDMA: Frequency Division Multiple Access
FeMBB: Further-Enhanced Mobile Broadband
FHSS: Frequency-Hopping Spread Spectrum
FR 1: Frequency Range 1
FR 2: Frequency Range 2
FSK: Frequency-Shift Keying
GaAs: Gallium Arsenide
GaN: Gallium Nitride
GFSK: Gaussian Frequency-Shift Keying
GPRS: General Packet Radio Service
GPS: Global Positioning System
HAPs: High-Altitude Platforms
HCS: Human-Centric Services
HBF: Holographic Beamforming
HD: High Definition
HSR: High-Availability Seamless Redundancy
HVAC: Heating, Ventilation, and Air Conditioning
IEEE: Institute of Electrical and Electronics Engineers
IIoT: Industrial Internet of Things
IMT-2020: International Mobile Communications – 2020
InP: Indium Phosphide
IoE: Internet of Everything
IoT: Internet of Things
IPv6: Internet Protocol version 6
ISA100.11a: International Society of Automation
ISM: Industrial, Scientific, and Medical
ITU: International Telecommunication Union
LAPs: Low-Altitude Platforms
LDHMC: Long-Distance and High-Mobility Communications
LDPC: Low-Density Parity-Check Codes
LEOs: Low Earth Orbits
LIS: Large Intelligent Surfaces
LoRa CCS: Long-Range Chirp Spread Spectrum
LoRa/LoRaWAN: Long-Range Wide-Area Network
LoS: Line-Of-Sight
LPWAN: Low-Power Wide-Area Network
LSAS: Large-Scale Antenna Systems
LTE: Long-Term Evolution
LTE-A: Long-Term Evolution-Advanced
LTE-M: Long-Term Evolution-Machines
LTE-U: Long-Term Evolution-Unlicensed
MAC: Medium Access Control

MBBLL: Mobile Broadband Bandwidth & Low Latency
mBBMT: Massive Broad Bandwidth Machine Type
MBRLLC: Mobile Broadband Reliable Low Latency Communication
mHEMTs: Metamorphic High Electron Mobility Transistors
MIMO: Multiple Input, Multiple Output
MiWi: Microchip Wireless
ML: Machine Learning
mLLMT: Massive Low Latency Machine Type
MMICs: Monolithic Microwave Integrated Circuits
mMTC: Massive Machine Type Communications
mmWave: Millimeter Wave
MPS: Multi-Purpose 3CLS and Energy Services
M-QAM: M-ary Quadrature Amplitude Modulation
MU-MIMO: Multi-User Multiple Input, Multiple Output
muRLLC: Massive Ultra-Reliable, Low Latency Communication
MW: Microwave
NFC: Near-Field Communication
NFV: Network Function Virtualization
NLoS: Non-Line-Of-Sight
NOMA: Non-Orthogonal Multiple Access
NB-IoT: Narrowband-Internet of Things
NFV: Network Function Virtualization
OAM: Orbital Angular Momentum
OFDMA: Orthogonal Frequency-Division Multiple Access
OOK: On-Off Keying
PCF: Point Coordination Function
PSK: Phase-Shift Keying
QAM: Quadrature Amplitude Modulation
QoS: Quality of Service
QPSK: Quadrature Phase-Shift Keying
QR: Quick Response
RF: Radio Frequencies
RFID: Radio Frequency Identification
RTLS: Real Time Location Systems
SDN: Software-Defined Networking
Si: Silicon
SNAP: Subnetwork Access Protocol
SNR: Signal-to-Noise Ratio
Sub-mmWave: Sub-Millimeter Wave
SURLLC: Secure Ultra-Reliable Low-Latency Communications
TDD: Time Division Duplex
TDMA: Time Division Multiple Access
THz: Terahertz

TRAN: Terrestrial Radio Access Networks
UAV: Unmanned Aerial Vehicle
UCDC: Unconventional Data Communications
uHDD: Ultra-High Data Density
uHSLLC: Ultra-High Speed with Low Latency Communications
umMTC: Ultra-Massive Machine Type Communications
UMTS: Universal Mobile Telecommunications System
uMUB: Ubiquitous Mobile Ultra-Broadband
uRLLC: Ultra-reliable and Low-latency Communications
UWB: Ultra-Wideband
V2X: Vehicle-to-Everything
VLC: Visible Light Communication
VR: Virtual Reality
Wi-Fi: Wireless Fidelity
Wi-Gig: Wireless Gigabit Alliance
Wi-Sun: Wireless Smart Utility Network
WirelessHART: Wireless Highway Addressable Remote Transducer Protocol
WLAN: Wireless Local Area Network
WoS: Web of Science
WPAN: Wireless Personal Area Network
WWAN: Wireless Wide Area Network
XR: Extended Reality
Zigbee: Zonal Intercommunication Global-Standard

Chapter 11
IoT Blockchains for Digital Twins

Steven A. Wright
Georgia State University, USA

ABSTRACT

Digital twins (DTs) have emerged as a critical concept in cyberspace infrastructure. DTs are virtual representations of physical things including model smart structures or environments, manufacturing processes, humans, and a variety of other things. The value provided by DTs relies on their fidelity in representation. Blockchains provide trust assurance mechanisms, particularly where multiple parties are involved. The expected life cycle operations of the IoT, blockchain, and DT need to be considered to develop economically useful blockchain digital twin (BDT) models. BDTs do not exist in isolation, but rather within a DT environment (DTE). A DTE may include multiple DTs of different objects to enable interactions between these objects to be evaluated in both virtual reality and mixed reality cases. To populate DTEs with multiple DTs requires industrialized tooling to support the rapid creation of DTs. The industrialization of DT creation requires frameworks, architectures, and standards to enable interoperability between DTs and DTEs.

INTRODUCTION

Digital Twins (DTs) are digital representations of living or non-living physical objects. DTs have been widely considered in the context of manufacturing as a conceptual model in the product lifecycle management process (Grieves, 2019). Digital modelling in manufacturing combines computer aided design techniques (e.g., 3D models) with additive or subtractive manufacturing processes. ISO defines a DT in manufacturing as *a fit for purpose digital representation of an observable manufacturing element with a means to enable convergence between the element and its digital representation at an appropriate rate of synchronization* (ISO 2021). A DT can also be considered as a projection of physical objects into digital spaces e.g., virtual reality. DTs can also be used to optimize asset performance through monitoring, diagnostics, or prognostics (Tao et al., 2018). In the context of the built environments, DTs have been used to capture spatial data capturing the building, smart city, etc. (Deng et al., 2021). DT

DOI: 10.4018/978-1-7998-9266-3.ch011

technologies can also model living organisms, including humans. Human DTs (HDTs) are emerging for healthcare (Croatti et al., 2020) and social interaction. A broad range of applications for DTs and related technologies has led to a broad range of definitions for DTs (See, e.g., Minerva et al., 2020 and Voas et al., 2021). All DTs rely, to a greater or lesser extent on sensing operations (typically based on IoT devices) for the creation and operation of the DT. Blockchain Digital Twins (BDTs) are a subset of the DTs that incorporate blockchains to provide additional trust-based features.

The connectivity between the physical object and its DT is one of the main characteristics of DTs. A *static* digital twin only requires connectivity with the physical world (1) when the digital twin is created as a digital model of the physical object, or (2) when the DT is used to drive some physical process (e.g., manufacturing replicas of a scanned physical object- model based manufacturing). A *dynamic* DT maintains a digital representation of the current state of the physical object. The current state of the physical object is typically characterized using IoT sensors. This DT - Physical object connectivity, whether for model creation or state maintenance, is a form of machine- machine communication. 5G and 6G networks provide additional capabilities to support machine-machine communication. The ITU-T recognized DTs as a use case driving additional requirements for 6G features (ITU-T, 2020). Maintaining the linkage between the DT and physical world requires continuous connectivity between the DT and the Sensors monitoring the physical object's status. For a movable physical object, these sensors either need to be attached to the object or the range of motion needs to be constrained. The integrity and trustworthiness of the DTs as representation of the physical objects' current state relies on the authenticity of the data as well as the modelling process.

The connectivity patterns of digital twins differ from typical human communications or web browsing. The IoT sensors are clustered sensing the physical object and its physical environment in a specific locality. The DT and the Digital Twin Environment (DTE) where the DT executes may be localized in some manufacturing 4.0 scenarios to reduce latency from the modelled physical objects. Other use cases may provide significantly greater value if the DT can be executed in a remote DTE. Early approaches to the DT concepts, systems, and technologies, typically leveraged the prevalent centralized computing system architectures. As DTs start to move from research concept and implementation prototype to commercial deployments, they are being to be applied to problems that matter to people. As commercial users of DTs come to rely on those models, it becomes more important to ensure the data provenance, audit, and traceability in the creation of the DT model as well as its operation (Hasan et al., 2020). Not only the data from the original physical object, but also the data from the DT – transactions, logs and history - need to be secure and tamper-proof. Decentralized connectivity patterns and blockchain architectures can provide lower risks than centralized architectures. Commercially successful DTs are likely to be operated by multiple independent users that trust in the proper operation of the DTs. Such users' objectives may also require the trustworthy operation of more than one DT, where some DTs come from different sources to execute in the same DTE.

For users of DTs to benefit from this digital representation, they must trust that it provides an adequate representation for their purposes. For a static DT, this requires trust in both the modelling process and the underlying data; dynamic DTs obviously add temporal considerations. Many early DTs were developed within a single organization. As the technology becomes more widely deployed, multiple parties are likely to be required. DT models of components in a supply chain context may be independently developed by the respective component manufacturers but required for use by the downstream users of those components. Privacy is also a concern for commercial uses of DTs. The physical objects modelled by the DT may include intellectual property (e.g., trade secrets). The owners of the physical objects may be concerned about inadvertent disclosures through the DTs. The users of DTs may similarly be concerned about the confidentiality of their uses of the DTs (e.g., in the modelling of their trade secret processes). HDTs raise privacy concerns as they are likely constructed using personally identifiable biometric information, where additional protective regulations may apply. Mechanisms to support privacy and trust often build on cryptographic methods and architectures. Mechanisms and design patterns to support privacy and trust in IoT blockchains are starting to emerge (Wright, 2019) as are trustworthy digital twins (Suhail et al., 2021).

BDTs are an emerging category of cyber-physical systems that have been receiving attention recently (See, e.g., Raj, 2021 and Leng et al., 2021). IoT sensors combined with 5G and 6G communications services and blockchain technologies seem to provide an infrastructure to support the industrialization and deployment of BDTs on a larger scale. Live BDTs have requirements for specific machine-machine communications services and architectures to operate effectively. BDTs in industry verticals from manufacturing, the built environment and healthcare, seem poised for wider deployment over 5G and 6G networks.

MOTIVATIONS FOR DIGITAL TWINS

A DT in operation can be used to configuration, monitoring, diagnostics, and prognostics of the corresponding physical object. A *live* DT of a *specific* airplane engine, for example, could be used optimize the preventative maintenance schedule of that engine based on monitoring and diagnostics. Similarly, a DT of a machine used in manufacturing, for example, could be used to optimize production processes. The manufacturing implications of DTs are significant, with potential for gains from DTs of both products and production processes throughout the product life cycle. The motivation for using a DT in a manufacturing context may be the product or the production process, or both.

In the manufacturing of products, DTs can be used in the design, pre-production, and production phases (Söderberg et al., 2017). In a highly automated production factory for complex assembled products, several hundreds of robots could be organized for handling and joining operations. Geometry related problems, usually constitute a significant part of the total cost and delay for poor quality. The product's physical geometry, for example, could be optimized as a DT in the design phase to withstand expected manufacturing tolerances. While the DT can then be used as a model for production, industrializing the process of creating DTs enables mass customization in production. Based on the DT geometry and functionality, the preproduction planning for process flow and inspection can be developed. Inspection data during production can be compared with the DT to identify diagnose defects. Note that DTs are not restricted to modelling physical geometry (e.g., they could be used to model biological processes for biotech manufacturing processes and products). DT roles in manufacturing start with the DT of the product, then develops the DT of the production process, before realizing the production process, then operating that process to create the finished product.

Another motivation for DT use is the reduced costs in comparison with testing on some categories of physical objects. Testing of product quality is required at various points during supply chains to eliminate non-conformant products. Non-destructive evaluation and testing provide information about the physical product without causing damage. The use of non-destructive evaluation can significantly impact costs across supply chains, not just individual manufacturers (Kapustina et al., 2020). DTs provide a digital representation of the product. If the required testing can be performed on the DT, then this causes no damage to the physical object. A *live* DT has an ongoing data feed from the underlying physical object; and the physical object is not significantly damaged by the IoT sensors generating this data feed. While testing is a common economic rationale for DTs, in some cases there may be additional requirements – e.g., the testing results need to be witnessed by an independent party for regulatory purposes or evaluated by other authorities to issue compliance certifications. These other parties then need to trust the authenticity of the test results from the DT.

Composition of DTs into larger, more complex system models is a motivation for the use of DTs. Modern society is constructing and using systems with greater complexity that ever before. The interactions of the components of complex systems can be unplanned, unfamiliar and unexpected resulting in system level behavior that is unanticipated, incomprehensible and potentially disastrous. Where those components learn or are composed with random components, the systems are changing over time and can exhibit emergent behavior. That system behavior can be classified as to whether it was predicted or not, and whether it was beneficial or not. The use of DTs to model the components of complex systems allows online computation to identify anomalous emergent behavior (Grieves & Vickers, 2017). The cost of DT development and operation can be less than the losses incurred in the event of destructive emergent behavior in complex systems.

Another motivation for use of a DT may be where this both cheaper and more convenient than the use of the original object. Education and training are important factors in developing the skills necessary to support the production of complex goods and services. DTs can be as a tool for training and education (Chhetri et al., 2004). Learners can develop skills using a DT rather than the underlying physical object. This is particularly helpful when the underlying objects are expensive or rare (e.g., complex machinery). Skill development using a DT is also preferable where practice using the physical object could be dangerous (e.g., surgical skills, explosive devices, aircraft). Skill certification based on simulations is already used in several fields (Granda & Salik, 2020). Blockchains are also proposed for verifying academic credentials (Caldarelli & Ellul, 2021). Trustworthy evidence of skill performance based on the users' simulation results is a motivation for the deployment of BDTs for both academic and industrial training.

DTS as a shared abstraction in collaborative settings is another motivation. Modelling has long been a technique used in the development and operation of telecommunications services to assure various aspects system performance. DTs, however, correspond to physical objects and have communication flowing between the DT and the physical object. DTs are now being proposed in telecommunications (Ahmadi et al., 2021; Seilov et al., 2021) for modelling network configurations prior to deployment. In the past decade, network operators have been moving towards virtualizing network functions (ETSI, 2012) in a process some refer to as softwarization. The DT, in this sense, could be considered a softwarization of the physical object. The scale of telecommunications networks is a significant characteristic. Deployment of DTs at scale requires a consistent DTE to enable easy automation and management of large numbers of DTs in response to unexpected changes in demand. The economic efficiency from automation and management of DTs in a consistent DTE is a motivation for the deployment of further DTs.

The built environment is an area with multiple applications for DTs. Architectural models (even digital ones) of new buildings have been used for some time and some DTs have been proposed as a form of Building Information Model (BIM). BIM is a collaborative system that helps industry professionals in designing, delivering and maintaining assets throughout the entire lifecycle by outlining the characteristics of a building facility. BIM facilitates knowledge sharing among stakeholders from conception to demolition. BIM maturity levels range from 0 (no collaboration) to 3 (collaboration and data sharing using open standards. The construction of large complex buildings often involves multiple parties. This hinders information sharing leading to errors and consequential costs and delays. DTs that model the building information (including feeds from IoT sensors traceable data communication between the parties) provide a trusted framework for all the parties (Lee et al., 2021). BDTs provide a BIM maturity level 3 context for collaboration in construction (Götz et al., 2020). The trusted, shared collaborative environment required for large building projects involving multiple parties over the life cycle of the building provides a motivation for the deployment of BDTs.

With a broad range of applications for DTs, and an even broader range of users, the range of motivations will vary. The potential value delivered by a DT needs to be balanced by the costs associated with the technology. DTs that are customized towards specific operational niches may provide greater value for their users. Lower complexity DTs that are targeted at broader applications and composable into larger systems may be developed at lower cost. Note that the adoption costs for new technologies should also consider the maturity and readiness of the organization to adopt the new technology. The formal or *defacto* standardization of DTEs, and other tooling to simplify the adoption, configuration and operation of DTs reduces the barriers to their deployment. Reduced barriers to entry enable a broader range of organizations to develop the motivations and business cases to move from awareness of DT technologies towards deployments.

Objectives for Blockchains in Support of Digital Twins

Blockchains have been proposed for assurance of authenticity of the DT (Hasan et al., 2020; Hemdan & Mahmoud, 2021). These approaches use cryptographically signed transactions recorded on blockchains to provide assurances of the provenance of the data used in developing and maintaining the state of the DT. Some products (e.g., aircraft) already have regulatory requirements constraining which components can be used in the aircraft, and blockchains have been proposed to establish provenance (Mandolla et al., 2019). Assurance of the integrity of a live DT requires trust in the data feed to that DT. Beyond the data feeds, trust in the authenticity of the DT also requires assurances regarding the modeling and simulation provided by the DT. In some cases, the mechanisms to build these assurances may require notarization of transactions by independent nodes on the blockchain, or certifications by other parties in off chain transactions.

Blockchains can be used for assurance of authenticity of actions by DT. If DTs are deployed for a commercial purpose, the results of their use have commercial value, and hence the authenticity of those results needs to be assured. For example, the owner of a manufacturing facility makes available a DT of the facility so that potential customers can verify the manufacturability of their products using that facility. Users of this DT receive a result signed by the DT to predict a certain level of manufacturing performance given the input specifications and configurations. Such users may need such assurances of manufacturability in order to obtain the financing necessary to proceed. In this case, assurance that the DT provides an adequate representation of the manufacturing facility is insufficient. It is the authenticity of the results produced by the DT that are required.

Blockchains for settlement operations is a common use case. There are settlement or payment operations associated with the commercial use of DTs. In this use case, the DT is a digital asset. Access to this DT is conditioned on some commercial contract. The commercial contract here being at least partially automated via a smart contract executing in a blockchain. Commercial contracts are legal documents, and there is a legal distinction in the role of the DT in the smart contract that may be significant. In cases where the DT is a party to the smart contract, it is typically acting as an agent for some other legal entity.

There is some precedent, however, for recognition of blockchain software as a legal entity (Wright, 2021) which may be attractive for some use cases.

BDT Trust

Trust models follow from the objectives driving the deployment of the blockchain associated with the digital twin. The trust models may be targeted at assurances of the integrity of DT, the results of the DTs operation, or commercial settlement operations around the use of DTs. The users of DTs need to consider the degree of trust required in comparison with the consequences and likelihood of plausible threat models. System architectures have incremental costs and complexity associated with adding trust features. Additional system complexity may also impact ease of use and other technology adoption hurdles. A degree of assurance in the integrity is required for all DTs because of the level of investment required to create and operate them.

Users trust that DT technology will provide the desired operational functionality with an acceptable level of quality. Trust can be viewed as a level of confidence. From this perspective trust is the probability that the intended behavior and the actual behavior are equivalent given a fixed context, fixed environment, and fixed point in time. Trust can be considered at several levels: Functionally equivalence of the DT & physical object; Integration & composition with other DTs; Sufficiency of available information about the environment and context of the physical object; Generalizability of DT technology across multiple fields of use. (Voas et al., 2021) identified 14 trust considerations to enhance the usefulness of DT technology:

- DT Creation Ordering
- Temporal
- Environment
- Manufacturing Defects
- Functional Equivalence
- Composability and Complexity
- Implementation and Monitoring
- Heterogeneity of Standards
- Non-Functional Requirements
- DT Accuracy
- DT Testing
- DT Certification
- DT Error Propagation
- DT Counterfeiting

IoT devices designed for low cost and low power consumption can be used as trusted data-sources for blockchains. Blockchain operations require as little as 6% of energy in the IoT device (Pincheira et al., 2021). Using trusted data feeds helps assure the integrity of the development or operation of the DT using those data feeds. IoT has also developed into many areas of critical national infrastructure such as transport, hospitals, and power distribution grids. The trust, security, and immutability of blockchains can improve the protection of such critical infrastructures and any DTs of those infrastructures (Hammoudeh, 2020). Blockchain can be seen as a fundamental component for trustless security of connected critical infrastructures. The DTs connected to those infrastructures are also important, and depending on their role, may be essential for the proper operation of the infrastructure. For these applications, the implementation costs of BDTs appear modest in comparison with the trust benefits they enable.

BDT Lifecycle

Creation of DTs requires sensors to measure the physical object to be modelled. The parameters to be measured depend on the purpose of the digital model (e.g., static physical geometry vs range of motion). The digital model may need to incorporate physical object properties other than geometries (e.g., electromagnetic properties for a DT of an electric motor, or biological, chemical, physiological properties for DTs intended for healthcare applications). The sensor arrangement could be configured to scan multiple smaller physical objects either in a fixed location like a photo booth or larger physical objects e.g., built structures, landscapes. The software capturing the data from these sensors, converts in into the digital model. Such digital modelling software needs to understand the constraints of the sensors as well as the digital model characteristics and the environment where the DT will be operating. The creation of a DT can be a one-time event; the DT can be continuously updating its status based on the physical object; or the DT can be learning and improving the fidelity of its model over time. A given physical object may have multiple types of DTs created for different purposes. A DT created at one point in time could be allowed to evolve independently of the original object.

A given physical object may have multiple instances of its DT created for operation by different parties. DT Environments (DTEs) provide a context within which to operate the DT. DTEs can be customized for a specific DT, or more generalized to support simultaneous operation of multiple DTs. A DTE for evaluating smart city operations may include independently developed DTs of multiple buildings or other structures in the smart city. Portability of DTs across DTEs will be required when the operators of separate DTEs require access to the same DT.

Operation of a DT instance may be done by the developer of the DT, but in a commercial context it is more likely a different user. The operational controls will depend on the specific objectives of the user. A common industrial objective for DTs is in product lifecycle management, e.g., as an information system to maintain a product biography (Barata et al., 2020). Smart contracts can be used to govern and track transactions initiated by participants involved in the operation of BDTs in the product life cycle (Hasan et al., 2020).

Termination of a DT instance ceases its operation and relinquishes computing resources. For commercial use of DTs, termination of a DT instance may impact charging in some commercial contexts (e.g., DT as a Service). Care should also be taken to relinquish the appropriate communication resources. This may be a little more complex if the data feeds are multicast between the physical object and multiple instances.

More complex operations may be required to support some DT roles. For example, it may be necessary to migrate an operational DT to execute on new infrastructure. DTs as a software entity can also be expected to follow a software lifecycle where additional releases become available periodically. Many software products reuse other software components from open source or other software repositories. A Software Bill of Materials (SBOM) provides a transparency mechanism for securing software product supply chains (Carmody et al., 2021). Blockchains and other tools provide convenient methods to establish software provenance (Barclay et al., 2019). Software provenance of DTs is useful not only from a commercial licensing perspective, but also for maintenance. Many software components release functionality updates on a 6-month cycle, with urgent security patches on interim releases. Depending on the operational use case, the complexity of the DT, and the number and severity of upstream software components' security patches and functionality releases, operating DTs may need to support in service updates.

DTEs can be considered as a platform for to manage a DT's life cycle (Minerva et al., 2020). The DTE becomes the context within which multiple DTs are instantiated, composed into larger systems and operated to perform their business purpose. In this context, a standard format for ingestion of new DTs from different DT developers is necessary to scale the variety of DTs available on the DTE. The variety of DTs available within the DTE is one factor in the market attractiveness of such a platform. The DTE is, of course, also software that has its own lifecycle. This lifecycle can be masked from users to a certain extent if the DTE is made available on an "as a Service" basis. Several major cloud service providers already provide blockchains on such a basis. The DT itself could also be delivered as a service (Aheleroff et al., 2021).

The lifecycle of the DT should also be considered in comparison to the lifecycle of the underlying physical object. A DT can be developed in a manufacturing context before a product exists. A DT may also remain functional and useful after the underlying physical object has reached its end of life or been destroyed. Some categories of physical objects may be very long lived. Buildings, for example, are commonly expected to have lifetimes measured in the decades.

The building lifecycle includes the design, construction, operation, maintenance, and end-of-life stages (Yitmen et al., 2021). Managing the building's lifecycle directly and indirectly, affects many aspects, such as buildings or infrastructures' operation and efficiency, operational risks, the environmental impact of buildings, people's quality of life, safety, and businesses (Yousefnezhad et al., 2020). To maintain a DT over such a long life-cycle requires additional considerations as typical software lifetimes follow much shorter technology timeframes. For example, the DT may need to be designed to accommodate changes in archival storage technologies over these timeframes.

Frameworks, Architectures, and Standards for BDTs

One approach to framing the evolution of DTs comes from some of the terminology various authors have used in classifying DTs. From the perspective of the integration between physical and digital objects, (Fuller et al., 2020) (and others) distinguish a Digital Twin, Digital Shadow, Digital Model. Using this terminology, a Digital Model has only manual data flow in either direction between the digital and physical object while a Digital Twin has automatic data flow in both directions. A Digital Shadow has automated data flow in one direction, but only manual data flow in the other direction (Sepasgozar, 2021). This distinction in the degree of integration constrains the functionality. The Digital Model, in this terminology corresponds to the notion of a *static* DT introduced earlier; and does not reflect the current state of the physical object. Automated connectivity between the physical and digital object provides greater functionality; but requires additional resources and infrastructure. This terminology might be useful for distinguishing capability levels of the type of DTs. For example, it would be easier to develop and bring to market static Digital models than live DTs, though they would be less functional.

DTs can be classified in terms of the value they create and/or the data resources that they use. (Barth et al., 2020) developed such an ontology for DTs through literature review. This ontology distinguishes between value created for external users of services delivered by the DT and internal value delivered by the operation of a DT within the organization creating it. The data resources associated with a DT come from different data sources, in different data formats and may be categorized differently. These perspectives on the data resources and value delivered also provide a framework to evaluate the DT. This framework enhances an individual's understanding of DTs, facilitates efficient conveyance of system details between stakeholders, provides a point of reference for system designers to extract specifications, documents the DT for future reference, and provides a means for collaboration. Consider the ratio of value delivered to data input. A DT with a higher ratio value could be considered as more data efficient. Looking more closely at the value created and data resources could also lead to other improvements in the DT. From a value chain perspective, four points of value aggregation can be identified: the *Physical Objects*, i.e., the products sold or leased to the final customers; the *Virtualization Platform*, i.e., the set of functions, data and representations of objects that behave in synergy with the associated physical objects; the *interfaces and views* on DTs that can be used in order to create services; and the *services and applications*, that use the DT's functionality. These can all be sold or offered to final customers (Minerva et al., 2020).

External value from the DT is delivered to other actors in the ecosystem at different levels – e.g., as a connected product level, a system level, a system of systems level. DTs as smart connected products can provide features such as autonomy, self-optimization, or controllability that deliver value to external actors in the ecosystem. DTs deliver a service within a constrained scope of external ecosystem actors. This service scope may be constrained by functionality, but also availability, performance and quality. A DT creates internal value by connecting product generations through analyzing past generations, optimizing current generations and innovating future generations. The internal value created by a DT may be operationalized as product management targeted at product instances, product masters or product types. Internal value can also be delivered by improvements in the product lifecycle at various stages e.g., Beginning of Life, Middle of Life, End of Life. Value creation in DTs can be improved by optimizing the value created by the DT for its intended user. For an internal user this might be a matter of designing the DT for better fit with internal organizational processes. For an external user, this might be a matter of better fit with that targeted user's processes and business needs. From the value creation perspective, the DT can be seen delivering value as an integrator of the product and customer lifecycles. BDTs (in comparison to non-blockchain DTs) can deliver greater external value to the extent that improve the trustworthiness of the system.

Data sources feeding the DT may be from measurements of the product itself, from other internal enterprise systems with product information (e.g., design studies), or third parties (e.g., component suppliers, specialized testing and certification agencies). Data formats may be structured, unstructured or interpreted. Structured and unstructured data must be processed; creating interpreted or inferred information which can then be used for automated decision- and sense-making. An information architecture (e.g., Dreyer et al., 2017) is required for the storage and processing of the data, the creation and delivery of services based on the data, and the presentation of data and services, so that value can be delivered to the users of the DT.

Connectivity, particularly wireless connectivity, is becoming both more ubiquitous, and in greater bandwidth. IoT and other smart devices are significantly increasing the data generated and changing the nature of the connectivity to support machine-machine interactions, rather than human interactions. Cloud computing architectures have thrived by aggregating scarce data to centralized locations for efficient computation. The rising volume of data at the edge requires one or more edge layers between the IoT sensors creating the data and centralized clouds. With multiple IoT devices associated with a specific environment of physical object, peer to peer connectivity between those sensors is often used to develop an aggregate view of the behavior of the space or object being sensed. A Blockchain of Things (BoT) has been proposed to provide greater security and trust in the data generated from IoT sensors

(Abadi et al., 2018). While IoT devices are typically optimized for low cost and low power, a BoT is a more heterogeneous computing environment that can include edge or fog computing architectures with nodes that have more computing resources. These *Edge Computing* architectures enable more computation intensive processing of the data – e.g., Artificial Intelligence processing of raw data feeds for event detection, or access controls for the data, or off chain storage and computation (Lu et al., 2020). This could enable data trading between the sources of the data and other parties (Nawaz et al., 2020). One application of such trading would be the secure enablement of data feeds for multiple instances of DTs.

In their survey of DT features, scenarios, and architectural models, Minerva et al. (2020) positioned DTs with related concepts from augmented and virtual reality (AR/VR), multiagent systems and virtualization. From this broader context, the considered a DT as having the following essential features: A DT strictly refers to a physical object; A DT contains all the information needed to fully characterize a physical object and its intended or predicted behavior; and since the DT is framed in a lifecycle composed of different steps, it can encompass data and information that describe the 'history' of the physical object.

Augmented Reality is associated with 'augmenting' the amount of information associated to a physical object with additional data. The physical object and the 'augmentation' information are represented as a single entity. DTs of the physical objects in the field of view can be used to aggregate additional information about those objects for display to the user of the AR system. The DT could be said to be augmented with additional information not available at the physical object. Virtual Reality, in comparison, creates a virtual environment in which a user can interact with virtual objects. Sometimes these objects have a physical counterpart, and sometimes they do not (but they still reflect the expected behavior of physical objects). DTs in VR could be used to populate the virtual environment with virtual objects corresponding to physical objects. In a virtualized world, an *Avatar* is a virtual representation of an object, usually a person. The Avatar can behave as a substitute of the physical individual – a form of DT.

Minerva et al. (2020) identified twelve dimensions for characterization of DTs including: Representativeness and Contextualization, Reflection, Replication, Entanglement, Persistency, Memorization, Composability, Accountability/Manageability, Augmentation, Ownership, Servitization, and Predictability. Another technological trend that has a relationship with the Digital Twin concept is Virtualization. It is the ability to virtualize entire systems by means of software and to execute them on general-purpose machines. Virtualization offers to DT infrastructures the ability to deal with extensive virtualization of functions, e.g., a logical object, its orchestration, and the chaining into different services. These are enabling capabilities for an effective DT platform. Multi-agent Software Architectures are systems based on the implementation of agents that act on behalf of another entity and explore and collect data in several environments. Multi-agent Systems offer several interesting properties that DTs may embody:

- As a software agent, a DT represents an external entity that wants to operate in a specific environment – e.g., a DTE.
- DTs and Agents can be of different types, e.g., passive, active or cognitive, depending on the use case.
- Agents typically operate with other software components (e.g., other DTs) in complex environments (e.g., DTEs) that are difficult to model. Many agents (and DTs), especially cognitive ones, are autonomous and operate in a decentralized manner.
- Agents (and DTs) can collect a large amount of data to improve their knowledge of the environment in which they operate.

The industrialization and commercialization of BDTs requires the development and implementation of multiple standards to enable interoperability in a broad ecosystem. ISO TC 184 SC 4 has developed a series of standards for product data representation and exchange. ISO also has in process a Digital Twin Manufacturing Framework (ISO, 2021) and a report on human visualization of product design data from the perspective of the digital twin (ISO, 2020). The ITU has also positioned DTs as a use case for 6G networks (ITU-T, 2020). While the development of detailed technical standards takes time, other industry bodies and open-source communities can enable some DT related markets on shorter timeframes.

The Digital Twin Consortium (DTC, 2021) was founded to accelerate the development, adoption, interoperability, and security of digital twins and enabling technologies. This consortium has working groups consider DT applications in multiple industry verticals including:

aerospace & defence, fintech, healthcare & life sciences, infrastructure, manufacturing, natural resources. The Industrial Internet Consortium (IIC) considers that DTs can be deployed in any Industrial IoT tier, realized with the available technology choices, and synchronization between digital twins is the only communication between tiers. Data replicated into a digital twin looks like ingest and triggers the associated published notification events (Harper et al., 2019). They proposed 6 operations for DTs:

- DTs are discoverable in an interoperable fashion
- DTs maintain an information model that can be queried
- DTs exchange data with the underlying physical objects
- DTs support a publish/subscribe model for transactions
- DTs securely synchronize between connected tiers
- DT properties are independently configurable and managable

Harper et al. (2019) also developed a set of nine architectural criteria for DTs:

- App store deployment of configuration
- Integrated information model
- Flexible classification of types, properties, and instances
- Encrypted data at rest and in transfer
- Role based access control configured for authenticated users
- Data ingest configuration for each column store
- CRUD data exchange with cascading side effects based on role
- Publish and subscribe notification of CRUD transactions
- Filtered synchronization between tiers.

BDT Implementations

Several implementations used a blockchain-based framework to address the issues of data management and security. Once the data collected from trustworthy sources and recorded via the blockchain, product lifecycle events can be fed into data-driven systems for process monitoring, diagnostics, and optimized control. DTs that can then draw intelligent conclusions from data by identifying the faults and recommending precautionary measures (Suhail et al., 2021). The block is typically used to store transaction logs required to demonstrate provenance when managing a product life cycle with a BDT. The volume of data may require offline storage, with the block just retaining hashes of the stored files. The DTs in this product lifecycle information system can also be stored and shared via IPFS (Hasan et al., 2020) or Swarm (Putz et al., 2021). Some implementations separate lower level blockchain and upper-level digital twin models to balance between holistic planning and local execution in manufacturing systems (Leng et al., 2020).

Ethereum is a widely used open source blockchain. Ethereum smart contract code can be used for tracing signoffs by the various personnel responsible for creating DTs (e.g., design manager, quality Manager, testing Manager, Delivery Manager). The smart contract code (e.g., solidity) can be made publicly available through repositories such as GitHub. Another open-source prototype "EtherTwin" (Putz et al., 2021) provides role-based access controls through the underlying asset life cycle from manufacturing through operation and end of existence. Blockchains operate to ensure trustworthy transactions on tokens. For IoT blockchains, tokens are associated with IoT devices or data. Integrating physical assets with decentralized blockchains can be challenging. (Mazzei et al., 2020) developed tokenization software for use with the Ethereum blockchain. (Nielson et al., 2020) also proposed the use of ERC721 non-fungible tokens for tracking product information.

While the Ethereum blockchain is widely used for decentralized financial applications, it may not be optimal for IoT applications. (Bai et al., 2019) developed a lightweight blockchain network architecture for IIOT applications that moved the complex data processing and storage off chain. "Twinchain" is another blockchain designed for use with IoT devices. It offers significant reduction in the time to confirm transactions compared to traditional blockchains like Ethereum and is also based on has based digital signatures to be resilient to quantum computing attacks (Khan et al., 2020).

BDT Challenges

As DTs move from concept to initial custom deployments, and then widespread commercial deployments, the users and deployers of DTs cannot be assumed to have deep expertise in the creation and deployment of the many underlying technologies. Moving to widespread commercial deployment implies not only an effortless integration of the data flowing in either direction between the DT and the physical object it models, but also the international standardization necessary to enable global markets (Fuller et al., 2020). DTs require the integration of multiple technologies – IOT devices, connectivity, big data analytics, AI, blockchain etc. as well as expertise in the specific domain to be modelled. Enabling widespread markets for DT commercialization implies acceptance of a certain degree of heterogeneity in the components and other infrastructure components.

Scalable standardized IT infrastructures are required that can enable widespread DT enablement. While cloud computing resources are scalable, edge computing architectures are needed to manage the reduced latency requirements. In addition, these IT architectures need to provide the necessary assurances of security, privacy and trust for the various parties engaged in DT commerce. DTs are fundamentally models of the objects they represent. DTs are proposed for a wide variety of applications. Domain expertise is required to develop and validate models for each of those domains. The generalizability to different industries and products requires further study.

DTs emerged in the manufacturing realm. Three relevant issues may arise in the application of the DT in large open systems (Minerva et al., 2020): *knowledge of the physical world*: it is can be difficult to describe the models, laws and effects of the real world in a DT; *System complexity:* Large complex systems and products may require multiple DTs to interoperate; *Interoperability:* if proprietary and closed interfaces are used, it will be extremely difficult to create interoperable ecosystems of DTs.

CONCLUSION

DTs have emerged as a critical concept in cyberspace infrastructure. DTs are virtual representations of physical things. DTs can be used to model smart structures or environments, manufacturing processes, humans, and a variety of other things. The value provided by DTs relies on their fidelity in representation. Blockchains provide trust assurance mechanisms, particularly where multiple parties are involved. The expected life cycle operations of the IoT, Blockchain and DT need to be considered to develop economically useful BDT models. BDTs do not exist in isolation, but rather within a DTE. A DTE may include multiple DTs of different objects, to enable interactions between these objects to be evaluated in both virtual reality and mixed reality cases. To populate DTEs with multiple DTs requires industrialized tooling to support the rapid creation of DTs. The industrialization of DT creation requires frameworks, architectures and standards to enable interoperability between DTs and DTEs. While blockchains developed from fintech applications, BDT applications will have different requirements for blockchain features and performance. Beyond the systems design of BDTs, their implementation and operation will create additional challenges.

REFERENCES

Abadi, F. A., Ellul, J., & Azzopardi, G. (2018). The blockchain of things, beyond bitcoin: A systematic review. In *International Conference on Internet of Things (iThings) and IEEE Green Computing and Communications (GreenCom) and IEEE Cyber, Physical and Social Computing (CPSCom) and IEEE Smart Data (SmartData)* (pp. 1666-1672). IEEE. 10.1109/Cybermatics_2018.2018.00278

Aheleroff, S., Xu, X., Zhong, R. Y., & Lu, Y. (2021). Digital twin as a service (DTaaS) in industry 4.0: An architecture reference model. *Advanced Engineering Informatics*, *47*, 101225. doi:10.1016/j.aei.2020.101225

Ahmadi, H., Nag, A., Khan, Z., Sayrafian, K., & Rahadrja, S. (2021). *Networked Twins and Twins of Networks: an Overview on the Relationship Between Digital Twins and 6G*. https://arxiv.org/abs/2108.05781 doi:10.1016/j.aei.2020.101225

Bai, L., Hu, M., Liu, M., & Wang, J. (2019). BPIIoT: A light-weighted blockchain-based platform for industrial IoT. *IEEE Access : Practical Innovations, Open Solutions*, *7*, 58381–58393. doi:10.1016/j.aei.2020.101225

Barata, J., Pereira, V., & Coelho, M. (2020). Product Biography Information System: A Lifecycle Approach to Digital Twins. In *International Conference on Systems, Man, and Cybernetics (SMC)* (pp. 899-904). IEEE. 10.1109/SMC42975.2020.9283061

Barclay, I., Preece, A., Taylor, I., & Verma, D. (2019). *Towards Traceability in Data Ecosystems using a Bill of Materials Model*. https://arxiv.org/abs/1904.04253 doi:10.1016/j.aei.2020.101225

Barth, L., Ehrat, M., Fuchs, R., & Haarmann, J. (2020). Systematization of digital twins: ontology and conceptual framework. In *Proceedings of the 3rd International Conference on Information Science and System* (pp. 13-23). 10.1145/3388176.3388209

Caldarelli, G., & Ellul, J. (2021). Trusted Academic Transcripts on the Blockchain: A Systematic Literature Review. *Applied Sciences (Basel, Switzerland)*, *11*(4), 1842. doi:10.3390/app11041842

Carmody, S., Coravos, A., Fahs, G., Hatch, A., Medina, J., Woods, B., & Corman, J. (2021). Building resilient medical technology supply chains with a software bill of materials. *NPJ Digital Medicine*, *4*(1), 1–6. doi:10.103841746-021-00403-w PMID:33623135

Chhetri, M. B., Krishnaswamy, S., & Loke, S. W. (2004) Smart virtual counterparts for learning communities. *International Conference on Web Information Systems Engineering*, 125–134.

Croatti, A., Gabellini, M., Montagna, S., & Ricci, A. (2020). On the integration of agents and digital twins in healthcare. *Journal of Medical Systems*, *44*(9), 1–8. doi:10.100710916-020-01623-5 PMID:32748066

Deng, T., Zhang, K., & Shen, Z.J.M. (2021). A Systematic Review of a Digital Twin City: A New Pattern of Urban Governance toward Smart Cities. *Journal of Management Science and Engineering*.

Dreyer, S., Olivotti, D., Lebek, B., & Breitner, M. H. (2017). Towards a Smart Services Enabling Information Architecture for Installed Base Management in Manufacturing. *Wirtschaftsinformatik*.

DTC. (2021). https://www.digitaltwinconsortium.org/about-us/index.htm

ETSI. (2012). *Network Functions Virtualisation (NFV): Network Operator Perspectives on Industry Progress*. Paper presentation at SDN and OpenFlow World Congress. https://portal.etsi.org/NFV/NFV_White_Paper.pdf

Fuller, A., Fan, Z., Day, C., & Barlow, C. (2020). Digital twin: Enabling technologies, challenges and open research. *IEEE Access : Practical Innovations, Open Solutions*, *8*, 108952–108971. doi:10.1109/ACCESS.2020.2998358

Götz, C. S., Karlsson, P., & Yitmen, I. (2020). *Exploring applicability, interoperability and integrability of Blockchain-based digital twins for asset life cycle management. In Smart and Sustainable Built Environment*. Emerald Group Publishing Ltd., doi:10.1108/SASBE-08-2020-0115

Granda, F. A. P., & Salik, I. (2020). Simulation Training and Skill Assessment in Critical Care. *StatPearls*. https://www.ncbi.nlm.nih.gov/books/NBK549895/

Grieves, M. (2019). Virtually Intelligent Product Systems: Digital and Physical Twins. In *Complex Systems Engineering: Theory and Practice* (pp. 175–200). American Institute of Aeronautics and Astronautics. doi:10.2514/5.9781624105654.0175.0200

Grieves, M., & Vickers, J. (2017). Digital twin: Mitigating unpredictable, undesirable emergent behavior in complex systems. In *Transdisciplinary Perspectives on Complex Systems* (pp. 85–113). Springer. doi:10.1007/978-3-319-38756-7_4

Hammoudeh, M. (2020, November). Blockchain, Internet of Things and Digital Twins in Trustless Security of Critical National Infrastructure. *The 4th International Conference on Future Networks and Distributed Systems (ICFNDS)*.

Harper, K. E., Ganz, C., & Malakuti, S. (2019). Digital twin architecture and standards. *IIC Journal of Innovation*, *12*, 72–83.

Hasan, H. R., Salah, K., Jayaraman, R., Omar, M., Yaqoob, I., Pesic, S., Taylor, T., & Boscovic, D. (2020). A blockchain-based approach for the creation of digital twins. *IEEE Access : Practical Innovations, Open Solutions*, *8*, 34113–34126. doi:10.1109/ACCESS.2020.2974810

Hemdan, E. E. D., & Mahmoud, A. S. A. (2021). BlockTwins: A Blockchain-Based Digital Twins Framework. In Blockchain Applications in IoT Ecosystem (pp. 177-186). Springer International Publishing. doi:10.1007/978-3-030-65691-1_12

ISO. (2020). *Automation systems and integration — Industrial data — Visualization elements of digital twins*. ISO/TR 24464.

ISO. (2021). *Automation systems and integration — Digital Twin framework for manufacturing — Part 1: Overview and general principles*. ISO/DIS 23247-1.

ITU-T (2020). *Y.3000-series - Representative use cases and key network requirements for Network 2030.* Series Y Supplement 67 (07/2020).

Kapustina, I., Kalinina, O., Ovchinnikova, A., & Barykin, S. (2020). The logistics network digital twin in view of concept of the non-destructive quality control methods. In *E3S Web of Conferences* (Vol. 157, p. 05001). EDP Sciences. doi:10.1051/e3sconf/202015705001

Khan, A., Shahid, F., Maple, C., Ahmad, A., & Jeon, G. (2020). Towards Smart Manufacturing Using Spiral Digital Twin Framework and Twinchain. *IEEE Transactions on Industrial Informatics*. Advance online publication. doi:10.1109/TII.2020.3047840

Lee, D., Lee, S. H., Masoud, N., Krishnan, M. S., & Li, V. C. (2021). Integrated digital twin and blockchain framework to support accountable information sharing in construction projects. *Automation in Construction*, *127*, 103688. doi:10.1016/j.autcon.2021.103688

Leng, J., Wang, D., Shen, W., Li, X., Liu, Q., & Chen, X. (2021). Digital twins-based smart manufacturing system design in Industry 4.0: A review. *Journal of Manufacturing Systems*, *60*, 119–137. doi:10.1016/j.jmsy.2021.05.011

Leng, J., Yan, D., Liu, Q., Xu, K., Zhao, J. L., Shi, R., Wei, L., Zhang, D., & Chen, X. (2020). Manuchain: Combining permissioned blockchain with a holistic optimization model as bi-level intelligence for smart manufacturing. *IEEE Transactions on Systems, Man, and Cybernetics. Systems*, *50*(1), 182–192. doi:10.1109/TSMC.2019.2930418

Lu, Y., Huang, X., Zhang, K., Maharjan, S., & Zhang, Y. (2020). Low-latency federated learning and blockchain for edge association in digital twin empowered 6G networks. *IEEE Transactions on Industrial Informatics*, *17*(7), 5098–5107. doi:10.1109/TII.2020.3017668

Mandolla, C., Petruzzelli, A. M., Percoco, G., & Urbinati, A. (2019). Building a digital twin for additive manufacturing through the exploitation of blockchain: A case analysis of the aircraft industry. *Computers in Industry*, *109*, 134–152. doi:10.1016/j.compind.2019.04.011

Mazzei, D., Baldi, G., Fantoni, G., Montelisciani, G., Pitasi, A., Ricci, L., & Rizzello, L. (2020). A Blockchain Tokenizer for Industrial IOT trustless applications. *Future Generation Computer Systems*, *105*, 432–445. doi:10.1016/j.future.2019.12.020

Minerva, R., Lee, G. M., & Crespi, N. (2020). Digital twin in the IoT context: A survey on technical features, scenarios, and architectural models. *Proceedings of the IEEE*, *108*(10), 1785–1824. doi:10.1109/JPROC.2020.2998530

Nawaz, A., Peña Queralta, J., Guan, J., Awais, M., Gia, T. N., Bashir, A. K., Kan, H., & Westerlund, T. (2020). Edge computing to secure iot data ownership and trade with the ethereum blockchain. *Sensors (Basel)*, *20*(14), 3965. doi:10.339020143965 PMID:32708807

Nielsen, C. P., da Silva, E. R., & Yu, F. (2020). Digital Twins and Blockchain–Proof of Concept. *Procedia CIRP*, *93*, 251–255. doi:10.1016/j.procir.2020.04.104

Pincheira, M., Vecchio, M., Giaffreda, R., & Kanhere, S. S. (2021). Cost-effective IoT devices as trustworthy data sources for a blockchain-based water management system in precision agriculture. *Computers and Electronics in Agriculture*, *180*, 105889. doi:10.1016/j.compag.2020.105889

Putz, B., Dietz, M., Empl, P., & Pernul, G. (2021). Ethertwin: Blockchain-based secure digital twin information management. *Information Processing & Management*, *58*(1), 102425. doi:10.1016/j.ipm.2020.102425

Raj, P. (2021). Empowering digital twins with blockchain. *Advances in Computers*, *121*, 267–283. doi:10.1016/bs.adcom.2020.08.013

Seilov, S., Kuzbayev, A. T., Seilov, A. A., Shyngisov, D. S., Goikhman, V. Y., Levakov, A. K., Sokolov, N. A., & Zhursinbek, Y. S. (2021). The Concept of Building a Network of Digital Twins to Increase the Efficiency of Complex Telecommunication Systems. *Complexity*, *2021*, 2021. doi:10.1155/2021/9480235

Sepasgozar, S. M. (2021). Differentiating digital twin from digital shadow: Elucidating a paradigm shift to expedite a smart, sustainable built environment. *Buildings*, *11*(4), 151. doi:10.3390/buildings11040151

Söderberg, R., Wärmefjord, K., Carlson, J. S., & Lindkvist, L. (2017). Toward a digital twin for real-time geometry assurance in individualized production. *CIRP Annals*, *66*(1), 137–140. doi:10.1016/j.cirp.2017.04.038

Suhail, S., Hussain, R., Jurdak, R., & Hong, C. S. (2021). Trustworthy digital twins in the industrial internet of things with blockchain. In *Internet Computing*. IEEE

Tao, F., Zhang, M., Liu, Y., & Nee, A. (2018). Digital twin driven prognostics and health management for complex equipment. *CIRP Annals*, *67*(1), 169–172. doi:10.1016/j.cirp.2018.04.055

Voas, J., Mell, P., & Piroumin, V. (2021). *Considerations for Digital Twin Technology and Emerging Standards, draft NISTR 8356*. https://nvlpubs.nist.gov/nistpubs/ir/2021/NIST.IR.8356-draft.pdf

Wright, S. A. (2019). Privacy in IoT Blockchains: with big data comes big responsibility. *Proceedings of the International Conference on Big Data* (pp. 5282-5291). IEEE. 10.1109/BigData47090.2019.9006341

Wright, S. A. (2021) Identity. *Proceedings of 27th Int'l Conf. on Parallel & Distributed Processing Techniques & Applications (PDPTA'21).*

Yitmen, I., Alizadehsalehi, S., Akıner, İ., & Akıner, M. E. (2021). An Adapted Model of Cognitive Digital Twins for Building Lifecycle Management. *Applied Sciences (Basel, Switzerland)*, *11*(9), 4276. doi:10.3390/app11094276

Yousefnezhad, N., Malhi, A., Kinnunen, T., Huotari, M., & Främling, K. (2020, July). Product Lifecycle Information Management with Digital Twin: A Case Study. In *18th International Conference on Industrial Informatics (INDIN)* (Vol. 1, pp. 321-326). IEEE. 10.1109/INDIN45582.2020.9442208

KEY TERMS AND DEFINITIONS

Blockchain Digital Twin: A type of Digital Twins that incorporates blockchains to provide additional trust-based features.

Digital Model: A Digital Model has only manual data flow in either direction between the digital and physical object.

Digital Shadow: A Digital Shadow has automated data flow in one direction, but only manual data flow in the other direction.

Digital Twin: A digital model of a physical object. A *static* Digital Twin has characteristics that do not change with time. A *dynamic* Digital Twin captures behavioral responses of the physical object. A *live* Digital Twin represents the current state of the physical object.

Digital Twin Environment: A context within which a Digital Twin may be operated.

Human Digital Twin: A digital model of a human.

Compilation of References

(2016). Optical performance monitoring for fiber-optic communication networks. InZhou, X., & Xie, C. (Eds.), *Enabling Technologies for High Spectral-Efficiency Coherent Optical Communication Networks*. John Wiley & Sons.

(2020). Indoor Visible Light Communication Networks for Camera-Based Mobile Sensing. InShen, X. S., Lin, X., & Zhang, K. (Eds.), *Encyclopedia of Wireless Networks* (pp. 1–7). Springer International Publishing. doi:10.1007/978-3-319-32903-1_270-1

GPP TR 21.916. (2020a). *Services and System Aspects*. Release 16, v0.5.0.

GPP TR 23.745. (2021). *Study on application layer support for Factories of the Future in 5G network*. Release 17, v17.0.0.

GPP TS 23.222. (2021). *Common API Framework for 3GPP Northbound APIs*. Release 17, v17.4.0.

GPP TS 23.286. (2021). *Application layer support for Vehicle-to-Everything (V2X) services*. Release 17, v17.1.0.

GPP TS 23.434. (2019). *Service Enabler Architecture Layer for Verticals (SEAL); Functional architecture and information flows*. Release 16, v1.1.0.

GPP TS 23.434. (2021). *Service Enabler Architecture Layer for Verticals (SEAL)*. Release 17, v17.1.0.

GPP TS 23.501. (2021). *System architecture for the 5G System (5GS)*. Release 17, v17.0.0.

GPP TS 33.122. (2020). *Security aspects of Common API Framework (CAPIF) for 3GPP northbound APIs*. Release 16, v16.3.0.

GPPP. (2016). *5G Vision, The 5G Infrastructure Public Private Partnership: the next generation of communication networks and services* [Whitepaper]. https://espas.secure.europarl.europa.eu/orbis/document/5g-vision-5g-infrastructure-public-private-partnership-next-generation-communication

GPPP. (2021). *View on 5G Architecture* [White paper]. https://5gppp.eu/wp-content/uploads/2020/02/5G-PPP-5G-Architecture-White-Paper_final.pdf

Aazam, M., Zeadally, S., & Harras, K. (2018). Deploying fog computing in industrial internet of things and industry 4.0. IEEE Transactions on Industrial Informatics.

Abadi, F. A., Ellul, J., & Azzopardi, G. (2018). The blockchain of things, beyond bitcoin: A systematic review. In *International Conference on Internet of Things (iThings) and IEEE Green Computing and Communications (GreenCom) and IEEE Cyber, Physical and Social Computing (CPSCom) and IEEE Smart Data (SmartData)* (pp. 1666-1672). IEEE. 10.1109/Cybermatics_2018.2018.00278

Abashidze, I., & Dąbrowski, M. (2016). *Internet of Things in marketing: opportunities and security issues*. Management Systems in Production Engineering.

Abdulkader, O., Bamhdi, A. M., Thayananthan, V., Elbouraey, F., & Al-Ghamdi, B. (2019). A lightweight blockchain based cybersecurity for IoT environments. *Proc. 6th IEEE Int. Conf. Cyber Secur. Cloud Comput. (CSCloud)/5th IEEE Int. Conf. Edge Comput. Scalable Cloud (Edge-Com)*, 139–144. 10.1109/CSCloud/EdgeCom.2019.000-5

Abomhara, M., & Køien, G. M. (2014). Security and privacy in the internet of things: Current status and open issues. *2014 International Conference on Privacy and Security in Mobile Systems (PRISMS),* 1–8. 10.1109/PRISMS.2014.6970594

Abuhasel, K. A., & Khan, M. A. (2020). A secure industrial internet of things (iiot) framework for resource management in smart manufacturing. *IEEE Access : Practical Innovations, Open Solutions*, *8*, 117354–117364. doi:10.1109/ACCESS.2020.3004711

Admin. (2020). Beamforming in 5G mmWave radios. *5G mmWave*. https://www.5gmmwave.com/5g-mmwave/beamforming-in-5g-mmwave-radios/

Admin. (2021). Economics of 5G mmWave:is it cost effective? *5G mmWave*. https://www.5gmmwave.com/5g-mmwave/economics-of-5g-mmwave-is-it-cost-effective/

Agiwal, M., Roy, A., & Saxena, N. (2016). Next generation 5G wireless networks: A comprehensive survey. *IEEE Communications Surveys and Tutorials*, *18*(3), 1617–1655. doi:10.1109/COMST.2016.2532458

Aheleroff, S., Xu, X., Zhong, R. Y., & Lu, Y. (2021). Digital twin as a service (DTaaS) in industry 4.0: An architecture reference model. *Advanced Engineering Informatics*, *47*, 101225. doi:10.1016/j.aei.2020.101225

Ahmed, E., Yaqoob, I., Gani, A., Imran, M., & Guizani, M. (2016). Internet-Of-Things-Based Smart Environments: State of the Art, Taxonomy, and Open Research Challenges. *IEEE Wireless Communications*, *23*(5), 10–16. doi:10.1109/MWC.2016.7721736

Ahmed, U., Lin, J. C.-W., & Srivastava, G. (2022). Privacy-Preserving Active Learning on the Internet of 5G Connected Artificial Intelligence of Things. *IEEE Internet of Things Magazine*, *5*(1), 126–129. doi:10.1109/IOTM.001.2100205

Akhtar, M. W., Hassan, S. A., Ghaffar, R., Jung, H. J., Garg, S., & Hossain, M. S. (2020). The shift to 6G communications: Vision and requirements. *Human-centric Computing and Information Sciences*, *10*(53), 1–27. doi:10.118613673-020-00258-2

Akyildiz, Kak, & Nie. (2020). 6G and beyond: The future of wireless communications systems. *IEEE Access*, *8*, 133995-134030.

Akyildiz, I. F., Kak, A., & Nie, S. (2020). 6G and beyond: The future of wireless communications systems. *IEEE Access : Practical Innovations, Open Solutions*, *8*, 44983–44998. doi:10.1109/ACCESS.2020.3010896

Alahakoon, D., Nawaratne, R., Xu, Y., De Silva, D., Sivarajah, U., & Gupta, B. (2020). Self-Building Artificial Intelligence and Machine Learning to Empower Big Data Analytics in Smart Cities. *Information Systems Frontiers*, 1–20. doi:10.100710796-020-10056-x

Alavi, A. H., Jiao, P., Buttlar, W. G., & Lajnef, N. (2018). Internet of Things-enabled smart cities: State-of-the-art and future trends. *Measurement*, *129*, 589–606. doi:10.1016/j.measurement.2018.07.067

Alazab, M., Manogaran, G., & Montenegro-Marin, C. E. (2022). Trust management for internet of things using cloud computing and security in smart cities. *Cluster Computing*, *25*(3), 1765–1777. doi:10.100710586-021-03427-9

Albino, V., Berardi, U., & Dangelico, R. M. (2015). Smart Cities: Definitions, Dimensions, Performance, and Initiatives. *Journal of Urban Technology*, *22*(1), 3–21. doi:10.1080/10630732.2014.942092

Alec, J. (2020). *The 4 Stages of IoT Architecture.* https://www.digi.com/blog/post/the-4-stages-of-iot-architecture

Aleksic, S., Winkler, D., Franzl, G., Poppe, A., Schrenk, B., & Hipp, F. (2013). Quantum key distribution over optical access networks. *Proceedings of 2013 18th European Conference on Network and Optical Communications & 2013 8th Conference on Optical Cabling and Infrastructure (NOC-OC&I)*, 11–18.

Al-Hourani, A., Kandeepan, S., & Lardner, S. (2014). Optimal LAP altitude for maximum coverage. *IEEE Wireless Communications Letters*, *3*(6), 569–572. doi:10.1109/LWC.2014.2342736

Alladi, T., Chamola, V., Parizi, R., & Choo, K.-K. R. (2019). Blockchain applications for industry 4.0 and industrial iot: A review. IEEE Access.

Allam, Z., Bibri, S. E., Jones, D. S., Chabaud, D., & Moreno, C. (2022). Unpacking the '15-Minute City' via 6G, IoT, and Digital Twins: Towards a New Narrative for Increasing Urban Efficiency, Resilience, and Sustainability. *Sensors (Basel)*, *22*(4), 1369. doi:10.339022041369 PMID:35214271

Al-Otaibi, Y. D. (2021). Distributed multi-party security computation framework for heterogeneous internet of things (iot) devices. *Soft Computing*, *25*(18), 12131–12144. doi:10.100700500-021-05864-5

Alpesh, P. (2021). *The Internet of Things and Its Impact on Digital Marketing.* https://theseoplatform.co.uk/blog/the-internet-of-things-and-its-impact-on-digital-marketing/

Alsharif, M. H., Kelechi, A. H., Albreem, M. A., Chaudhry, S. A., Zia, M. S., & Kim, S. (2020). Sixth generation (6G) wireless networks: Vision, research activities, challenges and potential solutions. *Symmetry*, *12*(4), 1–27. doi:10.3390ym12040676

Altexsoft. (2020). *IoT Architecture: the Pathway from Physical Signals to Business Decisions.* https://www.altexsoft.com/blog/iot-architecture-layers-components/#:~:text=IoT%20Architecture%3A%20the%20Pathway%20from%20Physical%20Signals%20to%20Business%20Decisions&text=IoT%20solutions%20have%20become%20a,enterprises%2C%20connected%20devices%20are%20everywhere.&text=This%20article%20describes%20IoT%20%E2%80%94%20the,its%20architecture%2C%20layer%20to%20layer

Al-Turjman, F., Zahmatkesh, H., & Shahroze, R. (2022). An overview of security and privacy in smart cities' IoT communications. *Transactions on Emerging Telecommunications Technologies*, *33*(3), e3677. doi:10.1002/ett.3677

Ambika, P. (2020). Machine learning and deep learning algorithms on the industrial internet of things (iiot). *Advances in Computers*, *117*(1), 321–338. doi:10.1016/bs.adcom.2019.10.007

Ameer, S., & Shah, M. A. (2018, August). Exploiting Big Data Analytics for Smart Urban Planning. In *2018 IEEE 88th Vehicular Technology Conference (VTC-Fall)* (pp. 1-5). IEEE. 10.1109/VTCFall.2018.8691036

Analyticsinsight.net. (2018). *How the Internet of Things Is Transforming Digital Marketing.* https://www.analyticsinsight.net/how-the-internet-of-things-is-transforming-digital-marketing/

Andreev, S., Petrov, V., Dohler, M., & Yanikomeroglu, H. (2019, June). Future of ultra-dense networks beyond 5G: Harnessing heterogeneous moving cells. *IEEE Communications Magazine*, *57*(6), 86–92. doi:10.1109/MCOM.2019.1800056

Andrews, G., & Zhang, J. (2014). What will 5G be? *IEEE Journal on Selected Areas in Communications*, *32*(6), 1065–1082. doi:10.1109/JSAC.2014.2328098

Astarloa, A., Bidarte, U., Jim'enez, J., Zuloaga, A., & L'azaro, J. (2016). Intelligent gateway for industry 4.0-compliant production. In *IECON 2016-42nd Annual Conference of the IEEE Industrial Electronics Society* (pp. 4902–4907). IEEE.

Atitallah, S. B., Driss, M., Boulila, W., & Ghézala, H. B. (2020). Leveraging Deep Learning and IoT Big Data Analytics to Support the Smart Cities Development: Review and Future Directions. *Computer Science Review*, *38*, 100303. doi:10.1016/j.cosrev.2020.100303

Atlam, H. F., Azad, M. A., Alzahrani, A. G., & Wills, G. (2020). A Review of Blockchain in Internet Of Things and AI. *Big Data and Cognitive Computing*, *4*(4), 28. doi:10.3390/bdcc4040028

Atwal, K. S., & Bassiouni, M. (2018). Deep SDN: connecting the dots towards self-driving networks. *Proceedings of* 2018 *IEEE 37th International Performance Computing and Communications Conference (IPCCC).* 10.1109/PCCC.2018.8711025

Atzori, L., Iera, A., Morabito, G., & Nitti, M. (2012). The social internet of things (siot) – when social networks meet the internet of things: Concept, architecture and network characterization. *Computer Networks*, *56*(16), 56. doi:10.1016/j.comnet.2012.07.010

Ayoub, O., Bovio, A., Musumeci, F., & Tornatore, M. (2020). Survivable Virtual Network Mapping in Filterless Optical Networks. *Proceedings of 2020 International Conference on Optical Network Design and Modeling (ONDM)*, 1-6. 10.23919/ONDM48393.2020.9133008

Bahalul Haque, A. K. M., Bhushan, B., Nawar, A., Talha, K. R., & Ayesha, S. J. (2022). Attacks and Countermeasures in IoT Based Smart Healthcare Applications. In *Recent Advances in Internet of Things and Machine Learning* (pp. 67–90). Springer. doi:10.1007/978-3-030-90119-6_6

Barakat, B., Taha, A., Samson, R., Steponenaite, A., Ansari, S., Langdon, P. M., Wassell, I. J., Abbasi, Q. H., Imran, M. A., & Keates, S. (2021). 6G opportunities arising from internet of things use cases: A review paper. *Future Internet*, *13*(6), 1–29. doi:10.3390/fi13060159

Barata, J., Pereira, V., & Coelho, M. (2020). Product Biography Information System: A Lifecycle Approach to Digital Twins. In *International Conference on Systems, Man, and Cybernetics (SMC)* (pp. 899-904). IEEE. 10.1109/SMC42975.2020.9283061

Barth, L., Ehrat, M., Fuchs, R., & Haarmann, J. (2020). Systematization of digital twins: ontology and conceptual framework. In *Proceedings of the 3rd International Conference on Information Science and System* (pp. 13-23). 10.1145/3388176.3388209

Bartholomew, C. (2020). China and 5G. *Issues in Science and Technology*, *36*(2), 50–57. https://www.jstor.org/stable/26949108

Bdigimarketer. (2019). *The growing Need for Ethics in Digital Marketing-Bdigimarketer*. https://medium.com/@bdigimarketer1/the-growing-need-for-ethics-in-digital-marketing-bdigimarketer-56da482c032a

Behrad, S., Bertin, E., Tuffin, S., & Crespi, N. (2020). A new scalable authentication and access control mechanism for 5G-based IoT. *Future Generation Computer Systems*, *108*, 46–61. doi:10.1016/j.future.2020.02.014

Behrendt, F. (2019). Cycling the smart and sustainable city: Analyzing ec policy documents on internet of things, mobility and transport, and smart cities. *Sustainability*, *11*(3), 763. doi:10.3390u11030763

Bhabad, M. A., & Bagade, S. T. (2015). Internet Of Things: Architecture, Security Issues and Countermeasures. *International Journal of Computers and Applications*, *125*(14).

Bhatia, B. B. (2022). 6G Spectrum studies for ITU. *ITU-APT Foundation of India*. https://www.itu-apt.org/system/static/uploads/mediap/6G%20Spectrum%20Studies%20for%20ITU%20(1).pdf

Bhushan, B., Khamparia, A., Sagayam, K. M., Sharma, S. K., Ahad, M. A., & Debnath, N. C. (2020). Blockchain for Smart Cities: A Review of Architectures, Integration Trends and Future Research Directions. *Sustainable Cities and Society*, *61*, 102360. doi:10.1016/j.scs.2020.102360

Bibri, S. E. (2018). The IoT for smart sustainable cities of the future: An analytical framework for sensor-based big data applications for environmental sustainability. *Sustainable Cities and Society*, *38*, 230–253. doi:10.1016/j.scs.2017.12.034

Bifulco, F., Tregua, M., Amitrano, C. C., & D'Auria, A. (2016). ICT and sustainability in smart cities management. *International Journal of Public Sector Management*, *29*(2), 132–147. doi:10.1108/IJPSM-07-2015-0132

Biswas, K., & Muthukkumarasamy, V. (2016, December). Securing Smart Cities Using Blockchain Technology. In *2016 IEEE 18th international conference on high performance computing and communications; IEEE 14th international conference on smart city; IEEE 2nd international conference on data science and systems (HPCC/SmartCity/DSS)* (pp. 1392-1393). IEEE. 10.1109/HPCC-SmartCity-DSS.2016.0198

Biswas, S., Li, F., Maharjan, S., Mohanty, S., Wang, Y., & Sharif, K. (2019). Pobt: A lightweight consensus algorithm for scalable iot business blockchain. *IEEE Internet of Things Journal*, *7*, 2327–4662.

Bonnetain, X., Naya-Plasencia, M., & Schrottenloher, A. (2019). Quantum Security Analysis of AES. *IACR Transactions on Symmetric Cryptology*, *2*, 55–93. doi:10.46586/tosc.v2019.i2.55-93

Borgia, E. (2014). The internet of things vision: Key features, applications and open issues. *Computer Communications*, *54*(12), 1–31. doi:10.1016/j.comcom.2014.09.008

Brenner, M. (2020). *Marketing and the Internet of Things*. https://marketinginsidergroup.com/strategy/marketing-internet-of-things/

Bu, L., Zhang, Y., Liu, H., Yuan, X., Guo, J., & Han, S. (2021). An iiot-driven and ai-enabled framework for smart manufacturing system based on three-terminal collaborative platform. *Advanced Engineering Informatics*, *50*, 101370. doi:10.1016/j.aei.2021.101370

Caldarelli, G., & Ellul, J. (2021). Trusted Academic Transcripts on the Blockchain: A Systematic Literature Review. *Applied Sciences (Basel, Switzerland)*, *11*(4), 1842. doi:10.3390/app11041842

Calvanese Strinati, Barbarossa, Gonzalez-Jimenez, Ktenas, Cassiau, Maret, & Dehos. (2019). 6G: The next frontier: From holographic messaging to artificial intelligence using subterahertz and visible light communication. *IEEE Veh. Technol. Mag.*, *14*(3), 42-50.

campaignsoftheworld.com. (2018). *How the Internet of things is Transforming Digital Marketing*. https://campaignsoftheworld.com/technology/how-the-internet-of-things-is-transforming-digital-marketing/

Caragliu, A., & Del Bo, C. F. (2019). Smart innovative cities: The impact of Smart City policies on urban innovation. *Technological Forecasting and Social Change*, *142*, 373–383. doi:10.1016/j.techfore.2018.07.022

Carey, L. D., Niu, J. G., Yang, P., Kankiewicz, J. A., Larson, V. E., & Haar, T. H. V. (2008). The vertical profile of liquid and ice water content in midlatitude mixed-phase altocumulus clouds. *Journal of Applied Meteorology and Climatology*, *47*(9), 2487–2495. doi:10.1175/2008JAMC1885.1

Carmody, S., Coravos, A., Fahs, G., Hatch, A., Medina, J., Woods, B., & Corman, J. (2021). Building resilient medical technology supply chains with a software bill of materials. *NPJ Digital Medicine*, *4*(1), 1–6. doi:10.103841746-021-00403-w PMID:33623135

Chakchouk, N. (2015). A survey on opportunistic routing in wireless communication networks. *IEEE Communications Surveys and Tutorials*, *17*(4), 2214–2241. doi:10.1109/COMST.2015.2411335

Chalapathi, G. S. S., Chamola, V., Vaish, A., & Buyya, R. (2021). Industrial internet of things (iiot) applications of edge and fog computing: A review and future directions. Fog/edge computing for security, privacy, and applications, 293–325.

Chauhan, D., Jain, J. K., & Sharma, S. (2016, December). An end-to-end header compression for multihop IPv6 tunnels with varying bandwidth. In *2016 Fifth international conference on eco-friendly computing and communication systems (ICECCS)* (pp. 84-88). IEEE. 10.1109/Eco-friendly.2016.7893247

Chauhan, D., Jain, J. K., & Bahad, P. (2021). Performance evaluation of 802.11 A/G wireless networks with IP6HC. *Journal of Management Information and Decision Sciences*, *24*, 1–7.

Chen, H., Yuan, L., & Jing, G. (2020, October). 5G Boosting Smart Cities Development. In *2020 2nd International conference on artificial intelligence and advanced manufacture (AIAM)* (pp. 154-157). IEEE. 10.1109/AIAM50918.2020.00038

Chen, B., Wan, J., Lan, Y., Imran, M., Li, D., & Guizani, N. (2019). Improving cognitive ability of edge intelligent iiot through machine learning. *IEEE Network*, *33*(5), 61–67. doi:10.1109/MNET.001.1800505

Chen, C. (2004). Searching for Intellectual Turning Points: Progressive Knowledge Domain Visualization. *Proceedings of the National Academy of Sciences of the United States of America*, *101*(suppl_1), 5303–5310. doi:10.1073/pnas.0307513100 PMID:14724295

Chen, D., Liu, Z., Wang, L., Dou, M., Chen, J., & Li, H. (2013). Natural disaster monitoring with wireless sensor networks: A case study of data-intensive applications upon low-cost scalable systems. *Mobile Networks and Applications*, *18*(5), 651–663. doi:10.100711036-013-0456-9

Chen, K., Lu, W., Xue, F., Tang, P., & Li, L. H. (2018). Automatic building information model reconstruction in high-density urban areas: Augmenting multi-source data with architectural knowledge. *Automation in Construction*, *93*, 22–34. doi:10.1016/j.autcon.2018.05.009

Chen, N., & Okada, M. (2021). Toward 6G internet of things and the convergence with RoF system. *IEEE Internet of Things Journal*, *8*(11), 8719–8733. doi:10.1109/JIOT.2020.3047613

Chettri, L., & Bera, R. (2020). A comprehensive survey on internet of things (IoT) toward 5G wireless systems. *IEEE Internet of Things Journal*, *7*(1), 16–32. doi:10.1109/JIOT.2019.2948888

Chhetri, M. B., Krishnaswamy, S., & Loke, S. W. (2004) Smart virtual counterparts for learning communities. *International Conference on Web Information Systems Engineering*, 125–134.

Chin, J., Callaghan, V., & Lam, I. (2017, June). Understanding and Personalising Smart City Services Using Machine Learning, the Internet-Of-Things and Big Data. In *2017 IEEE 26th international symposium on industrial electronics (ISIE)* (pp. 2050-2055). IEEE. 10.1109/ISIE.2017.8001570

China Academy of Information and Communication Research. (2018). China Academy of Information and Communication Research.

Cho, J. Y., Sergeev, A., & Zou, J. (2020). Securing Ethernet-based Optical Front haul for 5G Network. *Journal of Cyber Security and Mobility*, *9*(1), 91–110. doi:10.13052/jcsm2245-1439.913

Chowdhury, M., & Jang, Y. (2019). *6G wireless communication systems: Applications, requirements, technologies, challenges, and research directions.* https://arxiv.org/abs/1909.11315 doi:10.1109/ACCESS.2018.2844843

Christidis, K., & Devetsikiotis, M. (2016). Blockchains and Smart Contracts for the Internet of Things. *IEEE Access : Practical Innovations, Open Solutions*, *4*, 2292–2303. doi:10.1109/ACCESS.2016.2566339

Cirillo, F., Gómez, D., Diez, L., Maestro, I. E., Gilbert, T. B. J., & Akhavan, R. (2020). Smart City IoT Services Creation Through Large-Scale Collaboration. *IEEE Internet of Things Journal*, *7*(6), 5267–5275. doi:10.1109/JIOT.2020.2978770

Clarke. (2018). *How Internet of Things (IoT) Will Transform the Future of Marketing Businesses.* https://www.branex.ca/blog/how-iot-transform-future-marketing/

Cobo, M. J., López-Herrera, A. G., Herrera-Viedma, E., & Herrera, F. (2011). Science Mapping Software Tools: Review, Analysis, and Cooperative Study Among Tools. *Journal of the American Society for Information Science and Technology*, *62*(7), 1382–1402. doi:10.1002/asi.21525

Collaert, N., & Peeters, M. (2022). InP+CMOS heterogeneous integration for the next generation of wireless. *Microwave Journal*, *65*(11), 50–56.

Common Public Radio Interface. (n.d.). *ECPRI Interface Specification v1.2.* http://www.cpri.info/spec.html

Croatti, A., Gabellini, M., Montagna, S., & Ricci, A. (2020). On the integration of agents and digital twins in healthcare. *Journal of Medical Systems*, *44*(9), 1–8. doi:10.100710916-020-01623-5 PMID:32748066

Cvijetic, N. (2014). Optical network evolution for 5G mobile applications and SDN-based control. *Proceedings of 2014 16th International Telecommunications Network Strategy and Planning Symposium (Networks),* 1-5. 10.1109/NETWKS.2014.6958537

Cvijetic, N., Tanaka, A., Ji, P., Sethuraman, K., Murakami, S., & Wang, T. (2014). SDN and OpenFlow for dynamic flex-grid optical access and aggregation networks. *Journal of Lightwave Technology*, *32*(4), 864–870. doi:10.1109/JLT.2013.2274991

Cvijetic, N., Tanaka, A., Kanonakis, K., & Wang, T. (2014). SDN-controlled topology-reconfigurable optical mobile fronthaul architecture for bidirectional CoMP and low latency inter-cell D2D in the 5G mobile era. *Optics Express*, *22*(17), 20809. Advance online publication. doi:10.1364/OE.22.020809 PMID:25321284

Dahlberg, A., Oliveira Filho, J., Hanson, R., Wehner, S., Skrzypczyk, M., Coopmans, T., Wubben, L., Rozpedek, F., Pompili, M., Stolk, A., Pawełczak, P., & Knegjens, R. (2019). A link layer protocol for quantum networks. In *Proceedings of the ACM Special Interest Group on Data Communication.* ACM. 10.1145/3341302.3342070

Dan, S. (2020). *Ethical Marketing: 5 Examples of Companies with a Conscience.* https://www.wordstream.com/blog/ws/2017/09/20/ethical-marketing

Dang, S. P., Amin, O., Shihada, B., & Alouini, M. S. (2020). What should 6G be? *Nature Electronics*, *3*(1), 20–29. doi:10.103841928-019-0355-6

Dao, N. N., Pham, Q. V., Tu, N. H., Thanh, T. T., Bao, V. N. Q., Lakew, D. S., & Cho, S. (2021). Survey on aerial radio access networks: Toward a comprehensive 6G access infrastructure. *IEEE Communications Surveys and Tutorials*, *23*(2), 1193–1225. doi:10.1109/COMST.2021.3059644

Das, S. K., Mudi, R., Rahman, M. S., & Fapojuwo, A. O. (2022). (Preprint). Distributed learning for 6G–IoT networks: A comprehensive survey. *TechRxiv.* Advance online publication. doi:10.36227/techrxiv.20069051.v3

Das, S., & Saha, P. (2018). A review of some advanced sensors used for health diagnosis of civil engineering structures. *Measurement*, *129*, 68–90. doi:10.1016/j.measurement.2018.07.008

David & Berndt. (2018). 6G vision and requirements: Is there any need for beyond 5G? *IEEE Veh. Technol. Mag.*, *13*(3), 72-80.

De Alwis, C. (2021). Survey on 6G frontiers: Trends, applications, requirements, technologies and future research. *IEEE Open Journal of the Communications Society, 2*, 836–886. doi:10.1109/OJCOMS.2021.3071496

Deals Insight Team. (2020). *The Big Revolution in B2B Marketing-Internet of Things.* https://www.dealsinsight.com/the-big-revolution-in-b2b-marketing-internet-of-things-iot/

Deng, T., Zhang, K., & Shen, Z.J.M. (2021). A Systematic Review of a Digital Twin City: A New Pattern of Urban Governance toward Smart Cities. *Journal of Management Science and Engineering.*

Deshpande, I. (2021). *What Is IoT And Why Should Marketers Care About It?* https://www.spiceworks.com/tech/iot/articles/what-is-iot-for-marketers/

DeTomasi, S. (2019). Navigating the 5G NR standards. In P. Hindle (Ed.), *Advanced 5G Over the Air Testing (OTA) Challenges and Solutions* (pp. 10–13). Microwave Journal.

Díaz Zayas, A., Rivas Tocado, F. J., & Rodríguez, P. (2020). Evolution and testing of NB-IoT solutions. *Applied Sciences, 10*(21), 1–17.

Díaz Zayas, A., Caso, G., Alay, Ö., Merino, P., Brunstrom, A., Tsolkas, D., & Koumaras, H. (2020). A modular experimentation methodology for 5G deployments: The 5GENESIS approach. *Sensors (Basel), 20*(22), 6652. doi:10.339020226652 PMID:33233691

Digital, S. (2022). *What Is The Role Of Data Science In Digital Marketing?* https://www.stellardigital.in/blog/what-is-the-role-of-data-science-in-digital-marketing/

Diogo, S. (2021). *5 Ways in which IoT will Impact Digital Marketing.* https://www.youlead.agency/blog/5-ways-in-which-iot-will-impact-digital-marketing

Dong, F., Li, Y., Li, K., Zhu, J., & Zheng, L. (2022). Can smart city construction improve urban ecological total factor energy efficiency in China? Fresh evidence from generalized synthetic control method. *Energy, 241*, 122909. doi:10.1016/j.energy.2021.122909

Doudou, M., Djenouri, D., & Badache, N. (2012). Survey on latency issues of asynchronous MAC protocols in delay-sensitive wireless sensor networks. *IEEE Communications Surveys and Tutorials, 15*(2), 528–550. doi:10.1109/SURV.2012.040412.00075

Douglas, K. (2019). *10 Modern Technologies That Are Enhancing Digital Marketing.*https://martech.zone/modern-technologies-impacting-digital-marketing

Dreyer, S., Olivotti, D., Lebek, B., & Breitner, M. H. (2017). Towards a Smart Services Enabling Information Architecture for Installed Base Management in Manufacturing. *Wirtschaftsinformatik.*

DTC. (2021). https://www.digitaltwinconsortium.org/about-us/index.htm

Dynes, J. F., Kindness, S. J., Tam, S. W. B., Plews, A., Sharpe, A. W., Lucamarini, M., Fröhlich, B., Yuan, Z. L., Penty, R. V., & Shields, A. J. (2016). Quantum key distribution over multicore fiber. *Optics Express, 24*(8), 8081–8087. doi:10.1364/OE.24.008081 PMID:27137247

Eichler, T., & Ziegler, R. (2022). *Fundamentals of THz Technology for 6G.* Rohde & Schwarz.

El Meouche, R., Hijazi, I., Poncet, P. A., Abunemeh, M., & Rezoug, M. (2016). UAV photogrammetry implementation to enhance land surveying, comparisons and possibilities. *The International Archives of the Photogrammetry, Remote Sensing and Spatial Information Sciences, 42*(W2), 107–114. doi:10.5194/isprs-archives-XLII-2-W2-107-2016

El Rachkidy, N., Guitton, A., & Misson, M. (2012). Pivot routing improves wireless sensor networks performance. *Journal of Networks*.

Elayan, H., Amin, O., Shihada, B., Shubair, R. M., & Slim Alouini, M. (2020). Terahertz band: The last piece of RF spectrum puzzle for communication systems. *IEEE Open Journal of the Communications Society, 1*, 1–32. doi:10.1109/OJCOMS.2019.2953633

Elliott, C. (2002). Building the quantum network. *New Journal of Physics, 4*, 46. doi:10.1088/1367-2630/4/1/346

Ericsson. (2019). *Network Programmability, a new frontier in 5G* [White paper]. https://www.ericsson.com/en/blog/2019/1/network-programmability---in-5g-an-invisible-goldmine-for-service-providers-and-industry

Eriksson, T. A., Hirano, T., Puttnam, B. J., Rademacher, G., Luís, R. S., Fujiwara, M., Namiki, R., Awaji, Y., Takeoka, M., Wada, N., & Sasaki, M. (2019). Wavelength division multiplexing of continuous variable quantum key distribution and 18.3 Tbit/s data channels. *Communications on Physics, 2*(1), 1–8. doi:10.103842005-018-0105-5

Essiambre, R. J., & Tkach, R. W. (2012). Capacity Trends and Limits of Optical Communication Networks. *Proceedings of the IEEE, 100*(5), 1035–1055. doi:10.1109/JPROC.2012.2182970

ETSI. (2012). *Network Functions Virtualisation (NFV): Network Operator Perspectives on Industry Progress.* Paper presentation at SDN and OpenFlow World Congress. https://portal.etsi.org/NFV/NFV_White_Paper.pdf

Farooq, M., Waseem, M., Khairi, A., & Mazhar, S. (2015). A critical analysis on the security concerns of internet of things (iot). *International Journal of Computers and Applications, 111*, 1–6. doi:10.5120/19547-1280

Fiftyfiveandfive.com. (2020). *6 Digital Marketing Technologies to Help You Raise Your Game.* https://www.fiftyfiveandfive.com/6-digital-marketing-technologies/

Fortify, H. (2015). *Internet of Things security study: Smartwatches.* Hewlett-Packard.

Fortino, G., & Trunfio, P. (2014). *Internet of things based on smart objects: Technology, middleware and applications.* Springer. doi:10.1007/978-3-319-00491-4

Fragkos, D., Makropoulos, G., Sarantos, P., Koumaras, H., Charismiadis, A. S., & Tsolkas, D. (2021, September). 5G Vertical Application Enablers Implementation Challenges and Perspectives. In *2021 IEEE International Mediterranean Conference on Communications and Networking (MeditCom)* (pp. 117-122). IEEE. 10.1109/MeditCom49071.2021.9647460

Frizzo-Barker, J., Chow-White, P. A., Adams, P. R., Mentanko, J., Ha, D., & Green, S. (2020). Blockchain as a disruptive technology for business: A systematic review. *International Journal of Information Management, 51*, 102029. doi:10.1016/j.ijinfomgt.2019.10.014

Frustaci, M., Pace, P., Aloi, G., & Fortino, G. (2018). Evaluating critical security issues of the IoT world: Present and future challenges. *IEEE Internet of Things Journal, 5*(4), 2483–2495. doi:10.1109/JIOT.2017.2767291

Fuller, A., Fan, Z., Day, C., & Barlow, C. (2020). Digital twin: Enabling technologies, challenges and open research. *IEEE Access : Practical Innovations, Open Solutions, 8*, 108952–108971. doi:10.1109/ACCESS.2020.2998358

Gandhewar, R., Gaurav, A., Kokate, K., Khetan, H., & Kamat, H. (2019). Cloud based framework for iiot application with asset management. In *2019 3rd International conference on Electronics, Communication and Aerospace Technology (ICECA)* (pp. 920–925). IEEE. 10.1109/ICECA.2019.8821897

Gan, L., Shen, L., Tang, M., Xing, C., Li, Y., Ke, C., Tong, W., Li, B., Fu, B., & Liu, D. (2018). Investigation of channel model for weakly coupled multicore fiber. *Optics Express, 26*(5), 5182–5199. doi:10.1364/OE.26.005182 PMID:29529725

Gao, Z., Song, Y., Li, C., Zeng, F., & Wang, F. (2017). Research on the application of rapid surveying and mapping for large scare topographic map by UAV aerial photography system. *The International Archives of the Photogrammetry, Remote Sensing and Spatial Information Sciences*, *42*(W6), 121–125. doi:10.5194/isprs-archives-XLII-2-W6-121-2017

Giordani, M. (n.d.). *Towards 6G networks: use cases and technologies.* arXiv:1903.12216

Giordani, Polese, Mezzavilla, Rangan, & Zorzi. (2020). Towards 6G networks: Use cases and technologies. *IEEE Commun. Mag.*, *58*(3), 51-63.

Giordani, M., & Zorzi, M. (2021). Non-terrestrial networks in the 6G era: Challenges and opportunities. *IEEE Network*, *35*(2), 244–251. doi:10.1109/MNET.011.2000493

Glaessgen, E., & Stargel, D. (2012). *The digital twin paradigm for future NASA and U.S. air force vehicles*. Academic Press.

Gleim, A. V., Egorov, V. I., Nazarov, Yu. V., Smirnov, S. V., Chistyakov, V. V., Bannik, O. I., Anisimov, A. A., Kynev, S. M., Ivanova, A. E., Collins, R. J., Kozlov, S. A., & Buller, G. S. (2016). Secure Polarization-Independent Subcarrier Quantum Key Distribution in Optical Fiber Channel Using BB84 Protocol with a Strong Reference. *Optics Express*, *24*(3), 2619–2633. doi:10.1364/OE.24.002619 PMID:26906834

Gohar, A., & Nencioni, G. (2021). The Role of 5G Technologies in a Smart City: The Case for Intelligent Transportation System. *Sustainability*, *13*(9), 5188. doi:10.3390u13095188

Gordon, M., Jamshidi, D., Mahlke, S., Mao, Z., & Chen, X. (2012). Comet: Code offload by migrating execution transparently. Academic Press.

Götz, C. S., Karlsson, P., & Yitmen, I. (2020). *Exploring applicability, interoperability and integrability of Blockchain-based digital twins for asset life cycle management. In Smart and Sustainable Built Environment.* Emerald Group Publishing Ltd., doi:10.1108/SASBE-08-2020-0115

Goyal, S., Sharma, N., Kaushik, I., Bhushan, B., & Kumar, A. (2020). Precedence & issues of iot based on edge computing. *2020 IEEE 9th International Conference on Communication Systems and Network Technologies (CSNT)*, 72–77. 10.1109/CSNT48778.2020.9115789

Granda, F. A. P., & Salik, I. (2020). Simulation Training and Skill Assessment in Critical Care. *StatPearls*. https://www.ncbi.nlm.nih.gov/books/NBK549895/

Grieves, M. (2005). *Product Lifecycle Management: Driving the Next Generation of Lean Thinking*. Academic Press.

Grieves, M. (2019). Virtually Intelligent Product Systems: Digital and Physical Twins. In *Complex Systems Engineering: Theory and Practice* (pp. 175–200). American Institute of Aeronautics and Astronautics. doi:10.2514/5.9781624105654.0175.0200

Grieves, M., & Vickers, J. (2017). Digital Twin: Mitigating Unpredictable, Undesirable Emergent Behavior in Complex Systems. In F.-J. Kahlen, S. Flumerfelt, & A. Alves (Eds.), *Transdisciplinary Perspectives on Complex Systems: New Findings and Approaches* (pp. 85–113). Springer International Publishing. doi:10.1007/978-3-319-38756-7_4

Grimaldi, D., & Fernandez, V. (2019). Performance of an internet of things project in the public sector: The case of Nice smart city. *The Journal of High Technology Management Research*, *30*(1), 27–39. doi:10.1016/j.hitech.2018.12.003

Grizhnevich. (2021). *IoT Architecture: Building Blocks and How They Work.* https://www.scnsoft.com/blog/iot-architecture-in-a-nutshell-and-how-it-works

Grubesic, T. H., Matisziw, T. C., Murray, A. T., & Snediker, D. (2008). Comparative Approaches for Assessing Network Vulnerability. *International Regional Science Review*, *31*(1), 88–112. doi:10.1177/0160017607308679

Gubbi, J., Buyya, R., Marusic, S., & Palaniswami, M. (2013). Internet of Things (IoT): A vision, architectural elements, and future directions. *Future Generation Computer Systems*, *29*(7), 1645–1660. doi:10.1016/j.future.2013.01.010

Gudema. (2019). *7 Marketing Technologies Every Company Must Use*. https://hbr.org/2014/11/7-marketing-technologies-every-company-must-use

Guo, F. X., Richard Yu, F., Zhang, H. L., Ji, H., & Victor Leung, C. M. (2021). Enabling massive IoT toward 6G: A comprehensive survey. *IEEE Internet of Things Journal*, *8*(15), 11891–11915. doi:10.1109/JIOT.2021.3063686

Hammoudeh, M. (2020, November). Blockchain, Internet of Things and Digital Twins in Trustless Security of Critical National Infrastructure. *The 4th International Conference on Future Networks and Distributed Systems (ICFNDS).*

Han, B., Gopalakrishnan, V., Ji, L., & Lee, S. (2015). Network function virtualization: Challenges and opportunities for innovations. *IEEE Communications Magazine*, *53*(2), 90–97. doi:10.1109/MCOM.2015.7045396

Han, C., Wu, Y. Z., Chen, Z., & Wang, X. D. (2019). Terahertz communication (TeraCom): challenges and impact on 6G wireless systems. arXiv preprint, arXiv:1912.06040. doi:10.1109/ACCESS.2020.2977406

Hang, L., & Kim, D. H. (2019). Design and Implementation of an Integrated IoT Blockchain Platform for Sensing Data Integrity. *Sensors (Basel)*, *19*(10), 2228. doi:10.339019102228 PMID:31091799

Harper, K. E., Ganz, C., & Malakuti, S. (2019). Digital twin architecture and standards. *IIC Journal of Innovation*, *12*, 72–83.

Hasan, H. R., Salah, K., Jayaraman, R., Omar, M., Yaqoob, I., Pesic, S., Taylor, T., & Boscovic, D. (2020). A blockchain-based approach for the creation of digital twins. *IEEE Access : Practical Innovations, Open Solutions*, *8*, 34113–34126. doi:10.1109/ACCESS.2020.2974810

Hassan, Q. F., & Madani, S. A. (2017). *Internet of things: Challenges, advances, and applications*. Academic Press.

Hatim, J., Chaimae, S., & Habiba, C. (2022). Improved IOT/SDN Architecture with the Concept of NFV. *International Conference on Digital Technologies and Applications*, 294–301. 10.1007/978-3-031-01942-5_29

Hazra, A., Adhikari, M., Amgoth, T., & Srirama, S. N. (2021). A comprehensive survey on interoperability for iiot: taxonomy, standards, and future directions. *ACM Computing Surveys*, *55*(1), 1–35.

Hemdan, E. E. D., & Mahmoud, A. S. A. (2021). BlockTwins: A Blockchain-Based Digital Twins Framework. In Blockchain Applications in IoT Ecosystem (pp. 177-186). Springer International Publishing. doi:10.1007/978-3-030-65691-1_12

Hewa, Gür, Kalla, Ylianttila, Bracken, & Liyanage. (2020). The role of block chain in 6G: Challenges, opportunities and research directions. *Proc. 2nd 6G Wireless Summit (6G SUMMIT),* 1-5.

Himmler, A., & Alexander, R. N. (2021). *How 4D Imaging Radar Sensors Can Be Validated*. Autocar Professional. Available online: https://www.autocarpro.in/opinion-blogs/how-4d-imaging-radar-sensors-can-be-validated-79866

Hinton, G. E., Osindero, S., & Teh, Y. W. (2006). A fast learning algorithm for deep belief nets. *Neural Computation*, *18*(7), 1527–1554. doi:10.1162/neco.2006.18.7.1527 PMID:16764513

Holis, J., & Pechac, P. (2008). Elevation dependent shadowing model for mobile communications via high altitude platforms in built-up areas. *IEEE Transactions on Antennas and Propagation*, *56*(4), 1078–1084. doi:10.1109/TAP.2008.919209

Hong, Z., Chen, W., Huang, H., Guo, S., & Zheng, Z. (2019). Multi-hop cooperative computation offloading for industrial iot–edge–cloud computing environments. *IEEE Transactions on Parallel and Distributed Systems*, *30*(12), 2759–2774. doi:10.1109/TPDS.2019.2926979

Hosseinzadeh, M., Hemmati, A., & Rahmani, A. M. (2022). 6G-enabled internet of things: Vision, techniques, and open issues. *Computer Modeling in Engineering & Sciences, 133*(3), 509–556. doi:10.32604/cmes.2022.021094

Huang, Yang, Wu, Ma, Zhang, & Zhang. (2019). A survey on green 6G network: Architecture and technologies. *IEEE Access, 7*, 175758-175768.

Huang, J., Liu, Y., Wang, C. X., Sun, J., & Xiao, H. L. (2018). 5G millimetre-wave channel sounders, measurements, and models: Recent developments and future challenges. *IEEE Communications Magazine, 57*(1), 138–145. doi:10.1109/MCOM.2018.1701263

Huang, J., Meng, Y., Gong, X., Liu, Y., & Duan, Q. (2014). A novel deployment scheme for green internet of things. *IEEE Internet of Things Journal, 1*(2), 196–205. doi:10.1109/JIOT.2014.2301819

Huang, J., Zhang, C., & Zhang, J. (2020). A multi-queue approach of energy efficient task scheduling for sensor hubs. *Chinese Journal of Electronics, 29*(2), 242–247. doi:10.1049/cje.2020.02.001

Huang, K., Luo, W., Zhang, W., & Li, J. (2021). Characteristics and Problems of Smart City Development in China. *Smart Cities, 4*(4), 1403–1419. doi:10.3390martcities4040074

Huang, T. Y., Yang, W., Wu, J., Ma, J., Zhang, X., & Zhang, D. (2019). A survey on green 6G network: Architecture and technologies. *IEEE Access : Practical Innovations, Open Solutions, 7*, 175758–175768. doi:10.1109/ACCESS.2019.2957648

Huawei. (2017). *5G Spectrum Public Policy Position*. Huawei Technologies Co., Ltd.

Hugues-Salas, E., Alia, O., Wang, R., Rajkumar, K., Kanellos, G. T., Nejabati, R., & Simeonidou, D. (2020). 11.2 Tb/s Classical Channel Coexistence with DV-QKD over a 7-Core Multicore Fiber. *Journal of Lightwave Technology, 38*(18), 5064–5070. doi:10.1109/JLT.2020.2998053

Hussain, N., & Kim, N. (2022). Integrated Microwave and mm-Wave MIMO Antenna Module with 360∘ Pattern Diversity For 5G Internet-of-Things. *IEEE Internet of Things Journal, 9*(24), 24777–24789. doi:10.1109/JIOT.2022.3194676

IBM. (2023). *How does IoT work with blockchain?* IBM. https://www.ibm.com/my-en/topics/blockchain-iot

Ibrahim, M., Hanif, M., Ahmad, S., Jamil, F., Sehar, T., Lee, Y., & Kim, D. (2022). *SDN Based DDos Mitigating Approach Using Traffic Entropy for IoT Network*. Academic Press.

Ifeng. (2021). *Super Smart City 2.0, Artificial Intelligence Leads the Way to a New Trend*. https://www.ifeng.com/

Imoize, A. L., Adedeji, O., Tandiya, N., & Shetty, S. (2021). 6G enabled smart infrastructure for sustainable society: Opportunities, challenges, and research roadmap. *Sensors (Basel), 21*(5), 1–58. doi:10.339021051709 PMID:33801302

India. (2016). *The Internet of Things revolution and the transformation of Digital Marketing*. https://www.cyberclick.net/numericalblogen/the-internet-of-things-revolution-and-the-transformation-of-digital-marketing

INFOLYSiS. (2017). A Comparative CAPEX Techno-Economic Analysis of NFV Applicability on the Ground Segment of GEO/MEO/LEO Satellite Systems. INFOLYSIS P.C.

International Telecommunication Union. (1998). *Loss of signal (LOS), alarm indication signal (AIS) and remote defect indication (RDI) defect detection and clearance criteria for PDH signals*. ITU-T G.775.

International Telecommunication Union. (2005). *Specific Attenuation Model for Rain for Use in Prediction Methods*. Recommendation ITU-R P. 838-3.

International Telecommunication Union. (2012). *Series Y: Global Information Infrastructure, Internet Protocol Aspects and Next-Generation Networks*. Recommendation ITU-T Y.2060.

International Telecommunication Union. (2017a). *Characteristics of Precipitation for Propagation Modelling*. Recommendation ITU-R P. 837-7.

International Telecommunication Union. (2017b). *Reference Standard Atmospheres*. ITU-R P.

International Telecommunication Union. (2018a). *IoT Standards Part II: 3GPP Standards. Training on Planning Internet of Things (IoTs) Networks*. ITU Report.

International Telecommunication Union. (2018b). *Sharing between the Radio Astronomy Service and Active Services in the Frequency Range 275-3000 GHz*. Recommendation ITU-R Report RA.2189-1.

International Telecommunication Union. (2019a). *Attenuation Due to Clouds and Fog*. Recommendation ITU-R P. 840-8.

International Telecommunication Union. (2019b). *The Radio Refractive Index: Its Formula and Refractivity Data*. ITU-R P.

International Telecommunication Union. (2019c). *Attenuation by Atmospheric Gases and Related Effects*. ITU-R P.

International Telecommunication Union. (2021). *Propagation Data and Prediction Methods Required for the Design of Terrestrial Line-of-Sight Systems*. ITU-R P.

International Telecommunication Union. (2022). *Future Technology Trends of Terrestrial International Mobile Telecommunications Systems Towards 2030 and Beyond*. Report ITU-R M.2516-0.

IoT Dunia.com. (2020). *How does IoT work? - Explanation of IoT Architecture & layers*. https://iotdunia.com/iot-architecture/

IoT for all. (2021). *How IoT Data Can Improve Digital Marketing Outcomes*. https://www.iotforall.com/how-iot-consumer-data-affect-digital-marketing

IoT for all. (2021). What is the Internet of Things, or IoT? A simple explanation. *IoT For All*. https://www.iotforall.com/what-is-internet-of-things

Iqbal, W., Abbas, H., Daneshmand, M., Rauf, B., & Bangash, Y. A. (2020). An in-depth analysis of IoT security requirements, challenges, and their countermeasures via software-defined security. *IEEE Internet of Things Journal*, *7*(10), 10250–10276. doi:10.1109/JIOT.2020.2997651

Iscoop. (2020). *The Internet of Things in Marketing: The Integrated Marketing Opportunity*. https://www.i-scoop.eu/internet-of-things-iot/internet-things-marketing/

ISO. (2020). *Automation systems and integration — Industrial data — Visualization elements of digital twins*. ISO/TR 24464.

ISO. (2021). *Automation systems and integration — Digital Twin framework for manufacturing — Part 1: Overview and general principles*. ISO/DIS 23247-1.

ITU-T (2020). *Y.3000-series - Representative use cases and key network requirements for Network 2030*. Series Y Supplement 67 (07/2020).

Jaber, M., & Tukmanov, A. (2016). 5G backhaul challenges and emerging research directions: A survey. *IEEE Access : Practical Innovations, Open Solutions*, *4*, 1743–1766. doi:10.1109/ACCESS.2016.2556011

Jaidka, H., Sharma, N., & Singh, R. (2020). Evolution of iot to iiot: Applications & challenges. *Proceedings of the International Conference on Innovative Computing & Communications (ICICC)*.

Jain, J. K., Waoo, A. A., & Chauhan, D. (2022). *A Literature Review on Machine Learning for Cyber Security Issues*. Academic Press.

Jain, J. K., & Chauhan, D. (2021). An energy-efficient model for internet of things using compressive sensing. *Journal of Management Information and Decision Sciences, 24*, 1–7.

Jain, J. K., Dangi, C. S., & Chauhan, D. (2020, November). An Efficient Multipath Productive Routing Protocol for Mobile Ad-hoc Networks. In *2020 IEEE International Conference for Innovation in Technology (INOCON)* (pp. 1-5). IEEE. 10.1109/INOCON50539.2020.9298291

Jain, J. K., Jain, D. K., & Gupta, A. (2012). Performance analysis of node-disjoint multipath routing for mobile ad-hoc networks based on QOS. *International Journal of Computer Science and Information Technologies*, *3*(5), 5000–5004.

Jain, J. K., & Sharma, S. (2013). Performance Evaluation of Hybrid Multipath Progressive Routing Protocol for MANETs. *International Journal of Computers and Applications*, *71*(18).

Jain, J. K., & Sharma, S. (2017). A Novel Approach of Routing Protocol using Hybrid Analysis for MANETs. *International Journal of Computer Science and Information Security*, *15*(1), 582.

Jain, P., & Jain, J. K. (2022). First Order Control System Using Python Technology. In Innovations in Electrical and Electronic Engineering. *Proceedings of ICEEE*, *1*, 152–160.

Jain, S., Gupta, S., Sreelakshmi, K. K., & Rodrigues, J. J. (2022). Fog Computing In Enabling 5G-Driven Emerging Technologies for Development of Sustainable Smart City Infrastructures. *Cluster Computing*, *25*(2), 1–44. doi:10.100710586-021-03496-w

Javed, A. R., Hassan, M. A., Shahzad, F., Ahmed, W., Singh, S., Baker, T., & Gadekallu, T. R. (2022). Integration of blockchain technology and federated learning in vehicular (iot) networks: A comprehensive survey. *Sensors (Basel)*, *22*(12), 4394. doi:10.339022124394 PMID:35746176

Jha, M. (2022). Secure SDN Based IoT Network Through Blockchain for Smart Architectures. *2022 IEEE Region 10 Symposium (TENSYMP)*, 1–6.

Jiang, W., Han, B., Habibi, M. A., & Schotten, H. D. (2021). The road towards 6G: A comprehensive survey. *IEEE Open Journal of the Communications Society*, *2*, 334–366. doi:10.1109/OJCOMS.2021.3057679

Ji, B. F., Wang, Y., Song, K., Li, C., Wen, H., Menon, V. G., & Mumtaz, S. (2021). A survey of computational intelligence for 6G: Key technologies, applications and trends. *IEEE Transactions on Industrial Informatics*, *17*(10), 7145–7154. doi:10.1109/TII.2021.3052531

Ji, Y., & Chen, L. (2022). FedQNN: A Computation-Communication Efficient Federated Learning Framework for IoT with Low-bitwidth Neural Network Quantization. *IEEE Internet of Things Journal*.

Johan, F. (2021). *The Satellite IoT Communications Market*. Berg Insight.

Johannes, B. (2021). *Ways IoT Is Changing Digital Marketing*. https://iotmktg.com/ways-iot-is-changing-digital-marketing/

Jumira, O., Wolhuter, R., & Zeadally, S. (2013). Energy-efficient beaconless geographic routing in energy harvested wireless sensor networks. *Concurrency and Computation*, *25*(1), 58–84. doi:10.1002/cpe.2838

Kahn, J. M., You, R., Djahani, P., Weisbin, A. G., & Tang, A. (1998). Imaging diversity receivers for high-speed infrared wireless communication. *IEEE Communications Magazine*, *36*(12), 88–94. doi:10.1109/35.735884

Kantola, R. (2020). Trust networking for beyond 5G and 6G. Proc. 2nd 6G Wireless Summit (6G SUMMIT), 1-6. doi:10.1109/6GSUMMIT49458.2020.9083917

Kapustina, I., Kalinina, O., Ovchinnikova, A., & Barykin, S. (2020). The logistics network digital twin in view of concept of the non-destructive quality control methods. In *E3S Web of Conferences* (Vol. 157, p. 05001). EDP Sciences. doi:10.1051/e3sconf/202015705001

Karafiloski, E., & Mishev, A. (2017, July). Blockchain Solutions for Big Data Challenges: A Literature Review. In *IEEE EUROCON 2017-17th International Conference on Smart Technologies* (pp. 763-768). IEEE. 10.1109/EUROCON.2017.8011213

Karmakar, A., Dey, N., Baral, T., Chowdhury, M., & Rehan, M. (2019). Industrial internet of things. *RE:view*, 1–6.

Katz, M., Matinmikko-Blue, M., & Latva-Aho, M. (2018). 6 Genesis flagship program: Building the bridges towards 6G-enabled wireless smart society and ecosystem. *Proc. IEEE 10th Latin-American Conf. Commun. (LATINCOM)*, 1-9.

Kaur, K., Dhand, T., Kumar, N., & Zeadally, S. (2017). Container-as-a-service at the edge: Trade-off between energy efficiency and service availability at fog nano data centers. *IEEE Wireless Communications*, *24*(3), 48–56. doi:10.1109/MWC.2017.1600427

Kaur, K., Garg, S., Aujla, G. S., Kumar, N., Rodrigues, J. J. P. C., & Guizani, M. (2018). Edge computing in the industrial internet of things environment: Software-defined-networks-based edge-cloud interplay. *IEEE Communications Magazine*, *56*(2), 44–51. doi:10.1109/MCOM.2018.1700622

Kaur, M. J., Mishra, V. P., & Maheshwari, P. (2020). The Convergence of Digital Twin, IoT, and Machine Learning: Transforming Data into Action. In M. Farsi, A. Daneshkhah, A. Hosseinian-Far, & H. Jahankhani (Eds.), *Digital Twin Technologies and Smart Cities* (pp. 3–17). Springer International Publishing. doi:10.1007/978-3-030-18732-3_1

Keskinen, M. (2017). Mobile Network Evolution. University of Oulu.

Khajenasiri, I., Estebsari, A., Verhelst, M., & Gielen, G. (2017). A review on internet of things solutions for intelligent energy control in buildings for smart city applications. *Energy Procedia, 111*, 770–779. 10.1016/j.egypro.2017.03.239

Khan & Hong. (2020). 6G wireless systems: A vision, architectural elements, and future directions. *IEEE Access, 8*, 147029-147044.

Khan, A., Jhanjhi, N. Z., & Humayun, M. (2022). The Role of Cybersecurity in Smart Cities. In *Cyber Security Applications for Industry 4.0.* Chapman and Hall/CRC. doi:10.1201/9781003203087-9

Khan, A., Shahid, F., Maple, C., Ahmad, A., & Jeon, G. (2020). Towards Smart Manufacturing Using Spiral Digital Twin Framework and Twinchain. *IEEE Transactions on Industrial Informatics*. Advance online publication. doi:10.1109/TII.2020.3047840

Khodaei, A., & Deogun, J. (2021). *Optical MIMO Communication Using Holographic Spectral Multiplexing of Pulsed Ultrashort Laser*. doi:10.48550/arXiv.2106.13896

Kieu, K., Jones, R. J., & Peyghambarian, N. (2010). Generation of Few-Cycle Pulses From an Amplified Carbon Nanotube Mode-Locked Fiber Laser System. *IEEE Photonics Technology Letters*, *22*(20), 1521–1523. doi:10.1109/LPT.2010.2063423

Kim, J. H. (2021). 6G and Internet of Things: A survey. *Journal of Management Analytics*, *8*(2), 316–332. doi:10.1080/23270012.2021.1882350

Kim, T. H., Ramos, C., & Mohammed, S. (2017). Smart City and IoT. *Future Generation Computer Systems*, *76*, 159–162. doi:10.1016/j.future.2017.03.034

Kiran, B. (2020). *Top 7 Ways IoT is Transforming Digital Marketing in 2020 and Beyond.* https://medium.com/@kiran.bhatt270897/top-7-ways-iot-is-transforming-digital-marketing-in-2020-and-beyond-a855101b6607

Kong, W., Sun, Y., Cai, C., & Ji, Y. (2020). Impact of Modulation Formats and Bandwidth on Quantum Secured 5G Optical Fronthaul over Multicore Fiber. *Proceedings of the CLEO 2020.* 10.1364/CLEO_AT.2020.JTu2A.25

Kostakis, P., Charismiadis, A. S., Tsolkas, D., & Koumaras, H. (2021, May). An Experimentation Platform for Automated Assessment of Multimedia Services over Mobile Networks. In *IEEE INFOCOM 2021-IEEE Conference on Computer Communications Workshops (INFOCOM WKSHPS)* (pp. 1-2). IEEE. 10.1109/INFOCOMWKSHPS51825.2021.9484528

Koumaras, H., Makropoulos, G., Batistatos, M., Kolometsos, S., Gogos, A., Xilouris, G., Sarlas, A., & Kourtis, M. A. (2021). 5G-enabled UAVs with command-and-control software component at the edge for supporting energy efficient opportunistic networks. *Energies, 14*(5), 1480. doi:10.3390/en14051480

Kourtis, M. A., Anagnostopoulos, T., Kukliński, S., Wierzbicki, M., Oikonomakis, A., Xilouris, G., Chochliouros, I., Yi, N., Kostopoulos, A., Tomaszewski, L., Sarlas, T., & Koumaras, H. (2020, November). 5G network slicing enabling edge services. In *2020 IEEE Conference on Network Function Virtualization and Software Defined Networks (NFV-SDN)* (pp. 155-160). IEEE. 10.1109/NFV-SDN50289.2020.9289880

Ksentini, A., & Frangoudis, P. A. (2020). Towards Slicing-Enabled Multi-Access Edge Computing in 5G. *IEEE Network, 34*(2), 99–105. doi:10.1109/MNET.001.1900261

Kumar, S., & Singh, H. (2022). A Comprehensive Review of Metamaterials/Metasurface-Based MIMO Antenna Array for 5G Millimeter-Wave Applications. *Journal of Superconductivity and Novel Magnetism, 35*(11), 1–25. doi:10.100710948-022-06408-0

Kunsei, H., Hoole, P. R., Pirapaharan, K., & Hoole, S. R. H. (2022). Tracking Everyone and Everything in Smart Cities with an ANN Driven Smart Antenna. In *Machine Learning Techniques for Smart City Applications: Trends and Solutions* (pp. 75–94). Springer. doi:10.1007/978-3-031-08859-9_7

Lagkas, T., Klonidis, D., Sarigiannidis, P., & Tomkos, I. (2020). 5G/NGPON Evolution and Convergence: Developing on Spatial Multiplexing of Optical Fiber Links for 5G Infrastructures. *Fiber and Integrated Optics, 39*(1), 4–23. doi:10.1080/01468030.2020.1725184

Langendoen, K. (2008). Medium access control in wireless sensor networks. *Medium Access Control in Wireless Networks, 2*, 535-560.

Lee, D., Lee, S. H., Masoud, N., Krishnan, M. S., & Li, V. C. (2021). Integrated Digital Twin and Blockchain Framework to Support Accountable Information Sharing in Construction Projects. *Automation in Construction, 127*, 103688. doi:10.1016/j.autcon.2021.103688

Lee, I., & Lee, K. (2015). The Internet of Things (IoT): Applications, Investments, and Challenges for Enterprises. *Business Horizons, 58*(4), 431–440. doi:10.1016/j.bushor.2015.03.008

Lee, K., Romzi, P., Hanaysha, J., Alzoubi, H., & Alshurideh, M. (2022). Investigating the impact of benefits and challenges of iot adoption on supply chain performance and organizational performance: An empirical study in malaysia. *Uncertain Supply Chain Management, 10*(2), 537–550. doi:10.5267/j.uscm.2021.11.009

Lee, Y. L., Qin, D. H., Wang, L. C., & Sim, G. H. (2020). 6G massive radio access networks: Key applications, requirements and challenges. *IEEE Open Journal of Vehicular Technology, 2*, 54–66. doi:10.1109/OJVT.2020.3044569

Leng, J., Wang, D., Shen, W., Li, X., Liu, Q., & Chen, X. (2021). Digital twins-based smart manufacturing system design in Industry 4.0: A review. *Journal of Manufacturing Systems, 60*, 119–137. doi:10.1016/j.jmsy.2021.05.011

Leng, J., Yan, D., Liu, Q., Xu, K., Zhao, J. L., Shi, R., Wei, L., Zhang, D., & Chen, X. (2020). Manuchain: Combining permissioned blockchain with a holistic optimization model as bi-level intelligence for smart manufacturing. *IEEE Transactions on Systems, Man, and Cybernetics. Systems*, *50*(1), 182–192. doi:10.1109/TSMC.2019.2930418

Leo, M., Battisti, F., Carli, M., & Neri, A. (2014). A federated architecture approach for internet of things security. *2014 Euro Med Telco Conference (EMTC)*, 1–5. 10.1109/EMTC.2014.6996632

Letaief, Chen, Shi, Zhang, & Zhang. (2019). The roadmap to 6G: AI empowered wireless networks. *IEEE Commun. Mag.*, *57*(8), 84-90.

Letaief, K. B., Chen, W., Shi, Y. M., Zhang, J., & Angela Zhang, Y. J. (2019). The roadmap to 6G: AI empowered wireless networks. *IEEE Communications Magazine*, *57*(8), 84–90. doi:10.1109/MCOM.2019.1900271

Letterbug. (2016). *The Role of IoT in Digital Marketing*. https://suyati.com/blog/the-role-of-iot-in-digital-marketing/

Le, X. H., Lee, S., Butun, I., Khalid, M., Sankar, R., Kim, M., Han, M., Lee, Y.-K., & Lee, H. (2009). An energy-efficient access control scheme for wireless sensor networks based on elliptic curve cryptography. *Journal of Communications and Networks (Seoul)*, *11*(6), 599–606. doi:10.1109/JCN.2009.6388413

Liao, Y., Rocha Loures, E., & Deschamps, F. (2018). *Industrial internet of things: A systematic literature review and insights*. Academic Press.

Liebe, H., Hufford, G., & Cotton, M. (1993). Propagation modeling of moist air and suspended water/ice particles at frequencies below 1000 GHz. *Proceedings of AGARD*, 3-1–3-11.

Li, L., Cao, X., He, Q., Sun, J., Jia, B., & Dong, X. (2019). A new 3D laser-scanning and GPS combined measurement system. *Comptes Rendus Geoscience*, *351*(7), 508–516. doi:10.1016/j.crte.2019.09.004

Lin, J., Yu, W., Zhang, N., Yang, X., Zhang, H., & Zhao, W. (2017). A survey on internet of things: Architecture, enabling technologies, security, privacy, and applications. *IEEE Internet of Things Journal*, *4*(5), 1125–1142. doi:10.1109/JIOT.2017.2683200

Lin, R., Udalcovs, A., Ozolins, O., Tang, M., Fu, S., Popov, S., Ferreira da Silva, T., Xavier, G. B., & Chen, J. (2019). Embedding Quantum Key Distribution into Optical Telecom Communication Systems. *Proceedings of 2019 21st International Conference on Transparent Optical Networks (ICTON)*, 1-4. 10.1109/ICTON.2019.8840512

Linux Foundation. (n.d.). *Open Network Automation Platform (ONAP)*. https://www.onap.org/

Li, Q., Zhu, Q., Zheng, J. S., Liao, K. H., & Yang, G. S. (2014). Soil moisture response to rainfall in forestland and vegetable plot in Taihu Lake Basin, China. *Chinese Geographical Science*, *25*(4), 1–12.

Li, S. (2018, August). Application of Blockchain Technology in Smart City Infrastructure. In *2018 IEEE International Conference on Smart Internet of Things (SmartIoT)* (pp. 276-2766). IEEE. 10.1109/SmartIoT.2018.00056

Li, S., Da Xu, L., & Wang, X. (2012). Compressed sensing signal and data acquisition in wireless sensor networks and internet of things. *IEEE Transactions on Industrial Informatics*, *9*(4), 2177–2186. doi:10.1109/TII.2012.2189222

Liu, H., Shen, F., Liu, Z., Long, Y., Liu, Z., Sun, S., Tang, S., & Gu, D. (2019). A secure and practical blockchain scheme for IoT. *Proc. 18th IEEE Int. Conf. Trust, Secur. Privacy Comput. Commun./13th IEEE Int. Conf. Big Data Sci. Eng. (TrustCom / BigDataSE)*, 538–545. 10.1109/TrustCom/BigDataSE.2019.00078

Liu, S., Guo, C., Al-Turjman, F., Muhammad, K., & de Albuquerque, V. H. C. (2020). Reliability of response region: A novel mechanism in visual tracking by edge computing for iiot environments. *Mechanical Systems and Signal Processing*, *138*, 106537. doi:10.1016/j.ymssp.2019.106537

Li, X., Fong, P. S., Dai, S., & Li, Y. (2019). Towards Sustainable Smart Cities: An Empirical Comparative Assessment and Development Pattern Optimization in China. *Journal of Cleaner Production, 215*, 730–743. doi:10.1016/j.jclepro.2019.01.046

Li, Y., Yuan, A. N., Liu, X., Du, Y. B., Huang, T. X., & Cui, H. X. (2014). A weight cluster-based hybrid routing algorithm of ZigBee network. *International Journal of Future Generation Communication and Networking, 7*(2), 65–72. doi:10.14257/ijfgcn.2014.7.2.07

Lopez-Lopez, Matinmikko-Blue, Cardenas-Juarez, Stevens-Navarro, Aguilar-Gonzalez, & Katz. (2020). Spectrum challenges for beyond 5G: The case of Mexico. *Proc. 2nd 6G Wireless Summit (6G SUMMIT),* 1-5.

Lu & Zheng. (2020). 6G: A survey on technologies, scenarios, challenges, and the related issues. *J. Ind. Inf. Integr., 19*, 1-52.

Lu. (2020). Security in 6G: The prospects and the relevant technologies. *J. Ind. Integr. Manage., 5*, 1-24.

Luo, X., Yin, L., Li, C., Wang, C., Fang, F., Zhu, C., & Tian, Z. (2020). A lightweight privacy-preserving communication protocol for heterogeneous iot environment. *IEEE Access : Practical Innovations, Open Solutions, 8*, 67192–67204. doi:10.1109/ACCESS.2020.2978525

Lu, Y., Huang, X., Zhang, K., Maharjan, S., & Zhang, Y. (2020). Low-latency federated learning and blockchain for edge association in digital twin empowered 6G networks. *IEEE Transactions on Industrial Informatics, 17*(7), 5098–5107. doi:10.1109/TII.2020.3017668

Lynggaard, P., & Skouby, K. E. (2015). Deploying 5G-technologies in Smart City and Smart Home Wireless Sensor Networks with Interferences. *Wireless Personal Communications, 81*(4), 1399–1413. doi:10.100711277-015-2480-5

M¨uller, J. M., Kiel, D., & Voigt, K.-I. (2018). What drives the implementation of industry 4.0? the role of opportunities and challenges in the context of sustainability. *Sustainability, 10*(1), 247. doi:10.3390u10010247

Machuca, C. M., Wong, E., Furdek, M., & Wosinka, L. (2016). Energy consumption and reliability performance of survivable hybrid passive optical converged networks. *Advanced Photonics 2016.* doi:10.1364/NETWORKS.2016.NeTu3C.2

Mackenzie, B., Ferguson, R. I., & Bellekens, X. (2018, December). An Assessment of Blockchain Consensus Protocols for the Internet of Things. In *2018 International Conference on Internet of Things, Embedded Systems and Communications (IINTEC)* (pp. 183-190). IEEE. 10.1109/IINTEC.2018.8695298

Madarasz, T. (2019). Microwave will drive the development of 5G. In P. Hindle (Ed.), *Design Ideas and Tradeoffs for 5G Infrastructure* (pp. 10–13). Microwave Journal.

Mahanta, N. R., & Lele, S. (2022). Evolving Trends of Artificial Intelligence and Robotics in Smart City Applications: Crafting Humane Built Environment. *Trust-Based Communication Systems for Internet of Things Applications,* 195–241.

Mahbas, A. J., Zhu, H., & Wang, J. (2019, February). Impact of small cells overlapping on mobility management. *IEEE Transactions on Wireless Communications, 18*(2), 1054–1068. doi:10.1109/TWC.2018.2889465

Mahdi, M. N., Ahmad, A. R., Qassim, Q. S., Natiq, H., Subhi, M. A., & Mahmoud, M. (2021). From 5G to 6G technology: meets energy, internet-of-things and machine learning: a survey. *Applied Sciences (Basel, Switzerland), 11*(17), 8117. doi:10.3390/app11178117

Malakuti, S., & Gr¨uner, S. (2018). Architectural aspects of digital twins in iiot systems. *Proceedings of the 12th European Conference on Software Architecture,* 1–2. 10.1145/3241403.3241417

Malik, N., Alkhatib, K., Sun, Y., Knight, E., & Jararweh, Y. (2021). A comprehensive review of blockchain applications in industrial internet of things and supply chain systems. *Applied Stochastic Models in Business and Industry*, *37*(3), 391–412. doi:10.1002/asmb.2621

Mandolla, C., Petruzzelli, A. M., Percoco, G., & Urbinati, A. (2019). Building a digital twin for additive manufacturing through the exploitation of blockchain: A case analysis of the aircraft industry. *Computers in Industry*, *109*, 134–152. doi:10.1016/j.compind.2019.04.011

Marhoefer, M., Wimberger, I., & Poppe, A. (2007). Applicability of quantum cryptography for securing mobile communication networks. In Long-Term and Dynamical Aspects of Information Security: Emerging Trends in Information and Communication Security. Nova Science Publishers.

Ma, Z., & Liu, S. (2018). A review of 3D reconstruction techniques in civil engineering and their applications. *Advanced Engineering Informatics*, *37*, 163–174. doi:10.1016/j.aei.2018.05.005

Mazzei, D., Baldi, G., Fantoni, G., Montelisciani, G., Pitasi, A., Ricci, L., & Rizzello, L. (2020). A Blockchain Tokenizer for Industrial IOT trustless applications. *Future Generation Computer Systems*, *105*, 432–445. doi:10.1016/j.future.2019.12.020

Medvedev, A., Fedchenkov, P., Zaslavsky, A., Anagnostopoulos, T., & Khoruzhnikov, S. (2015). *Waste Management as an IoT-Enabled Service in Smart Cities.* Paper presented at the Internet of Things, Smart Spaces, and Next Generation Networks and Systems, Cham, Switzerland.

Mehic, M., Niemiec, M., Rass, S., Ma, J., Peev, M., Aguado, A., Martin, V., Schauer, S., Poppe, A., Pacher, C., & Voznak, M. (2020). Quantum Key Distribution: A Networking Perspective. *ACM Computing Surveys*, *53*(5), 41. doi:10.1145/3402192

Meneghello, F., Calore, M., Zucchetto, D., Polese, M., & Zanella, A. (2019). IoT: Internet of threats? a survey of practical security vulnerabilities in real IoT devices. *IEEE Internet of Things Journal*, *6*(5), 8182–8201. doi:10.1109/JIOT.2019.2935189

Meng, F. G., Jiang, B. C., & Wang, C. Y. (2012). An improvement of zigbee cluster-tree routing protocol. *Advanced Materials Research*, *588*, 1214–1217. doi:10.4028/www.scientific.net/AMR.588-589.1214

Menouar, H., Guvenc, I., Akkaya, K., Uluagac, A. S., Kadri, A., & Tuncer, A. (2017). UAV-Enabled Intelligent Transportation Systems for the Smart City: Applications and Challenges. *IEEE Communications Magazine*, *55*(3), 22–28. doi:10.1109/MCOM.2017.1600238CM

Mergel, I., Edelmann, N., & Haug, N. (2019). Defining digital transformation: Results from expert interviews. *Government Information Quarterly*, *36*(4), 101385. doi:10.1016/j.giq.2019.06.002

Mestres, A., Rodriguez-Natal, A., Carner, J., Barlet-Ros, P., Alarcón, E., Solé, M., Muntés-Mulero, V., Meyer, D., Barkai, S., Hibbett, M. J., Estrada, G., Máruf, K., Coras, F., Ermagan, V., Latapie, H., Cassar, C., Evans, J., Maino, F., Walrand, J., & Cabellos, A. (2017). Knowledge-defined networking. *Computer Communication Review*, *47*(3), 2–10. doi:10.1145/3138808.3138810

Meyer, O., Rauhoeft, G., Schel, D., & Stock, D. (2018). *Industrial internet of things: covering standardization gaps for the next generation of reconfigurable production systems*. Academic Press.

Michael, B. (2020). *Marketing and the Internet of Things.* https://marketinginsidergroup.com/strategy/marketing-internet-of-things/

Michailidis, E. T., Potirakis, S. M., & Kanatas, A. G. (2020). AI-inspired non-terrestrial networks for IIoT: Review on enabling technologies and applications. *IoT*, *1*(1), 21–48. doi:10.3390/iot1010003

Michalski, R. S., Carbonell, J. G., & Mitchell, T. M. (2013). *Machine learning: An artificial intelligence approach.* Springer Science & Business Media.

Microwave, F. (2021). 6G fantastic, yes. Fantasy? Not so much. *Microwaves & RF, 60*(1), 10–13.

Mili'c, S. D., & Babi'c, B. M. (2020). Toward the future—Upgrading existing remote monitoring concepts to iiot concepts. *IEEE Internet of Things Journal, 7*(12), 11693–11700. doi:10.1109/JIOT.2020.2999196

Minerva, R., Lee, G. M., & Crespi, N. (2020). Digital twin in the IoT context: A survey on technical features, scenarios, and architectural models. *Proceedings of the IEEE, 108*(10), 1785–1824. doi:10.1109/JPROC.2020.2998530

Minoli, D., & Occhiogrosso, B. (2019). Practical Aspects for the Integration of 5G Networks and IoT Applications in Smart Cities Environments. *Wireless Communications and Mobile Computing, 2019*, 1–30. Advance online publication. doi:10.1155/2019/5710834

Mohammadi, N., & Taylor, J. E. (2017). *Smart city digital twins.* Paper presented at the 2017 IEEE Symposium Series on Computational Intelligence (SSCI).

Monserrat, J. F., Martin-Sacristan, D., Bouchmal, F., Carrasco, O., de Valgas, J. F., & Cardona, N. (2020). Key technologies for the advent of the 6G. *Proc. IEEE Wireless Commun. Netw. Conf. Workshops (WCNCW)*, 1-6. 10.1109/WCNCW48565.2020.9124725

Morais, R. M. (2021). On the suitability, requisites, and challenges of machine learning. *Journal of Optical Communications and Networking, 13*(1), A1–A12. doi:10.1364/JOCN.401568

Mumtaz, S. (2017). Terahertz communication for vehicular networks. *IEEE Transactions on Vehicular Technology, 66*(7), 5617–5625. doi:10.1109/TVT.2017.2712878

Nakamoto, S. (2008). Bitcoin: A Peer-to-Peer Electronic Cash System. *Decentralized Business Review, 21260.* Advance online publication. doi:10.2139srn.3440802

Nakano, K., Motozuka, H., Wee, G. Y. H., Irie, M., Egami, A., Sakamoto, T., Takinami, K., & Takahashi, K. (2022). High-Capacity Data Collection Platform for Smart Cities Using IEEE 802.11 ad-Based Millimeter-Wave V2X Communication. *Wireless Communications and Mobile Computing*, 2022.

Nawaz, A., Peña Queralta, J., Guan, J., Awais, M., Gia, T. N., Bashir, A. K., Kan, H., & Westerlund, T. (2020). Edge computing to secure iot data ownership and trade with the ethereum blockchain. *Sensors (Basel), 20*(14), 3965. doi:10.339020143965 PMID:32708807

Nawaz, S., & Asaduzzaman, M. (2019). Quantum machine learning for 6G communication networks: State-of-the-art and vision for the future. *IEEE Access : Practical Innovations, Open Solutions, 7*, 46317–46350. doi:10.1109/ACCESS.2019.2909490

Neirotti, P., De Marco, A., Cagliano, A. C., Mangano, G., & Scorrano, F. (2014). Current trends in Smart City initiatives: Some stylised facts. *Cities (London, England), 38*, 25–36. doi:10.1016/j.cities.2013.12.010

Neocleous. (2019). *Digital Marketing and the Internet of Things.* https://www.baseelement.digital/en/digital-marketing-blog/digital-marketing-the-internet-of-things

Ng, I. C., & Wakenshaw, S. Y. (2017). The Internet-of-Things: Review and research directions. *International Journal of Research in Marketing, 34*(1), 3–21. doi:10.1016/j.ijresmar.2016.11.003

Ngo Hoang, T., & Lee, K. (2022). *Short-Packet URLLCs for Multihop MIMO Full-Duplex Relay Networks: Analytical and Deep-Learning-Based Real-Time Evaluation.* Academic Press.

Nguyen, Ding, Pubudu, Pathirana, Seneviratne, Li, Niyato, Dobre, & Poor. (2022). 6G internet of things: a comprehensive survey. *IEEE Internet of Things Journal, 9*(1), 836–886.

Nguyen, M.-S. V., Sur, S. N., & Do, D.-T. (2022). A Comprehensive Review on Physical Layer Design for Smart Cities. *IoT and IoE Driven Smart Cities*, 1–19.

Nguyen, T., Tran, N., Loven, L., Partala, J., Kechadi, M.-T., & Pirttikangas, S. (2020). Privacy-aware blockchain innovation for 6G: Challenges and opportunities. Proc. 2nd 6G Wireless Summit (6G SUM-MIT), 1-5. doi:10.1109/6GSUMMIT49458.2020.9083832

Nguyen, B., & Simkin, L. (2017). The Internet of Things (IoT) and marketing: The state of play, future trends and the implications for marketing. *Journal of Marketing Management, 33*(1-2), 1–6. doi:10.1080/0267257X.2016.1257542

Nguyen, H. P., Le, P. Q. H., Pham, V. V., Nguyen, X. P., Balasubramaniam, D., & Hoang, A. T. (2021). Application of the Internet of Things in 3E (efficiency, economy, and environment) factor-based energy management as smart and sustainable strategy. *Energy Sources. Part A, Recovery, Utilization, and Environmental Effects*, 1–23. doi:10.1080/15567036.2021.1954110

Nguyen, T. D., Khan, J. Y., & Ngo, D. T. (2017). Energy harvested roadside IEEE 802.15. 4 wireless sensor networks for IoT applications. *Ad Hoc Networks*, *56*, 109–121. doi:10.1016/j.adhoc.2016.12.003

Nielsen, C. P., da Silva, E. R., & Yu, F. (2020). Digital Twins and Blockchain–Proof of Concept. *Procedia CIRP*, *93*, 251–255. doi:10.1016/j.procir.2020.04.104

Noaman, M., Khan, M. S., Abrar, M. F., Ali, S., Alvi, A., & Saleem, M. A. (2022). Challenges in integration of heterogeneous internet of things. *Scientific Programming*, *2022*, 2022. doi:10.1155/2022/8626882

Nogales Dorado, B. (2022). *An NFV system to support adaptable multi-UAV service deployments*. Academic Press.

Notice of the State Council on the Issuance of “Made in China 2025”. (n.d.). https://www.gov.cn/

Nunes, B.A.A., Mendonca, M., Nguyen, X.N., Obraczka, & Turletti, T. (2014). A survey of software defined networking: past, present, and future of programmable networks. *IEEE Communications Survey & Tutorials, 16*(3), 1617–1634. doi:10.1109/SURV.2014.012214.00180

Ogbodo, E. U., Abu-Mahfouz, A. M., & Kurien, A. M. (2022). A Survey on 5G and LPWAN-IoT for Improved Smart Cities and Remote Area Applications: From the Aspect of Architecture and Security. *Sensors (Basel)*, *22*(16), 6313. doi:10.339022166313 PMID:36016078

Olsen, T. L., & Tomlin, B. (2019). Industry 4.0: Opportunities and Challenges for Operations Management. *Manufacturing & Service Operations Management*, *22*(1), 113–122. doi:10.1287/msom.2019.0796

Ordonez-Lucena, J., Chavarria, J. F., Contreras, L. M., & Pastor, A. (2019, October). The use of 5G Non-Public Networks to support Industry 4.0 scenarios. In *2019 IEEE Conference on Standards for Communications and Networking (CSCN)* (pp. 1-7). IEEE. 10.1109/CSCN.2019.8931325

Oza, T. (2020). Understanding the underlying sensor and wireless technologies in IIoT apps. *Microwaves & RF*, *9*, 18–22.

Padhi, P. K., & Charrua-Santos, F. (2021). 6G enabled industrial internet of everything: Towards a theoretical framework. *Applied System Innovation*, *4*(1), 11. doi:10.3390/asi4010011

Pajooh, H. H., Demidenko, S., Aslam, S., & Harris, M. (2022). Blockchain and 6G-enabled IoT. *Inventions (Basel, Switzerland)*, *7*(4), 109. doi:10.3390/inventions7040109

Palattella, M. R., Dohler, M., Grieco, A., Rizzo, G., Torsner, J., Engel, T., & Ladid, L. (2016). Internet of things in the 5G era: Enablers, architecture, and business models. *IEEE Journal on Selected Areas in Communications, 34*(3), 510–527. doi:10.1109/JSAC.2016.2525418

Palmere, T. (2019). *How IoT Data Can Improve Digital Marketing Outcomes.* https://www.iotforall.com/how-iot-consumer-data-affect-digital-marketing

Panarello, A., Tapas, N., Merlino, G., Longo, F., & Puliafito, A. (2018). Blockchain and iot integration: A systematic survey. *Sensors (Basel), 18*(8), 18. doi:10.339018082575 PMID:30082633

Pandey, G., Kumar, C. R., Kumar, M., Gupta, K., Jha, S. S., & Jha, J. (n.d.). IOT-BDA Architecture for Smart Cities. In *IoT and Big Data Analytics for Smart Cities* (pp. 57–74). Chapman and Hall/CRC.

Pan, T., Fan, Y., Shao, L., Chen, B., Chu, Y., He, G., Pan, Y., Wang, W., & Wu, Z. (2022). Multiple accounting and driving factors of water resources use: A case study of Shanghai. *Journal of Environmental Management, 313*, 114929. doi:10.1016/j.jenvman.2022.114929 PMID:35421695

Parameswaran, T., Reddy, Y. P., Nagaveni, V., & Sathiyaraj, R. (n.d.). Era of Computational Big Data Analytics and IoT Techniques in Smart City Applications. *IoT and Big Data Analytics for Smart Cities*, 1–22.

Parra, L., Sendra, S., Lloret, J., & Bosch, I. (2015). Development of a Conductivity Sensor for Monitoring Groundwater Resources to Optimize Water Management in Smart City Environments. *Sensors (Basel), 15*(9), 20990–21015. doi:10.3390150920990 PMID:26343653

Pattnaik, S. K., Samal, S. R., Bandopadhaya, S., Swain, K., Choudhury, S., Das, J. K., Mihovska, A., & Poulkov, V. (2022). Future wireless communication technology towards 6G IoT: An application-based analysis of IoT in real-time location monitoring of employees inside underground mines by using BLE. *Sensors (Basel), 22*(9), 3438. doi:10.339022093438 PMID:35591138

Paul, S. (2018). *4 Stages of IoT Architecture Explained In Simple Words.* https://medium.datadriveninvestor.com/4-stages-of-iot-architecture-explained-in-simple-words-b2ea8b4f777f

Peltonen, E., Leppanen, T., & Loven, L. (2020). EdgeAI: Edge-native distributed platform for arti_cial intelligence. Proc. 2nd 6G Wireless Summit (6G SUMMIT), 1-2.

Pincheira, M., Vecchio, M., Giaffreda, R., & Kanhere, S. S. (2021). Cost-effective IoT devices as trustworthy data sources for a blockchain-based water management system in precision agriculture. *Computers and Electronics in Agriculture, 180*, 105889. doi:10.1016/j.compag.2020.105889

Platenius-Mohr, M., Malakuti, S., Gr¨uner, S., Schmitt, J., & Goldschmidt, T. (2020). File-and api-based interoperability of digital twins by model transformation: An iiot case study using asset administration shell. *Future Generation Computer Systems, 113*, 94–105. doi:10.1016/j.future.2020.07.004

Pouttu, A., Burkhardt, F., Patachia, C., & Mendes, L. (2020). *6G white paper on validation and trials for verticals towards 2030.* 6GFlagship, 6G White Paper.

Power Digital. (2017). *The Internet of Things and The Future of Digital Marketing.* https://powerdigitalmarketing.com/blog/the-internet-of-things-and-the-future-of-digital-marketing/#gref

Pradhan, D., Dash, A., Tun, H. M., & Wah, N. K. S. (2022). Improvement of Capacity and QoE: Distributed Massive MIMO (DM-MIMO). *Technology.*

Prasanalakshmi, B., Murugan, K., Srinivasan, K., Shridevi, S., Shamsudheen, S., & Hu, Y.-C. (2022). Improved authentication and computation of medical data transmission in the secure IoT using hyperelliptic curve cryptography. *The Journal of Supercomputing*, *78*(1), 361–378. doi:10.100711227-021-03861-x

Premsankar, G., Francesco, M., & Taleb, T. (2018). Edge computing for the internet of things: A case study. IEEE Internet of Things Journal.

Putz, B., Dietz, M., Empl, P., & Pernul, G. (2021). Ethertwin: Blockchain-based secure digital twin information management. *Information Processing & Management*, *58*(1), 102425. doi:10.1016/j.ipm.2020.102425

Qadir, Z., Le, K. N., Saeed, N., & Munawar, H. S. (2022). Towards 6G internet of things: Recent advances, use cases, and open challenges. *ICT Express*. Advance online publication. doi:10.1016/j.icte.2022.06.006

Qiaoling, B., Tao, Y., Qiqing, H., & Ying, L. (2021). *Construction of a rule system for intelligent planning and construction management under the orientation of digital twin city: An example of BIM management platform for planning and construction of Xiongan New Area*. Academic Press.

Qorvo & RFMW. (2020). *Wi-Fi Convergence with IoT and 5G*. Microwave Journal.

Quaye, I., Amponsah, O., Azunre, G. A., Takyi, S. A., & Braimah, I. (2022). A Review of Experimental Informal Urbanism Initiatives and their Implications for Sub-Saharan Africa's Sustainable Cities' Agenda. *Sustainable Cities and Society*, *103938*, 103938. Advance online publication. doi:10.1016/j.scs.2022.103938

Rajanarthagi. (2020). *What is IoT? & Role of IoT in digital marketing*. https://gecdesigns.com/blog/role-of-iot-in-digital-marketing

RajathevaN. White paper on broadband connectivity in 6G. Available: http://arxiv. org/abs/2004.14247

Raj, P. (2021). Empowering digital twins with blockchain. *Advances in Computers*, *121*, 267–283. doi:10.1016/bs.adcom.2020.08.013

Rangsietti, A. K., & Kodali, S. S. P. (2022). SDN-Enabled Network Virtualization and Its Applications. *Software Defined Networks: Architecture and Applications*, 231–277.

Rao, S. K., & Prasad, R. (2018). Impact of 5G technologies on Smart City Implementation. *Wireless Personal Communications*, *100*(1), 161–176. doi:10.100711277-018-5618-4

Rappaport & Trichopoulos. (2019). Wireless communications and applications above 100 GHz: Opportunities and challenges for 6G and beyond. *IEEE Access, 7*, 78729-78757.

Rault, T., Bouabdallah, A., & Challal, Y. (2014). Energy efficiency in wireless sensor networks: A top-down survey. *Computer Networks*, *67*, 104–122. doi:10.1016/j.comnet.2014.03.027

Ren, D., Gui, X., & Zhang, K. (2022). Adaptive Request Scheduling and Service Caching for MEC-Assisted IoT Networks: An Online Learning Approach. *IEEE Internet of Things Journal*, *9*(18), 17372–17386. doi:10.1109/JIOT.2022.3157677

Reyna, A., Martín, C., Chen, J., Soler, E., & Díaz, M. (2018). On Blockchain and Its Integration with IoT. Challenges and Opportunities. *Future Generation Computer Systems*, *88*, 173–190. doi:10.1016/j.future.2018.05.046

Riahi Sfar, A., Natalizio, E., Challal, Y., & Chtourou, Z. (2018). A roadmap for security challenges in the internet of things. *Digital Communications and Networks*, *4*(2), 118–137. doi:10.1016/j.dcan.2017.04.003

Riazul Islam, S. M., & Kwak, D. (2015). The Internet of Things for health care: A comprehensive survey. *IEEE Access : Practical Innovations, Open Solutions*, *3*, 678–708. doi:10.1109/ACCESS.2015.2437951

Richardson, D. J., Fini, J. M., & Nelson, L. E. (2013). Space-division multiplexing in optical fibres. *Nature Photonics*, *7*(5), 354–362. doi:10.1038/nphoton.2013.94

Rinaldi, F., Maattanen, H.-L., Torsner, J., Pizzi, S., Andreev, S., Iera, A., Koucheryavy, Y., & Araniti, G. (2020). Non-terrestrial networks in 5G & beyond: A survey. *IEEE Access : Practical Innovations, Open Solutions*, *8*, 165178–165200. doi:10.1109/ACCESS.2020.3022981

Rivera, A. O. G., White, E. M., Acosta, J. C., & Tosh, D. (2022). Enabling Device Trustworthiness for SDN-Enabled Internet-of-Battlefield Things. *2022 IEEE Conference on Dependable and Secure Computing (DSC)*, 1–7. 10.1109/DSC54232.2022.9888903

Roman, R., Zhou, J., & Lopez, J. (2013). On the features and challenges of security and privacy in distributed internet of things. *Computer Networks*, *57*(10), 2266–2279. doi:10.1016/j.comnet.2012.12.018

Rommel, S., Perez-Galacho, D., Fabrega, J. M., Muñoz, R., Sales, S., & Tafur Monroy, I. (2019). High-Capacity 5G Front haul Networks Based on Optical Space Division Multiplexing. *IEEE Transactions on Broadcasting*, *65*(2), 434–443. doi:10.1109/TBC.2019.2901412

Saad, W., & Chen, M. (2020). *A vision of 6G wireless systems: applications, trends, technologies, and open research problems.* White Paper on 6G-the Next Hyper Connected Experience for All, Samsung, Seoul, South Korea.

Saad, W., Bennis, M., & Chen, M. Z. (2019). A vision of 6G wireless systems: Applications, trends, technologies, and open research problems. *IEEE Network*, *34*(3), 134–142. doi:10.1109/MNET.001.1900287

Saeed, M. M., Hasan, M. K., Obaid, A. J., Saeed, R. A., Mokhtar, R. A., Ali, E. S., Akhtaruzzaman, M., Amanlou, S., & Hossain, A. Z. (2022). A comprehensive review on the users' identity privacy for 5G networks. *IET Communications*, *16*(5), 384–399. doi:10.1049/cmu2.12327

Said, O., Al-Makhadmeh, Z., & Tolba, A. A. (2020). EMS: An energy management scheme for green IoT environments. *IEEE Access : Practical Innovations, Open Solutions*, *8*, 44983–44998. doi:10.1109/ACCESS.2020.2976641

Samdanis & Taleb. (2020). The road beyond 5G: A vision and insight of the key technologies. *IEEE Netw.*, *34*(2), 135-141.

Sari, A., Lekidis, A., & Butun, I. (2020). Industrial networks and iiot: Now and future trends. In *Industrial IoT* (pp. 3–55). Springer. doi:10.1007/978-3-030-42500-5_1

Sarigiannidis, P., Lagkas, T., Bibi, S., Ampatzoglou, A., & Bellavista, P. (2017). Hybrid 5G opticalwireless SDN-based networks, challenges and open issues. *IET Networks*, *6*(6), 141–148. doi:10.1049/iet-net.2017.0069

Schroeder, C. (2021). Early indications of 6G. *Microwave Journal*, *64*, 5–9.

Seilov, S., Kuzbayev, A. T., Seilov, A. A., Shyngisov, D. S., Goikhman, V. Y., Levakov, A. K., Sokolov, N. A., & Zhursinbek, Y. S. (2021). The Concept of Building a Network of Digital Twins to Increase the Efficiency of Complex Telecommunication Systems. *Complexity*, *2021*, 2021. doi:10.1155/2021/9480235

Sekaran, R., Patan, R., Raveendran, A., Al-Turjman, F., Ramachandran, M., & Mostarda, L. (2020). Survival study on blockchain based 6G-enabled mobile edge computation for IoT automation. *IEEE Access : Practical Innovations, Open Solutions*, *8*, 143453–143463. doi:10.1109/ACCESS.2020.3013946

Senger, S., & Malik, P. K. (2021). A Comprehensive Survey of Massive-MIMO Based on 5G Antennas. *International Journal of RF and Microwave Computer-Aided Engineering*, e23496.

Seo-alien.com. (2021). *The Role of IoT in Digital Marketing*. https://www.seo-alien.com/social-media-marketing/the-role-of-iot-in-digital-marketing/

Sepasgozar, S. M. (2021). Differentiating digital twin from digital shadow: Elucidating a paradigm shift to expedite a smart, sustainable built environment. *Buildings, 11*(4), 151. doi:10.3390/buildings11040151

Sergiou, Lestas, Antoniou, Liaskos, & Pitsillides. (2020). Complex systems: A Communication networks perspective towards 6G. *IEEE Access, 8*, 89007-89030.

Shafik, W., Matinkhah, M., & Sanda, M. N. (2020). Network resource management drives machine learning: A survey and future research direction. Journal of Communications Technology. *Electronics and Computer Science, 2020*, 1–15.

Shafi, M., Molisch, A. F., Smith, P. J., Haustein, T., Zhu, P., De Silva, P., Tufvesson, F., Benjebbour, A., & Wunder, G. (2017). 5G: a tutorial overview of standards, trials, challenges, deployment, and practice. *IEEE Journal on Selected Areas in Communications, 35*(6), 1201–1221. doi:10.1109/JSAC.2017.2692307

Shahriar, N., Taeb, S., Chowdhury, S. R., Zulfiqar, M., Tornatore, M., Boutaba, R., Mitra, J., & Hemmati, M. (2019). Reliable Slicing of 5G Transport Networks with Dedicated Protection. *Proceedings of 15th International Conference on Network and Service Management (CNSM)*, 1-9. 10.23919/CNSM46954.2019.9012711

Shammar, E. A., Zahary, A. T., & Al-Shargabi, A. A. (2021). A survey of IoT and blockchain integration: Security perspective. *IEEE Access : Practical Innovations, Open Solutions, 9*, 156114–156150. doi:10.1109/ACCESS.2021.3129697

Shamshuddin, K., & Jayalaxmi, G. N. (2022). Privacy Preserving Scheme for Smart Transportation in 5G Integrated IoT. In *ICT with Intelligent Applications* (pp. 59–67). Springer. doi:10.1007/978-981-16-4177-0_8

Sharma, H., Haque, A., & Blaabjerg, F. (2021). Machine learning in wireless sensor networks for smart cities: A survey. *Electronics (Basel), 10*(9), 1012. doi:10.3390/electronics10091012

Sharma, R., Mehta, K., & Sharma, O. (2021). Exploring Deep Learning to Determine the Optimal Environment for Stock Prediction Analysis. *2021 International Conference on Computational Performance Evaluation*, 148–152. 10.1109/ComPE53109.2021.9752138

Shatin, Liu, Chandrasekhar, Chen, Reed, & Zhang. (2019). Artificial intelligence-enabled cellular networks: A critical path to beyond-5G and 6G. *IEEE Wireless Commun., 27*(2), 212-217.

SheC.VuceticB. (2020). A tutorial on ultra-reliable and low-latency communications in 6G: Integrating domain knowledge into deep learning. Available: https://arxiv.org/abs/2009.06010

SheC.VuceticB. (2020). Deep learning for ultra-reliable and low-latency communications in 6G networks. Available: https://arxiv.org/abs/2002.11045

Shi, W., & Dustdar, S. (2016). The promise of edge computing. *Computer, 49*(5), 78–81. doi:10.1109/MC.2016.145

Sicari, S., Cappiello, C., De Pellegrini, F., Miorandi, D., & Coen-Porisini, A. (2016). A security-and quality-aware system architecture for internet of things. *Information Systems Frontiers, 18*(4), 665–677. doi:10.100710796-014-9538-x

Siegel, P. H. (2002). Terahertz technology. *IEEE Transactions on Microwave Theory and Techniques, 50*(3), 910–928. doi:10.1109/22.989974

Sikimić, M., Amović, M., Vujović, V., Suknović, B., & Manjak, D. (2020, March). An overview of wireless technologies for IoT network. In *2020 19th International Symposium INFOTEH-JAHORINA (INFOTEH)* (pp. 1-6). IEEE. 10.1109/INFOTEH48170.2020.9066337

Siles, G. A., Riera, J. M., & García-del-Pino, P. (2015). Atmospheric attenuation in wireless communication systems at millimeter and THz frequencies. *IEEE Antennas & Propagation Magazine, 57*(1), 48–61. doi:10.1109/MAP.2015.2401796

Singh, P., Dixit, S., Sammanit, D., & Krishnan, P. (2022). The automated farmlands of tomorrow: An IoT integration with farmlands. *IOP Conference Series. Materials Science and Engineering*, *1218*(1), 012048. doi:10.1088/1757-899X/1218/1/012048

Singh, S., Sharma, P. K., Yoon, B., Shojafar, M., Cho, G. H., & Ra, I. H. (2020). Convergence of Blockchain and Artificial Intelligence in IoT Network for the Sustainable Smart City. *Sustainable Cities and Society*, *63*, 102364. doi:10.1016/j.scs.2020.102364

Sisinni, E., Saifullah, A., Han, S., Jennehag, U., & Gidlund, M. (2018a). Industrial internet of things: Challenges, opportunities, and directions. *IEEE Transactions on Industrial Informatics*, *14*(11), 4724–4734. doi:10.1109/TII.2018.2852491

Skelia. (2020). *10 Ways IoT is Changing Digital Marketing in 2020*. https://skelia.com/articles/10-ways-iot-is-changing-digital-marketing-in-2020/

Skouby, K. E., & Lynggaard, P. (2014, November). Smart Home and Smart City Solutions Enabled by 5G, IoT, AAI and CoT Services. In *2014 International Conference on Contemporary Computing and Informatics (IC3I)* (pp. 874-878). IEEE. 10.1109/IC3I.2014.7019822

Sliwa, Falkenberg, & Wietfeld. (2020). Towards cooperative data rate prediction for future mobile and vehicular 6G networks. *Proc. 2nd 6G Wireless Summit (6G SUMMIT)*, 1-5.

Smstudy.com. (2018). *How IoT will help Digital Marketers in near future?* https://www.smstudy.com/article/how-iot-will-help-digital-marketers-in-near-future

Söderberg, R., Wärmefjord, K., Carlson, J. S., & Lindkvist, L. (2017). Toward a digital twin for real-time geometry assurance in individualized production. *CIRP Annals*, *66*(1), 137–140. doi:10.1016/j.cirp.2017.04.038

Soderi. (2020). Enhancing security in 6G visible light communications. *Proc. 6G Wireless Summit*, 1-5.

Soto-Valle, G., Hu, K., Holda, M., Cui, Y., & Tentzeris, M. (2022). Novel Additive Manufacturing-Enabled RF Devices for 5G/mmWave, IoT, Smart Skins, and Wireless Sensing Applications. *International Journal of High Speed Electronics and Systems, 31*(1-4), 2240017.

Spyridis, Y., Lagkas, T., Sarigiannidis, P., Argyriou, V., Sarigiannidis, A., Eleftherakis, G., & Zhang, J. (2021). Towards 6G IoT: Tracing mobile sensor nodes with deep learning clustering in UAV networks. *Sensors (Basel)*, *21*(11), 1–16. doi:10.339021113936 PMID:34200449

Statistical Bulletin of the People's Republic of China on National Economic and Social Development in 2021. (2022). Chinese Government.

StoicaR.AbreuF. (2019). 6G: The wireless communications network for collaborative and AI applications. Available: https://arxiv.org/abs/1904.03413

Strinati & Dehos. (2019). 6G: The next frontier: From holographic messaging to artificial intelligence using subterahertz and visible light communication. *IEEE Veh. Technol. Mag., 14*(3), 42-50.

Strinati, E. C., Barbarossa, S., Gonzalez-Jimenez, J. L., Kténas, D., Cassiau, N., Maret, L., & Dehos, C. (2019). 6G: The next frontier –from holographic messaging to artificial intelligence using subterahertz and visible light communication. *IEEE Vehicular Technology Magazine*, *14*(3), 42–50. doi:10.1109/MVT.2019.2921162

Suhail, S., Hussain, R., Jurdak, R., & Hong, C. S. (2021). Trustworthy digital twins in the industrial internet of things with blockchain. In *Internet Computing*. IEEE

Sumathi, A. C., Ahalawat, A., & Rameshkumar, A. (2022). Early detection of DDoS attack using integrated SDN-Blockchain architecture for IoT. *2022 International Conference on Innovative Computing, Intelligent Communication and Smart Electrical Systems (ICSES),* 1–5. 10.1109/ICSES55317.2022.9914202

Sun, N., Li, T., Song, G., & Xia, H. (2021). Network security technology of intelligent information terminal based on mobile internet of things. *Mobile Information Systems.*

Sun, G., Liu, G., Zhang, H., & Tan, W. (2013). Architecture on mobility management in OpenFlow-based radio access networks. *Proceedings of IEEE Global High Tech Congress on Electronics,* 88–92. 10.1109/GHTCE.2013.6767247

Sun, G., & Xu, B. (2010). Dynamic routing algorithm for priority guarantee in low duty-cycled wireless sensor networks. *Wireless Algorithms, Systems, and Applications: 5th International Conference, WASA 2010, Beijing, China, August 15-17, 2010 Proceedings*, *5*, 146–156.

Sz.gov.cn. (2022). *The 14th Five-Year Plan for digital government and smart city of Shenzhen was released.* http://www.sz.gov.cn/

Szatkowski, J. M., Li, Y., & Du, L. (2022). Enabling Reconfigurable Naval SCADA Network through Software-Defined Networking. *2022 IEEE Transportation Electrification Conference & Expo (ITEC),* 214–218. 10.1109/ITEC53557.2022.9813872

Tahir, A. (2016). *7 Fundamental Ethics of Social Media Marketing.* https://www.business2community.com/social-media/7-fundamental-ethics-social-media-marketing-01571504

Tangudu, N. D., Gupta, N., Shah, S. P., Pattan, B. J., & Chitturi, S. (2020, September). Common framework for 5G northbound APIs. In *2020 IEEE 3rd 5G World Forum (5GWF)* (pp. 275-280). IEEE.

Tanwar, S., Popat, A., Bhattacharya, P., Gupta, R., & Kumar, N. (2022). A taxonomy of energy optimization techniques for smart cities: Architecture and future directions. *Expert Systems: International Journal of Knowledge Engineering and Neural Networks*, *39*(5), e12703. doi:10.1111/exsy.12703

Tao, F., Sui, F., Liu, A., Qi, Q., Zhang, M., Song, B., Guo, Z., Lu, S. C.-Y., & Nee, A. Y. C. (2019). Digital twin-driven product design framework. *International Journal of Production Research*, *57*(12), 3935–3953. doi:10.1080/00207543.2018.1443229

Tao, F., Zhang, M., Liu, Y., & Nee, A. (2018). Digital twin driven prognostics and health management for complex equipment. *CIRP Annals*, *67*(1), 169–172. doi:10.1016/j.cirp.2018.04.055

Tariq, F. (n.d.). *A Speculative Study on 6G.* arXiv:1902.06700.

TatariaH.TufvessonF. (2020). 6G wireless systems: Vision, requirements, challenges, insights, and opportunities. Available: https://arxiv.org/abs/2008.03213

Taylor, M., Reilly, D., & Wren, C. (2020). Internet of things support for marketing activities. *Journal of Strategic Marketing*, *28*(2), 149–160. doi:10.1080/0965254X.2018.1493523

Technologies, K. (2018). *Smart Cows and Tips for Designing Mission-Critical IoT Products.* Keysight Technologies.

Technologies, K. (2019). *How to Ensure IoT Devices Work in Their Intended Environment (Locate and Identify Interference).* Keysight Technologies.

Technologies, K. (2020a). *Unleash the Power of IoT.* Keysight Technologies.

Technologies, K. (2020b). *The 5C's of IoT.* Keysight Technologies.

Tencent. (2021). *Smart City Policy in China and 31 Provinces and Cities in 2021*. https://news.qq.com/

The State Council of The People's Republic of China. (2015). *Notice of the State Council on Printing and Distributing "Made in China 2025"*. www.gov.cn/zhengce/content/2015-05/19/content_9784.htm

Thomas, W., & Weil, P. (Eds.). (2007). Lecture Notes in Computer Science: Vol. 4393. *Quantum network coding*. Springer Berlin. doi:10.1007/978-3-540-70918-3

Tian, K., Chai, H., Liu, Y., & Liu, B. (2022). Edge Intelligence Empowered Dynamic Offloading and Resource Management of MEC for Smart City Internet of Things. *Electronics (Basel), 11*(6), 879. doi:10.3390/electronics11060879

Tilson, D., Lyytinen, K., & Sørensen, C. (2010). Research Commentary—Digital Infrastructures: The Missing IS Research Agenda. *Information Systems Research, 21*(4), 748–759. doi:10.1287/isre.1100.0318

Trivedi, S., Mehta, K., & Sharma, R. (2021). Systematic Literature Review on Application of Blockchain Technology in E-Finance and Financial Services. *Journal of Technology Management & Innovation, 16*(3), 90–102. doi:10.4067/S0718-27242021000300089

Tsolkas, D., & Koumaras, H. (2022). On the development and provisioning of vertical applications in the beyond 5G era. *IEEE Networking Letters, 4*(1), 43–47. doi:10.1109/LNET.2022.3142088

Tuan, N. Y. (2020). Small cells help keep 5G connected. In B. Wong (Ed.), *Focus on: 5G technology and challenges* (pp. 2–4). Microwaves & RF.

Uddin, M. A., Stranieri, A., Gondal, I., & Balasubramanian, V. (2021). A survey on the adoption of blockchain in iot: Challenges and solutions. Blockchain. *Research and Applications*, *2*(2), 100006. doi:10.1016/j.bcra.2021.100006

Uikey, R., & Sharma, S. (2013). Zigbee cluster tree performance improvement technique. *International Journal of Computers and Applications*, *62*(19).

Ullah, R., Abbas, A. W., Ullah, M., Khan, R. U., Khan, I. U., Aslam, N., & Aljameel, S. S. (2021). EEWMP: An IoT-based energy-efficient water management platform for smart irrigation. *Scientific Programming, 2021*, 1–9. doi:10.1155/2021/5536884

ur Rehman, M. H., Yaqoob, I., Salah, K., Imran, M., Jayaraman, P. P., & Perera, C. (2019). The role of big data analytics in industrial internet of things. *Future Generation Computer Systems, 99*, 247–259.

Vandita, G. (2018). *5 Ways IoT is Changing Digital Marketing*. https://www.martechadvisor.com/articles/iot/5-ways-iot-is-changing-digital-marketing/

Venables, M. (2019). *Optimizing Digital Marketing with Data Science*. https://towardsdatascience.com/how-data-science-is-shaping-digital-marketing-5a149443f90

Voas, J., Mell, P., & Piroumin, V. (2021). *Considerations for Digital Twin Technology and Emerging Standards, draft NISTR 8356*. https://nvlpubs.nist.gov/nistpubs/ir/2021/NIST.IR.8356-draft.pdf

Waliwander, T. (2022). THz – To be or not to be in 6G. *Microwave Journal, 65*(5), 54–66.

Wang, B., Liu, X., & Zhang, Y. (2022). Future Prospects of BDS and IOT Integration Application. In *Internet of Things and BDS Application* (pp. 287–291). Springer. doi:10.1007/978-981-16-9194-2_7

Wang, C. X., Huang, J., Wang, H. M., Gao, X. Q., You, X. H., & Hao, Y. (2020). 6G wireless channel measurements and models. *IEEE Vehicular Technology Magazine, 15*(4), 22–32. doi:10.1109/MVT.2020.3018436

Wang, J., & Deng, K. (2022). Impact and mechanism analysis of smart city policy on urban innovation: Evidence from China. *Economic Analysis and Policy*, *73*, 574–587. doi:10.1016/j.eap.2021.12.006

Wang, K., Lin, C., & Wu, C. (2020). Trends and Planning Choices After the 60% Urbanization Rate in China. *City Planning Review*, *12*, 9–17.

Wang, Y., Guo, C. H., Chen, X. J., Jia, L. Q., Guo, X. N., Chen, R. S., ... Wang, H. D. (2021). Carbon Peak and Carbon Neutrality in China: Goals, Implementation Path and Prospects. *China Geology*, *4*(4), 720–746. doi:10.31035/cg2021083

Wang, Y., Ren, H., Dong, L., Park, H. S., Zhang, Y., & Xu, Y. (2019). Smart Solutions Shape for Sustainable Low-Carbon Future: A Review on Smart Cities and Industrial Parks in China. *Technological Forecasting and Social Change*, *144*, 103–117. doi:10.1016/j.techfore.2019.04.014

Wątróbski, J., Bączkiewicz, A., Ziemba, E., & Sałabun, W. (2022). Sustainable Cities and Communities Assessment Using the DARIA-TOPSIS Method. *Sustainable Cities and Society*, *103926*. Advance online publication. doi:10.1016/j.scs.2022.103926

Weir-McCall, D. (2020). *51 World Created a Digital Twin of Entire Shanghai*. Retrieved from https://www.unrealengine.com/en-US/spotlights/51world-creates-digital-twin-of-the-entire-city-of-shanghai

Weissberger, A. (2021). ITU-R future report: high altitude platform stations as IMT base stations (HIBS). *IEEE Communication Society Technology Blog*. https://techblog.comsoc.org/2021/02/17/itu-r-future-report-high-altitude-platform-stations-as-imt-base-stations-hibs/

Wikipedia. (2021). Internet of things. In *Wikipedia, the free encyclopedia*. https://en.wikipedia.org/wiki/Internet_of_things

Wilkins, G., & Stiff, A. (2019). Hem Realities: Augmenting Urbanism Through Tacit and Immersive Feedback. *Architecture and Culture*, *7*(3), 505–521. doi:10.1080/20507828.2019.1673545

William, P., Shrivastava, A., Chauhan, H., Nagpal, P., & Singh, P. (2022). Framework for Intelligent Smart City Deployment via Artificial Intelligence Software Networking. *2022 3rd International Conference on Intelligent Engineering and Management (ICIEM)*, 455–460.

Winzer, P. J., Neilson, D. T., & Chraplyvy, A. R. (2018). Fiber-optic transmission and networking: The previous 20 and the next 20 years. *Optics Express*, *26*(18), 24190–24239. doi:10.1364/OE.26.024190 PMID:30184909

Witten, I. H., & Frank, E. (2002). Data mining: Practical machine learning tools and techniques with java implementations. *SIGMOD Record*, *31*(1), 76–77. doi:10.1145/507338.507355

Wong, E., Grigoreva, E., Wosinska, L., & Machuca, C. M. (2017). Enhancing the survivability and power savings of 5G transport networks based on DWDM rings. *Journal of Optical Communications and Networking*, *9*(9), D74–D85. doi:10.1364/JOCN.9.000D74

Wong, E., Machuca, C. M., & Wosinska, L. (2016). Survivable architectures for power-savings capable converged access networks. *Proceedings of 2016 IEEE International Conference on Communications*, 1-7. 10.1109/ICC.2016.7510719

Wong, E., Machuca, C. M., & Wosinska, L. (2016). Survivable hybrid passive optical converged network architectures based on reflective monitoring. *Journal of Lightwave Technology*, *34*(18), 4317–4328. doi:10.1109/JLT.2016.2593481

World Economic Forum. (2019). *Fourth Industrial Revolution Beacons of Technology and Innovation in Manufacturing* [White paper]. https://www3.weforum.org/docs/WEF_4IR_Beacons_of_Technology_and_Innovation_in_Manufacturing_report_2019.pdf

World Economic Forum. (2022). *Digital Twin Cities: Framework and Global Practice*. Author.

Wright, S. A. (2021) Identity. *Proceedings of 27th Int'l Conf. on Parallel & Distributed Processing Techniques & Applications (PDPTA'21).*

Wright, S. A. (2019). Privacy in IoT Blockchains: with big data comes big responsibility. *Proceedings of the International Conference on Big Data* (pp. 5282-5291). IEEE. 10.1109/BigData47090.2019.9006341

Wu, S., Niu, J., Chou, W., & Guizani, M. (2016). Delay-aware energy optimization for flooding in duty-cycled wireless sensor networks. *IEEE Transactions on Wireless Communications, 15*(12), 8449–8462. doi:10.1109/TWC.2016.2615296

Xie, J., Tang, H., Huang, T., Yu, F. R., Xie, R., Liu, J., & Liu, Y. (2019). A Survey of Blockchain Technology Applied to Smart Cities: Research Issues and Challenges. *IEEE Communications Surveys and Tutorials, 21*(3), 2794–2830. doi:10.1109/COMST.2019.2899617

Xue, F., Chen, K., Lu, W., Niu, Y., & Huang, G. Q. (2018). Linking radio-frequency identification to Building Information Modeling: Status quo, development trajectory and guidelines for practitioners. *Automation in Construction, 93*, 241–251. doi:10.1016/j.autcon.2018.05.023

Xu, Y., & Jia, S. (2021, April). Research on Application of 5G to Smart City. *IOP Conference Series: Earth and Environmental Science, 760(1), 012014.* doi:10.1088/1755-1315/760/1/012014

YangH.AlphonesA.XiongZ.NiyatoD.ZhaoJ.WuK. (2019). Artificial intelligence-enabled intelligent 6G networks. Available: https://arxiv.org/abs/1912.05744

Yao, H., Qin, R., & Chen, X. (2019). Unmanned Aerial Vehicle for Remote Sensing Applications. *RE:view, 11*(12), 1443.

Yao, Y., Cao, Q., & Vasilakos, A. V. (2014). EDAL: An energy-efficient, delay-aware, and lifetime-balancing data collection protocol for heterogeneous wireless sensor networks. *IEEE/ACM Transactions on Networking, 23*(3), 810–823. doi:10.1109/TNET.2014.2306592

Yeh, L.-Y., Shen, N.-X., & Hwang, R.-H. (2022). Blockchain-Based Privacy-Preserving and Sustainable Data Query Service Over 5G-VANETs. *IEEE Transactions on Intelligent Transportation Systems.*

Yick, J., Mukherjee, B., & Ghosal, D. (2008). Wireless sensor network survey. *Computer Networks, 52*(12), 2292–2330. doi:10.1016/j.comnet.2008.04.002

Yitmen, I., Alizadehsalehi, S., Akıner, İ., & Akıner, M. E. (2021). An Adapted Model of Cognitive Digital Twins for Building Lifecycle Management. *Applied Sciences (Basel, Switzerland), 11*(9), 4276. doi:10.3390/app11094276

You, X. H. (2021). Towards 6G wireless communication networks: vision, enabling technologies, and new paradigm shifts. *Science China Information Sciences, 64*, 110301:1–110301:74.

You, K. Y. (2022a). Propagation channel modelling for low-altitude platform non-terrestrial networks from 275 GHz to 3 THz. *International Journal Wireless and Microwave Technologies, 3*(3), 1–17. doi:10.5815/ijwmt.2022.03.01

You, K. Y. (2022b). Survey on 5G and future 6G access networks for IoT applications. *International Journal Wireless and Microwave Technologies, 4*(4), 26–47. doi:10.5815/ijwmt.2022.04.03

Yousaf, F. Z., Bredel, M., Schaller, S., & Schneider, F. (2017). NFV and SDN - Key Technology Enablers for 5G Networks. *IEEE Journal on Selected Areas in Communications, 35*(11), 2468–2478. doi:10.1109/JSAC.2017.2760418

Yousefnezhad, N., Malhi, A., Kinnunen, T., Huotari, M., & Främling, K. (2020, July). Product Lifecycle Information Management with Digital Twin: A Case Study. In *18th International Conference on Industrial Informatics (INDIN)* (Vol. 1, pp. 321-326). IEEE. 10.1109/INDIN45582.2020.9442208 ytcstudios.com. (2019). *Digital Marketing in the age of IoT.* https://www.businessinsider.in/advertising/ad-agencies/article/digital-marketing-in-the-age-of-iot/articleshow/72055063.cms

ytcstudios.com. (2020). *The Convergence of Marketing & the Internet of Things.* https://www.ytcstudios.com/blog/2020/8/6/the-convergence-of-marketing-amp-the-internet-of-things-iot

Yu, P., Yu, M., & Sampat, M. (2022). Smart Management for Digital Transformation in China. In B. Barbosa, S. Filipe, & C. Santos (Eds.), Handbook of Research on Smart Management for Digital Transformation (pp. 411-438). IGI Global. doi:10.4018/978-1-7998-9008-9.ch019

Yu, Q., & Zhang, W. (2020). A cybertwin based network architecture for 6G. Proc. 2nd 6G Wireless Summit (6G SUMMIT), 1-5. doi:10.1109/6GSUMMIT49458.2020.9083808

Yu, A., Yang, H., Yao, Q., Li, Y., Guo, H., Peng, T., Li, H., & Zhang, J. (2019). Accurate Fault Location Using Deep Belief Network for Optical Front haul Networks in 5G and Beyond. *IEEE Access : Practical Innovations, Open Solutions, 7*, 77932–77943. doi:10.1109/ACCESS.2019.2921329

Yuan, Zhao, Zong, & Parolari. (2020). Potential key technologies for 6G mobile communications. *Sci. China Inf. Sci., 63*(8), 1-19.

YuanX.ZhangY.-J. A.ShiY.YanW.LiuH. (2020). Reconfigurable intelligent- surface empowered wireless communications: Challenges and opportunities. Available: https://arxiv.org/abs/2001.00364

Yu, J., Gutterman, C., Minakhmetov, A., Sherman, M., Chen, T., Zhu, S., Zussman, G., Seskar, I., & Kilper, D. (2020). Dual Use SDN Controller for Management and Experimentation in a Field Deployed Test bed. *Proceedings of 2020 Optical Fiber Communications Conference and Exhibition (OFC),* 1-3. 10.1364/OFC.2020.T3J.3

Yu, P., Chang, X., & Mandizvidza, K. (2023). Development of New Energy Vehicles in Entrepreneurial Ecosystem Under the Carbon Neutrality Policy in China. In B. Marco-Lajara, J. Martínez-Falcó, & L. Millán-Tudela (Eds.), *Corporate Sustainability as a Tool for Improving Economic, Social, and Environmental Performance* (pp. 55–84). IGI Global. doi:10.4018/978-1-6684-7422-8.ch004

Yu, P., Chen, D., & Ahuja, A. (2022). Smart and Sustainable Economy: How COVID-19 Has Acted as a Catalyst for China's Digital Transformation. In S. Kautish & G. Kaur (Eds.), *AI-Enabled Agile Internet of Things for Sustainable FinTech Ecosystems* (pp. 106–146). IGI Global. doi:10.4018/978-1-6684-4176-3.ch006

Yu, P., Chen, J., Sampat, M., & Misuko, N. (2022). The Digital Transformation of Rural Agricultural Business Management: A Case Study of China. In S. Bilgaiyan, J. Singh, & H. Das (Eds.), *Empirical Research for Futuristic E-Commerce Systems: Foundations and Applications* (pp. 23–52). IGI Global. doi:10.4018/978-1-6684-4969-1.ch002

Yu, P., Ge, Y., Mandizvidza, K., & Mulli, J. (2023). How Can Small and Medium Enterprises in the Chinese Market Achieve Sustainable Development Goals Through Blockchain? In D. Taleb, M. Abdelli, A. Khalil, & A. Sghaier (Eds.), *Examining the Vital Financial Role of SMEs in Achieving the Sustainable Development Goals* (pp. 52–85). IGI Global. doi:10.4018/978-1-6684-4834-2.ch004

Yu, P., Gong, R., & Sampat, M. (2022). Blockchain Technology in China's Digital Economy: Balancing Regulation and Innovation. In P. Tehrani (Ed.), *Regulatory Aspects of Artificial Intelligence on Blockchain* (pp. 132–157). IGI Global. doi:10.4018/978-1-7998-7927-5.ch007

Yu, P., Gu, H., Zhao, Y., & Ahuja, A. (2022). Digital Transformation Driven by Internet Data Center: Case Studies on China. In D. Piaggesi, H. Landazuri, & B. Jia (Eds.), *Cases on Applying Knowledge Economy Principles for Economic Growth in Developing Nations* (pp. 203–230). IGI Global. doi:10.4018/978-1-7998-8417-0.ch011

Yu, P., Jiao, A., & Sampat, M. (2022). The Effect of Chinese Green Transformation on Competitiveness and the Environment. In P. Ordóñez de Pablos, X. Zhang, & M. Almunawar (Eds.), *Handbook of Research on Green, Circular, and Digital Economies as Tools for Recovery and Sustainability* (pp. 257–279). IGI Global. doi:10.4018/978-1-7998-9664-7.ch014

Yu, P., Liu, X., Mahendran, R., & Lu, S. (2022). Analysis and Comparison of Business Models of Leading Enterprises in the Chinese Hydrogen Energy Industry. In R. Felseghi, N. Cobîrzan, & M. Raboaca (Eds.), *Clean Technologies and Sustainable Development in Civil Engineering* (pp. 179–216). IGI Global. doi:10.4018/978-1-7998-9810-8.ch008

Yu, P., Liu, Z., & Hanes, E. (2022). Supply Chain Resiliency, Efficiency, and Visibility in the Post-Pandemic Era in China: Case Studies of MeiTuan Waimai, and Ele.me. In Y. Ramakrishna (Ed.), *Handbook of Research on Supply Chain Resiliency, Efficiency, and Visibility in the Post-Pandemic Era* (pp. 195–225). IGI Global. doi:10.4018/978-1-7998-9506-0.ch011

Yu, P., Liu, Z., Hanes, E., & Mumtaz, J. (2022). Integration of IoT and Blockchain for Smart and Secured Supply Chain Management: Case Studies of China. In S. Goyal, N. Pradeep, P. Shukla, M. Ghonge, & R. Ravi (Eds.), *Utilizing Blockchain Technologies in Manufacturing and Logistics Management* (pp. 179–207). IGI Global. doi:10.4018/978-1-7998-8697-6.ch010

Yu, P., Liu, Z., & Sampat, M. (2023). Enhancing the Resilience of Food Cold Chain Logistics Through Digital Transformation: A Case Study of China. In I. Masudin, M. Almunawar, D. Restuputri, & P. Sud-On (Eds.), *Handbook of Research on Promoting Logistics and Supply Chain Resilience Through Digital Transformation* (pp. 200–224). IGI Global. doi:10.4018/978-1-6684-5882-2.ch014

Yu, P., Lu, S., Hanes, E., & Chen, Y. (2022). The Role of Blockchain Technology in Harnessing the Sustainability of Chinese Digital Finance. In P. Swarnalatha & S. Prabu (Eds.), *Blockchain Technologies for Sustainable Development in Smart Cities* (pp. 155–186). IGI Global. doi:10.4018/978-1-7998-9274-8.ch009

Yu, P., Lu, S., Sampat, M., Li, R., & Ahuja, A. (2022). How AI-Enabled Agile Internet of Things Can Enhance the Business Efficiency of China's FinTech Ecosystem. In S. Kautish & G. Kaur (Eds.), *AI-Enabled Agile Internet of Things for Sustainable FinTech Ecosystems* (pp. 190–223). IGI Global. doi:10.4018/978-1-6684-4176-3.ch009

Yu, P., Shen, X., & Hanes, E. (2023). Promoting Responsible Research and Innovation in China's Hi-Tech Zones: Based on Case Studies of Zizhu Hi-Tech Zone, East Lake Hi-Tech Zone, and Guangzhou Hi-Tech Zone. In B. Marco-Lajara, J. Martínez-Falcó, & L. Millán-Tudela (Eds.), *Corporate Sustainability as a Tool for Improving Economic, Social, and Environmental Performance* (pp. 222–245). IGI Global. doi:10.4018/978-1-6684-7422-8.ch012

Yu, P., Weng, Y., & Ahuja, A. (2022). Carbon Financing and the Sustainable Development Mechanism: The Case of China. In A. Rafay (Ed.), *Handbook of Research on Energy and Environmental Finance 4.0* (pp. 301–332). IGI Global. doi:10.4018/978-1-7998-8210-7.ch012

Yu, P., Xue, W., & Mahendran, R. (2022). The Development and Impact of China's Digital Transformation in the Medical Industry. In M. Rodrigues & J. Proença (Eds.), *Impact of Digital Transformation on the Development of New Business Models and Consumer Experience* (pp. 97–128). IGI Global. doi:10.4018/978-1-7998-9179-6.ch006

Yu, P., Xu, S., Cheng, Z., & Sampat, M. (2023). Does the Development of New Energy Vehicles Promote Carbon Neutralization?: Case Studies in China. In A. Pego (Ed.), *Climate Change, World Consequences, and the Sustainable Development Goals for 2030* (pp. 109–131). IGI Global. doi:10.4018/978-1-6684-4829-8.ch006

Yu, P., Zhang, Y., Sampat, M., & Chen, Y. (2023). Research on Cross-Industry Digital Transformation Under the New Normal: A Case Study of China. In B. Marco-Lajara, J. Martínez-Falcó, & L. Millán-Tudela (Eds.), *Corporate Sustainability as a Tool for Improving Economic, Social, and Environmental Performance* (pp. 246–277). IGI Global. doi:10.4018/978-1-6684-7422-8.ch013

Yu, P., Zhao, Z., & Sampat, M. (2023). How Digital Twin Technology Promotes the Development of Smart Cities: Case Studies in China. In I. Vasiliu-Feltes (Ed.), *Impact of Digital Twins in Smart Cities Development* (pp. 198–227). IGI Global. doi:10.4018/978-1-6684-3833-6.ch008

YuQ.ZhangW. (2019). Cybertwin: An origin of next generation network architecture. Available: https://arxiv.org/abs/1904.11313

Yu, Y., Li, Y., Tian, J., & Liu, J. (2018). Blockchain-based solutions to security and privacy issues in the internet of things. *IEEE Wireless Communications*, *25*(6), 12–18. doi:10.1109/MWC.2017.1800116

Zadid Shifat, A. S. M. (2018, January). Game-based approach for QoS provisioning and interference management in heterogeneous networks. *IEEE Access : Practical Innovations, Open Solutions*, *6*, 10208–10220. doi:10.1109/ACCESS.2017.2704094

Zanella, Filgueiras, Valério, Dartora, Mariano, & Cerqueira. (2020). Nano-antenna modelling based on plasmonic charge distribution for THz-based 6G applications. *Proc. 2nd 6G Wireless Summit (6G SUMMIT)*, 1-4.

Zavitsanos, D., Ntanos, A., Giannoulis, G., & Avramopoulos, H. (2020). On the QKD integration in converged fiber/wireless topologies for secured, low-latency 5G/B5G front haul. *Applied Sciences (Basel, Switzerland)*, *10*(15), 5193. doi:10.3390/app10155193

Zelbst, P. J., Green, K. W., Sower, V. E., & Bond, P. L. (2019). The impact of rfid, iiot, and blockchain technologies on supply chain transparency. *Journal of Manufacturing Technology Management*, *31*(3), 441–457. doi:10.1108/JMTM-03-2019-0118

Zhang, D., & Sato, T. (2016). One integrated energy efficiency proposal for 5G IoT communications. *IEEE Internet of Things Journal*, *3*(6), 1346–1354. doi:10.1109/JIOT.2016.2599852

Zhang, L., Liang, Y. C., & Niyato, D. (2019a). 6G visions: Mobile ultra-broadband, super internet-of-things, and artificial intelligence. *China Communications*, *16*(8), 1–14. doi:10.23919/JCC.2019.08.001

Zhang, P., Wang, C., Jiang, C., & Han, Z. (2021). Deep reinforcement learning assisted federated learning algorithm for data management of iiot. *IEEE Transactions on Industrial Informatics*, *17*(12), 8475–8484. doi:10.1109/TII.2021.3064351

Zhang, Y., Liu, W., Lou, W., & Fang, Y. (2006). Location-based compromise-tolerant security mechanisms for wireless sensor networks. *IEEE Journal on Selected Areas in Communications*, *24*(2), 247–260. doi:10.1109/JSAC.2005.861382

Zhang, Z. Q., Xiao, Y., Ma, Z., Xiao, M., Ding, Z., Lei, X., Karagiannidis, G. K., & Fan, P. (2019b). 6G wireless networks vision, requirements, architecture, and key technologies. *IEEE Vehicular Technology Magazine*, *14*(3), 28–41. doi:10.1109/MVT.2019.2921208

Zhao, K., & Ge, L. (2013). A survey on the internet of things security. *2013 Ninth International Conference on Computational Intelligence and Security*, 663–667. 10.1109/CIS.2013.145

Zhao, L., Zhu, D., Shafik, W., Matinkhah, S. M., Ahmad, Z., Sharif, L., & Craig, A. (2022). Artificial intelligence analysis in cyber domain: A review. *International Journal of Distributed Sensor Networks*, *18*(4). doi:10.1177/15501329221084882

Zhao, T. F., Wang, H., & Ma, Q. W. (2020). The coverage method of unmanned aerial vehicle mounted base station sensor network based on relative distance. *International Journal of Distributed Sensor Networks*, *16*(5), 1–12. doi:10.1177/1550147720920220

Zhao, Y., Cao, Y., Wang, W., Wang, H., Yu, X., Zhang, J., Tornatore, M., Wu, Y., & Mukherjee, B. (2018). Resource Allocation in Optical Networks Secured by Quantum Key Distribution. *IEEE Communications Magazine*, *56*(8), 130–137. doi:10.1109/MCOM.2018.1700656

Zheng, C., Yuan, J., Zhu, L., Zhang, Y., & Shao, Q. (2020). From digital to sustainable: A scientometric review of smart city literature between 1990 and 2019. *Journal of Cleaner Production*, *258*, 120689. doi:10.1016/j.jclepro.2020.120689

Zheng, Y., Yang, S., & Cheng, H. (2019). An application framework of digital twin and its case study. *Journal of Ambient Intelligence and Humanized Computing*, *10*(3), 1141–1153. doi:10.100712652-018-0911-3

Zhou, T., Jiang, N., Liu, Z., & Li, C. (2018). Joint cell activation and selection for green communications in ultra-dense heterogeneous networks. *IEEE Access : Practical Innovations, Open Solutions*, *6*, 1894–1904. doi:10.1109/ACCESS.2017.2780818

Zhou, Y., Wang, L., Love, P. E. D., Ding, L., & Zhou, C. (2019). Three-dimensional (3D) reconstruction of structures and landscapes: A new point-and-line fusion method. *Advanced Engineering Informatics*, *42*, 100961. doi:10.1016/j.aei.2019.100961

Zhu, C., Rodrigues, J., Shu, L., & Yang, L. (2018). Trust-based communication for the industrial internet of things. *IEEE Communications Magazine*, *56*(2), 16–22. doi:10.1109/MCOM.2018.1700592

Zj.gov.cn. (2022). *Taizhou City "Three Roles" to Build Intelligent Drainage Digital Twin Application System*. http://jst.zj.gov.cn/

Zong, B. Q., Fan, C., Wang, X., Duan, X., Wang, B., & Wang, J. (2019). 6G technologies: Key drivers, core requirements, system architectures, and enabling technologies. *IEEE Vehicular Technology Magazine*, *14*(3), 18–27. doi:10.1109/MVT.2019.2921398

About the Contributors

Poshan (Sam) Yu is a Lecturer in Accounting and Finance in the International Cooperative Education Program of Soochow University (China). He is also an External Professor of FinTech and Finance at SKEMA Business School (China), a Visiting Professor at Krirk University (Thailand), an EU-Asia Research Coordinator at the European Business University of Luxembourg and a Visiting Researcher at the Australian Studies Centre of Shanghai University (China). Sam leads FasterCapital (Dubai, UAE) as a Regional Partner (China) and serves as a Startup Mentor for AIC RAISE (Coimbatore, India) and University College London. His research interests include financial technology, regulatory technology, public-private partnerships, mergers and acquisitions, private equity, venture capital, start-ups, intellectual property, art finance, and China's "One Belt One Road" policy.

Nyaribo Wycliffe Misuko is a senior lecturer and currently the Dean School of Graduate Studies at KCA University Nairobi Kenya. Dr. Nyaribo holds a Ph.D. in Business Administration from MDS University, Ajmer, India. Dr. Nyaribo has been engaged in teaching and research at the university level for the last eighteen years. His area of expertise is management with a special focus on human resource management. Dr. Nyaribo has authored many articles published in peer-review journals. He is a member of professional bodies both locally and internationally In addition to his research work, he has mentored and supervised many master's students. He has also attended and presented papers at international conferences globally. He is also a member of the editorial board of academic journals

* * *

Srinivasan A. is now working as Associate Professor in SASTRA Deemed university in the Department of ECE and has teaching experience of 19 years. He obtained his Ph.D at SASTRA Deemed University. He has published more than 30 papers in national and international journals. He is having membership in IET, Institution of Engineers (India), Broadcasting society of India and HAM licence. Currently he is supervising 4 research scholars and one scholar awarded Ph.D.

Nancy Alonistioti (female) has a B.Sc. degree and a PhD degree in Informatics and Telecommunications (Dept. of Informatics and Telecommunications, NKUA). She has over 20 years of experience in numerous national and European R&D projects, including project/technical management experience. She is currently leading the SCAN group activities in the Dept. Informatics and Telecommunications in NKUA. She has served as member of the Future Internet Assembly Steering Committee. She is member of the ETSI Experts group and the Greek standardization group ELOT (5G, smart ctiy autonomic communications). She has over 100 publications in the area of new generation mobile networks, Smart Networks, IoT, SDN/NFV, AI enabled network management, autonomic communications and reconfigurable mobile systems. She is coauthor of 4 WO Patents and has more than 3000 citations.

Thangaraja Arumugam's area of interest is in the field of business and marketing analytics. He holds a Post Graduate Program in business analytics and business intelligence from the University of Texas. His research work has been published and presented in reputed journals and international conferences. He holds patents in the field of social mining, image processing and deep learning. He gained certifications in Marketing analytics, pricing analytics and sales forecasting from the Scandinavian Institute of business analytics. He holds several certificates such as marketing analytics: know your customers, Macquarie University, Australia. Applying data analytics in marketing, University of Illinois, USA. Artificial intelligence in marketing, university of Virginia, USA, etc.

Tina Babu, Assistant Professor. Department of Computer Science and Engineering, School of Engineering, Dayananda Sagar University, Bengaluru received her Doctor of Philosophy from Amrita University, Bengaluru. She received her B. Tech and M.Tech in Computer Science and Engineering from Mahatma Gandhi University, Kerala, India. Her area of interest include biomedical image processing, machine learning, pattern matching, computer vision.

Chandan Chavadi is the Dean & Professor of the Presidency Business School, Presidency College, Bengaluru. A PhD from Karnatak University, Dharwad, and an MBA (Mktg.) He holds a primary degree in B.E. (E&C). He has two years of corporate experience before his moving to academics. He has been in academics for the last 21 years. He has 32 papers & 2 book reviews to his credit, published in reputed journals and magazines such as ABDC-C journals, Web of Science, Scopus, UGC Care list and other indexed journals. Under Google scholar indices, he has total citations of 90, h-index of 6 & i10-index of 4. His paper has been accepted for publication in the IIM –A Vikalpa. Five of his research papers were recently published in the IIM Kozhikode Society And Management Review journal, IIM-Shillong Journal of Management Science, Business Perspective & Research journal of K J Somaiya Institute of Management, the MDI journal "Vision", and IIM-Lucknow journal "Metamorphosis". He is the recipient of Labdhi Bhandari Best Paper Award for the 7th IIM-A International Marketing Conference held on 11th to 13th Jan 2017. He is a recognized PhD Guide in Management for Bangalore City University.

Theoni Dounia is an undergraduate Electrical and Computer Engineer of the National Technical University of Athens, Greece. Mrs. Dounia has been actively involved as developer in the design and development of NLP/ROS-enriched apps, where she has applied various AI, ROS and NLP algorithms and methods. Mrs. Dounia joined INFOLYSiS in 2020 as R&D software engineer focusing mainly on the design and development of chatbot apps, supporting automated conversational flows enriched with DL/NLP technologies.

Dimitrios Fragkos received his BSc degree from the Department of Informatics and Telecommunications of the University of Peloponnese (UoP), Tripolis, Greece in September 2019. In October 2020 he received his MSc degree in Modern Wireless Communications from the same institute. Currently, he is a PhD candidate at Department of Informatics and Telecommunications, UoP with a thesis entitled

"Design and implementation of advanced 5G core network functions with capabilities of programmability, control and management of vertical applications, exploiting machine learning".

Jericho Galang is currently an Assistant Chief at the Public-Private Partnership Center of the Philippines under the National Economic and Development Authority, the country's premier socioeconomic planning body. He also serves as a lecturer at the Ateneo Department of Economics and formerly, at the University of the Philippines Los Baños (UPLB) Department of Economics. Before his current stint, he also worked as a consultant at the Development Academy of the Philippines. He earned his master's degree in development economics at the University of the Philippines Diliman and earned his bachelor's degree in agricultural economics in UPLB, earning the highest academic achievement in his College (Magna Cum Laude). Jericho is passionate in the field of human resource management and development economics. He is also soon to be married with his lovely fiance, Ash.

Emanuela Hanes is an independent researcher. Her cooperations include the Vienna University FH BFI Campus Wien, University of Graz, City University of Applied Sciences Bremen and University of Salzburg. Her research interests include China-EU business strategies, RegTech, FinTech, Cryptocurrencies, Geopolitics, Chinese Strategic Planning and Development Policies.

Usha Rani Janardhan received her B.E. Degree in CSE from VTU, Belagavi, India in June 2010. She acquired her master's degree IN Computer Networks Engineering from VTU, Belagavi, India in September 2015. She in pursuing Ph. D in from VTU, Belagavi, India. At present, she is working as an Assistant Professor in the department of CSE at the GSSS Institute of Engineering and Technology for Women. Mysuru. Her research areas include image processing and machine learning. She has published 03 papers in SCI and Scopus Indexed Journals.

Alexandros Kaloxylos has more than 25 years of experience in ICT. He is a professor at the department of informatics and telecommunications at the university of Peloponnese. He is also the executive director of the 6G-IA. For three years he was the head of the RAN department at Huawei research Center in Munich. He is the author of more than 150 publications and inventor of 10 5G patents. He has received more than 3000 citations to his work.

Harilaos Koumaras is a Research Assistant Professor at the Institute of Informatics and Telecommunications of NCSR "Demokritos", Greece. He has published more than 100 articles focused on 5G/B5G systems, network virtualization, SDN/NFV, MANO, and quality of service/experience in media delivery. Harilaos is the project coordinator of the H2020 5G-PPP 5GENESIS project and the technical manager of the H2020 5G-PPP EVOLVED-5G project.

Vaios Koumaras received his BSc degree in Computer Information Systems from the American College of Greece and his MBA in IT Project Management from the City University, USA. Since 1997, he has worked in several positions as Computer Analyst and Software Developer, participating in major IT projects. For more than 15 years, he holded the position of senior R&D software engineer and project manager, with participation and collaboration in R&D IT projects of numerous companies worldwide, mainly in marine software including the applicability of IoT in the shipping sector. Moreover, Mr. Koumaras is specialized in the management of IT projects concerning IoT applicability and data operability along with the design and development of chatbot apps both for commercial and research purposes (especially in IoT and 5G enabled industry environments). Since 2017, Mr. Koumaras is the Managing Director of INFOLYSiS and one of its co-founders.

Hongyu Lang is an independent researcher. His research interests include FinTech, finance, and data analytics.

George Makropoulos is a graduate in Electronics engineering from Technological Educational Institute of Athens and received his MSc degree in Advanced Informatics and Computing Systems from University of Piraeus in Greece. He is currently a PhD candidate (Resource management and slicing in 5G Networks) at Informatics and Telecommunications Department of National and Kapodistrian University of Athens. He has working experience as a Research Engineer and was involved in several EU-Funded projects in the fields of IoT and Telecommunications (fixed and mobile networks).

Sendhilkumar Manoharan is a hard-core researcher in the area of Marketing and organizational Behaviour. Experienced in Industry and Academics for more than 18 years with the educational qualifications of MSc (Psy)., MBA., Ph.D.

S. Meenakshi Sundaram is currently working as Professor and Head of the Department of CSE at GSSSIETW, Mysuru. He earned his BE in CSE in 1989 from Bharathidasan University in Tiruchirappalli, his M. Tech in 2006 from NIT in Tiruchirappalli, and his Ph.D. in CSE in 2014 from Anna University in Chennai. He has published 60 papers in refereed international journals, presented five papers at international conferences and delivered more than 50 seminars. He is a reviewer for Springer – Soft Computing Journal, International Journal of Ah Hoc Network Systems, Journal of Engineering Science and Technology, Taylor's University, Malaysia, and International Journal of Computational Science & Engineering, Inderscience Publishers, UK. He is a Senior Member of IEEE, a Life Member of IST, and a member of CSI. He has 32 years of teaching experience and 10 years of research experience. He has published 12 book chapters to his credit. Three research scholars have completed Ph.D. under his guidance and six research scholars are pursuing Ph. D from VTU Belagavi, India under his guidance. He received the "Research Excellence Award" from the RJS International Multidisciplinary Research Foundation in December 2021.

Tejaswini R. Murgod received her B.E. Degree in CSE from VTU, Belagavi, India in June 2008. She acquired her master's degree from VTU, Belagavi, India in Jan 2015. She completed her Ph. D in February 2022 from VTU, Belagavi, India. At present, she is working as an Associate Professor in the department of ISE at the NITTE Meenakshi Institute of Technology, Yelahanka, Bengaluru. Her research areas include underwater communication, optical networks, and wireless networks. She has published 10 papers in SCI and Scopus Indexed Journals. She has also published 9 book chapters to her credit. She received the "Best Researcher Award" from the RJS International Multidisciplinary Research Foundation in December 2021.

Rekha R. Nair, Assistant Professor, Department of Computer Applications, School of Engineering, Dayananda Sagar University, Bengaluru, received her Doctor of Philosophy from Amrita University,

Bengaluru. Her area of interest include biomedical image processing, machine learning, pattern matching, computer vision.

Madhu M. Nayak received her B.E. Degree in CSE from VTU, Belagavi, India in June 2004. She acquired her master's degree from VTU, Belagavi, India in September 2007. She in pursuing Ph. D from VTU, Belagavi, India. At present, she is working as an Assistant Professor in the department of CSE at the GSSS Institute of Engineering and Technology for Women. Mysuru. Her research areas include image processing and deep learning. She has published 05 papers in SCI and Scopus Indexed Journals.

Usha Obalanarasimhaiah has completed Bachelor of Science in Information Technology from KSOU, Mysuru in the year 2006. She is working as Programmer at GSSS Institute of Engineering and Technology for Women since 2006. She has gained sufficient knowledge of software installation, preventive and corrective maintenance of individual desktop systems and local area networks. Also she has enriched her knowledge in variety of Operating Systems and has configured multi boot Operating Systems in Computer Laboratories.

Shashikala Patil is an academician with two decades experience in teaching both post graduate and undergraduate programmes. Her area of expertise includes Marketing and Communication Studies. She has participated and presented papers at national and international conferences and has publications in Scopus and peer reviewed journals. She is an avid reader and has keen interest in CSR programmes and projects. Courses Taught: Marketing Management, Retail Management, Rural Marketing, Services Marketing, Research Methodology, Branding and Advertising, Business Communication, Soft Skills, Writing for Media Research Interests: Marketing – Retail, Rural and Services Marketing, Consumer Behaviour, Branding and Communication.

Mardeni Roslee is Deputy Director, Research Management Centre (RMC) & Chairman, Centre for Wireless Technology, Faculty of Engineering, Multimedia University. Advisor / Previous Past Chair, IEEE Malaysia, Communication Society and Vehicular Technology Society.

Baskaran S. received B. E. degree in Electronics and Communication Engineering from Bharathidasan University, India in 2003, and M. E. degree in Optical Communication from Anna University, India in 2005. He is a Ph. D. scholar of SASTRA Deemed University, India. From 2005 to 2016, he had worked as assistant professor in various engineering institutions in India. Since 2017, he has been working as an Assistant Professor, Department of Electronics and Communication Engineering at Srinivasa Ramanujan Centre, SASTRA Deemed University. His research interests include Optical Networks and WDM technologies.

Kishore S., Assistant Professor, Department of Computer Applications, School of Engineering, Dayananda Sagar University, Bengaluru, received post Graduation from CUSAT. His area of interest include biomedical image processing, machine learning, pattern matching, computer vision.

Christos Sakkas (male) received his BSc degree in Informatics and his MSc degree in Computer Science from the Department of Informatics of Athens University of Economics and Business (AUEB). Mr. Sakkas has participated in many research projects, collaborating with various universities, research centers and enterprises, with related presentations and publications at international conferences. Since 2017, Mr. Sakkas is INFOLYSiS CTO, coordinating and developing chatbot applications with emphasis on IoT and 5G use cases. His current research interests include SDN, NFV, DPI, Cloud Computing, IoT infrastructures, Monitoring systems and web/mobile applications.

Shouvik Sanyal is an Assistant Professor of Marketing at Dhofar University, Sultanate of Oman. He is an active researcher with over 40 published articles in journals of repute indexed in Scopus and Web of Science. He has presented his research at conferences in over 6 countries. His research interests include marketing, consumer behavior and new technologies in marketing.

Ajitha Savarimuthu is an enthusiastic and a motivated person with 13 + year of experience in academics and industry. A certified NLP practitioner, computer science engineer cum management professor specialized in Psychology currently working as an Associate Professor in Acharya Bangalore B School, Bangalore handling Management and Analytics subjects. Has been the resource person for various invited lectures and resource person for several FDP's and Webinars for both academic institutions and corporates. Published various papers in journals and presented numerous papers in conferences. Have written a book on Short Cases in Entrepreneurial Marketing. Was an adjunct Faculty in Learners Education, Sharjah, UAE teaching for Swiss School of management. A soft skills trainer and conducts workshop for students on Behavioural training and personality branding.

Wasswa Shafik (IEEE member) is pursuing his Ph.D. in Computer Science at the School of Digital Science, Universiti Brunei Darussalam, Brunei Darussalam. Received a master of engineering in Information Technology Engineering (MITE) from the Computer Engineering Department, Yazd University, Islamic Republic of Iran in in 2020. He received a bachelor of science in Information Technology Engineering with a minor in Mathematics from Ndejje University, Kampala, Uganda, in 2016. He is an associate researcher at the Computer Science Department, Intelligent Interconnectivity Laboratory at Yazd University, Islamic Republic of Iran. His areas of interest are Computer Vision, Anomaly Detection, Drones (UAVs), Machine/Deep Learning, AI-enabled IoT/IoMTs, IoT/IIoT/OT Security, Cyber Security, and Privacy. Shafik is the chair/co-chair/program chair of some Scopus/EI conferences. Also, academic editor/ associate editor for a set of indexed journals (Scopus journals' quartile ranking). He is the founder and lead investigator of the Digital Connectivity Research Laboratory (DCR-Lab) and the Managing Executive director of Asmaah Charity Organisation (ACO).

Dimitris Tsolkas leads activities related to research and development at FOGUS Innovations & Services P.C and currently he is responsible for network functions optimization as well as end-to-end network emulation and testing. Dimitris received a Ph.D degree from the National and Kapodistrian University of Athens (NKUA) in 2014. Since then, he has worked as senior research fellow at the department of Informatics and Telecommunications of NKUA and he has also served as PostDoc Marie Curie Fellow in MSCA RISE and ITN projects. He has long experience in ICT EU funded projects (e.g., EVOLVED-5G, 5GENESIS, and 5G-Monarch), serving as key technical person in some of them (e.g., EVOLVED5G and 5GENESIS). In the framework of ICT projects, he has participated in the 5GPPP Technical Board (TB) and 5GPPP/5GIA Working Groups (WGs), having contributed to key initiatives and white papers. In addition, Dimitris has published more than 50 papers in high quality journals, books, and conferences.

His current research interests include systems and protocols for wireless/mobile communication, radio access networks design and analysis, and user experience management.

Steven Wright, with over 50 patents awarded and multiple journal publications, is a worldwide-recognized professional and has been invited to share his knowledge in several international keynote presentations, forums, and conferences. Dr. Wright earned his MBA from Arizona State University (in marketing/finance), his PhD (in Computer Engineering) from North Carolina State University, and his JD at Georgia State University.

Xu Yifei is a researcher. Her research interests include public-private partnership, FinTech, smart cities and accounting. Yifei is conducting a part-time postgraduate study at Tongji University.

Kok Yeow You was born in 1977. He obtained his B.Sc. Physics (Honours) degree in Universiti Kebangsaan Malaysia (UKM) in 2001. He pursued his M.Sc. in Microwave at the Faculty of Science in 2003 and his Ph.D. in Wave Propagation at the Institute for Mathematical Research in 2006 in Universiti Putra Malaysia (UPM), Serdang, Selangor, Malaysia. Recently, he is a senior lecturer at the School of Electrical Engineering, Faculty of Engineering, Universiti Teknologi Malaysia (UTM), Skudai, Johor, Malaysia. His main personnel research interest is in the theory, simulation, and instrumentation of electromagnetic wave propagation at microwave frequencies focusing on the development of microwave passive devices and sensors for medical and agricultural applications.

Zixuan Zhao is an independent researcher. Her research interests include FinTech, Digital Twin and business analytics.

Index

R

S

T

U

V

W

Printed in the United States
by Baker & Taylor Publisher Services